PHYSICS

Oscillations, Waves and Optics

WITH LAB MANUAL

AF674589

A.B. Bhattacharya | Atanu Nag

KHANNA BOOK PUBLISHING CO. (P) LTD.
PUBLISHER OF ENGINEERING AND COMPUTER BOOKS
4C/4344, Ansari Road, Darya Ganj, New Delhi-110002
Phone: 011-23244447-48 **Mobile:** +91-99109 09320
E-mail: contact@khannabooks.com
Website: www.khannabooks.com

Dear Readers,

To prevent the piracy, this book is secured with HIGH SECURITY HOLOGRAM on the front title cover. In case you don't find the hologram on the front cover title, please write us to at contact@khannabooks.com or whatsapp us at +91-99109 09320 and avail special gift voucher for yourself.

Specimen of Hologram on front Cover title:

Moreover, there is a SPECIAL DISCOUNT COUPON for you with EVERY HOLOGRAM.

How to avail this SPECIAL DISCOUNT:

Step 1: Scratch the hologram
Step 2: Under the scratch area, your "coupon code" is available
Step 3: Logon to www.khannabooks.com
Step 4: Use your "coupon code" in the shopping cart and get your copy at a special discount
Step 5: Enjoy your reading!

ISBN: 978-93-91505-13-4

Physics (Oscillations, Waves and Optics) *by* A. B. Bhattacharya, Atanu Nag

First Edition: 2024

Published by:

Khanna Book Publishing Co. (P) Ltd.
Visit us at: **www.khannabooks.com**
Write us at: **contact@khannabooks.com**

CIN: U22110DL1998PTC095547

To view complete list of books, Please scan the QR Code:

KPH

Printed in Bharat.

अध्यक्ष
Prof. T. G. Sitharam
Chairman

सत्यमेव जयते

अखिल भारतीय तकनीकी शिक्षा परिषद्
(भारत सरकार का एक सांविधिक निकाय)
(शिक्षा मंत्रालय, भारत सरकार)
नेल्सन मंडेला मार्ग, वसंत कुंज, नई दिल्ली–110070
दूरभाष : 011–26131498
ई–मेल : chairman@aicte-india.org

ALL INDIA COUNCIL FOR TECHNICAL EDUCATION
(A STATUTORY BODY OF THE GOVT. OF INDIA)
(Ministry of Education, Govt. of India)
Nelson Mandela Marg, Vasant Kunj, New Delhi-110070
Phone : 011-26131498
E-mail : chairman@aicte-india.org

FOREWORD

Engineering has played a very significant role in the progress and expansion of mankind and society for centuries. Engineering ideas that originated in the Indian subcontinent have had a thoughtful impact on the world.

All India Council for Technical Education (AICTE) had always been at the forefront of assisting Technical students in every possible manner since its inception in 1987. The goal of AICTE has been to promote quality Technical Education and thereby take the industry to a greater heights and ultimately turn our dear motherland India into a Modern Developed Nation. It will not be inept to mention here that Engineers are the backbone of the modern society - better the engineers, better the industry, and better the industry, better the country.

NEP 2020 envisages education in regional languages to all, thereby ensuring that each and every student becomes capable and competent enough and is in a position to contribute towards the national growth and development.

One of the spheres where AICTE had been relentlessly working from last few years was to provide high-quality moderately priced books of International standard prepared in various regional languages to all it's Engineering students. These books are not only prepared keeping in mind it's easy language, real life examples, rich contents and but also the industry needs in this everyday changing world. These books are as per AICTE Model Curriculum of Engineering & Technology – 2018.

Eminent Professors from all over India with great knowledge and experience have written these books for the benefit of academic fraternity. AICTE is confident that these books with their rich contents will help technical students master the subjects with greater ease and quality.

AICTE appreciates the hard work of the original authors, coordinators and the translators for their endeavour in making these Engineering subjects more lucid.

Outcome Based Education

For the implementation of an outcome based education the first requirement is to develop an outcome based curriculum and incorporate an outcome based assessment in the education system. By going through outcome based assessments evaluators will be able to evaluate whether the students have achieved the outlined standard, specific and measurable outcomes. With the proper incorporation of outcome based education there will be a definite commitment to achieve a minimum standard for all learners without giving up at any level. At the end of the programme running with the aid of outcome based education, a student will be able to arrive at the following outcomes:

PO1. Engineering knowledge: Apply the knowledge of mathematics, science, engineering fundamentals, and an engineering specialization to the solution of complex engineering problems.

PO2. Problem analysis: Identify, formulate, review research literature, and analyze complex engineering problems reaching substantiated conclusions using first principles of mathematics, natural sciences, and engineering sciences.

PO3. Design/development of solutions: Design solutions for complex engineering problems and design system components or processes that meet the specified needs with appropriate consideration for the public health and safety, and the cultural, societal, and environmental considerations.

PO4. Conduct investigations of complex problems: Use research-based knowledge and research methods including design of experiments, analysis and interpretation of data, and synthesis of the information to provide valid conclusions.

PO5. Modern tool usage: Create, select, and apply appropriate techniques, resources, and modern engineering and IT tools including prediction and modeling to complex engineering activities with an understanding of the limitations.

PO6. The engineer and society: Apply reasoning informed by the contextual knowledge to assess societal, health, safety, legal and cultural issues and the consequent responsibilities relevant to the professional engineering practice.

PO7. Environment and sustainability: Understand the impact of the professional engineering solutions in societal and environmental contexts, and demonstrate the knowledge of, and need for sustainable development.

PO8. Ethics: Apply ethical principles and commit to professional ethics and responsibilities and norms of the engineering practice.

PO9. Individual and team work: Function effectively as an individual, and as a member or leader in diverse teams, and in multidisciplinary settings.

PO10. Communication: Communicate effectively on complex engineering activities with the engineering community and with society at large, such as, being able to comprehend and write effective reports and design documentation, make effective presentations, and give and receive clear instructions.

PO11. Project management and finance: Demonstrate knowledge and understanding of the engineering and management principles and apply these to one's own work, as a member and leader in a team, to manage projects and in multidisciplinary environments.

PO12. Life-long learning: Recognize the need for, and have the preparation and ability to engage in independent and life-long learning in the broadest context of technological change.

Course Outcomes

After the completion of the course the students will be able to:

CO-1: Explain different optical properties like reflection, refraction, diffraction, interference.

CO-2: Discuss the variation pattern of light intensity due to interference and diffraction.

CO-3: Apply wave equation to specified transverse and longitudinal vibration problems.

CO-4: Apply the concept of oscillations in different practical problems.

CO-5: Analyze working principle of different types of lasers to review their applications.

CO-6: Analyze real life problems involving waves and optics.

Course Outcomes	Expected Mapping with Programme Outcomes (1- Weak Correlation; 2- Medium Correlation; 3- Strong Correlation)											
	PO-1	PO-2	PO-3	PO-4	PO-5	PO-6	PO-7	PO-8	PO-9	PO-10	PO-11	PO-12
CO-1	2	-	-	-	-	-	-	-	-	-	-	-
CO-2	3	1	1	1	-	-	-	-	-	-	-	-
CO-3	2	2	2	1	-	-	-	-	-	-	-	-
CO-4	1	2	2	2	-	-	-	-	-	-	-	-
CO-5	3	-	-	1	-	-	-	-	-	-	-	-
CO-6	2	3	2	1	1	-	-	-	-	-	-	-

Abbreviations and Symbols

List of Abbreviations

General Terms			
Abbreviations	**Full form**	**Abbreviations**	**Full form**
CO	Course Outcome	PE	Potential Energy
CRO	Cathode Ray Oscilloscope	PO	Programme Outcome
He-Ne	Helium-Neon	RF	Radio Frequency
IR	Infra Red	RI	Refractive Index
KE	Kinetic Energy	RP	Resolving Power
LASER	Light Amplification by Stimulated Emission of Radiation	SHM	Simple Harmonic Motion
NA	Numerical Aperture	UO	Unit Outcome
Nd: YAG	Neodymium-doped Yttrium Aluminium Garnet	UV	Ultra Violet
Units Used			
Abbreviations	**Full form**	**Abbreviations**	**Full form**
cm	centimetre	mm	millimetre
eV	electron volt	ms	millisecond
Hz	hertz	mW	mill watt
kg	kilogram	nm	nanometre
kW	kilowatt	pm	pico metre
MW	megawatt	rad	radian

List of Symbols

Symbols	Description	Symbols	Description
Å	angstrom	I	intensity
Al	aluminium	L	inductance
Al_2O_3	aluminium oxide	l_c	coherence length
C	capacitance	m	magnification
c	speed of light	N	newton
CO_2	carbon dioxide	N_2	nitrogen
Cr	chromium	R	resistance
f	acceleration	T	time period
f_e	eyepiece focal length	v	velocity
f_o	objective focal length	v_g	group velocity

Symbols	Description	Symbols	Description
v_p	phase velocity	n	refractive index
Y	Young's modulus	ρ	density
Z	impedance	τ	relaxation time
β	fringe width	τ_c	coherence time
θ_c	critical angle	ω	angular velocity
θ_p	Brewster's angle	υ	frequency
λ	wavelength	$u(\upsilon)$	energy density
μ	permeability	ε	permittivity

List of Figures

Unit 1 Oscillations

Unit 2 Wave Propagation and Dispersion

Unit 3 The Propagation of Light and Geometric Optics

Unit 4 Wave Optics

Unit 5 Lasers

Guidelines for Teachers

To implement Outcome Based Education (OBE) knowledge level and skill set of the students should be enhanced. Teachers should take a major responsibility for the proper implementation of OBE. Some of the responsibilities (not limited to) for the teachers in OBE system may be as follows:

- Within reasonable constraint, they should manoeuvre time to the best advantage of all students.
- They should assess the students only upon certain defined criterion without considering any other potential ineligibility to discriminate them.
- They should try to grow the learning abilities of the students to a certain level before they leave the institute.
- They should try to ensure that all the students are equipped with the quality knowledge as well as competence after they finish their education.
- They should always encourage the students to develop their ultimate performance capabilities.
- They should facilitate and encourage group work and team work to consolidate newer approach.
- They should follow Blooms taxonomy in every part of the assessment.

Bloom's Taxonomy

Level	Teacher should check	Student should be able to	Possible mode of assessment
Create	Students ability to create	Design or Create	Mini project
Evaluate	Students ability to justify	Argue or Defend	Assignment
Analyse	Students ability to distinguish	Differentiate or Distinguish	Project/Lab Methodology
Apply	Students ability to use information	Operate or Demonstrate	Technical Presentation/ Demonstration
Understand	Students ability to explain the ideas	Explain or Classify	Presentation/Seminar
[illegible]	Students ability to recall (or remember)	Define or Recall	Quiz

Guidelines for Students

Students should take equal responsibility for implementing the OBE. Some of the responsibilities (not limited to) for the students in OBE system are as follows:

- Students should be well aware of each UO before the start of a unit in each and every course.
- Students should be well aware of each CO before the start of the course.
- Students should be well aware of each PO before the start of the programme.
- Students should think critically and reasonably with proper reflection and action.
- Learning of the students should be connected and integrated with practical and real life consequences.
- Students should be well aware of their competency at every level of OBE.

CONTENTS

1 Oscillations

UNIT SPECIFICS

This unit is emphasises the following main aspects:

- Mechanical and electrical simple harmonic oscillators
- Phasor representation of simple harmonic motion
- Damped harmonic oscillator – heavy, critical and light damping
- Energy decay in a damped harmonic oscillator
- Quality factor
- Forced mechanical oscillators
- Forced electrical oscillators
- Steady state motion of forced damped harmonic oscillator
- Power absorbed by oscillator

The practical applications of the topics are discussed for generating further curiosity and creativity as well as improving problem solving capacity.

Besides giving a large number of multiple choice questions as well as questions of short and long answer types marked in two categories following lower and higher order of Bloom's taxonomy, assignments through a number of numerical problems, a list of references and suggested readings are given in the unit so that one can go through them for practice. It is important to note that for getting more information on various topics of interest some QR codes have been provided in different sections which can be scanned for relevant supportive knowledge.

After the related practicals, based on the content, there is a "Know More" section. This section has been carefully designed so that the supplementary information provided in this part becomes beneficial for the users of the book. This section mainly highlights the initial activity, examples of some interesting facts, analogy, history of the development of the subject focusing the salient observations and finding, timelines starting from the development of the concerned topics up to the recent time, applications of the subject matter for our day-to-day real life or/and industrial applications on variety of aspects, case study related to environmental, sustainability, social and ethical issues whichever applicable, and finally inquisitiveness and curiosity topics of the unit.

RATIONALE

This fundamental unit on 'Oscillations' help students to get a primary idea about the simple harmonic motion as well as damped and forced simple harmonic oscillations. It has been illustrated how damped harmonic oscillators have non-conservative forces which dissipate their energy while critical damping returns the system to equilibrium as fast as possible without overshooting and an under damped system will oscillate through the equilibrium position. This unit clearly shows the difference between simple harmonic motion and damped harmonic motion as well as the difference between damped and forced oscillation.

A repeating motion is referred to as oscillatory motion where an object oscillates about an equilibrium position by a restoring force. This type of motion has importance for examining phenomena like alternating current circuits, electromagnetic waves etc. For the presence of resistive force, the periodic vibration of a body decreases its amplitude, known as the damped vibrations. On the other hand, the vibration of a body that takes place by the influence of any external periodic force is termed as the forced vibration.

PRE-REQUISITES

Mathematics: Differential equations (Class XII)

Physics: Simple Harmonic Motion (Class XII)

UNIT OUTCOMES

List of outcomes of this unit is as follows:

U1-O1: Analyzeize different physical characteristics of a harmonic motion

U1-O2: Interpret steady state solution of a forced harmonic oscillator

U1-O3: Demonstrate attenuation parameters of a damped harmonic motion

U1-O4: Illustrate the concepts of energy and sharpness of resonance in forced vibration

U1-O5: Apply the concept of Lissajous figures to categorize different types of simple harmonic oscillations

U1-O6: Classify underdamped, critically damped and overdamped vibratory motion

Unit-1 Outcome	EXPECTED MAPPING WITH COURSE OUTCOMES (1- Weak Correlation; 2- Medium Correlation; 3- Strong Correlation)					
	CO-1	CO-2	CO-3	CO-4	CO-5	CO-6
U1-O1	-	-	1	3	-	1
U1-O2	-	-	-	3	-	1
U1-O3	-	-	-	2	-	2
U1-O4	-	-	-	2	-	1
U1-O5	-	-	-	1	-	2
U1-O6	-	-	-	1	-	3

1.1 INTRODUCTION

Periodic motion is a motion which repeats its nature after a fixed time interval. Motion of a pendulum, oscillation of an object on a spring, earth's diurnal motion *etc* are examples of periodic motion. When a particle under periodic motion gets to-and-fro movement along the same path about equilibrium or the mean position, the motion is called a vibratory or oscillatory motion. All the oscillatory motions are periodic though all periodic motions are not oscillatory in nature.

Number of oscillation made by an oscillating body per unit time is termed as frequency (n) and the time required to complete one such oscillation is called the time period (T) of the motion. The relationship between these two is given by:

$$n = \frac{1}{T} \tag{1.1}$$

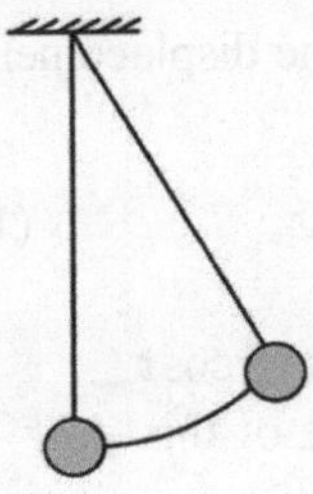

Fig. 1.1: Motion of a simple pendulum

When a particle moves to-and-fro in such a way that the acceleration of it is always directed towards a fixed point in its path of motion and is directly proportional to the displacement from the mean position, the motion of the particle is termed as simple harmonic motion (SHM). The distance traversed by a particle at any instant from the mean position is termed as the displacement of the particle and on either side the maximum displacement from the mean position is the amplitude of the motion.

If a particle of mass m executes SHM about a mean position such that at any time t the displacement from the mean position is x then the acceleration of the particle is given by:

$$f = -kx \tag{1.2}$$

Here k is a constant of proportionality.

The force acting on the particle at any instant is thus given by,

$$F = mf = -mkx = -\mu x \tag{1.3}$$

μ is proportionality constant. It is termed as the restoring force per unit displacement of the particle.

Characteristics of SHM: The characteristics of SHM are the following:

(a) The motion is oscillatory.

(b) The motion is linear.

(c) The acceleration of the body is proportional to the displacement with respect to the mean position.

(d) The acceleration is always directed away from the mean position.

1.2 ANALOGY OF SHM WITH CIRCULAR MOTION

Simple harmonic motion is a special case of circular motion. To understand this; let us consider a particle is moving around a circle of radius a with centre at O along the anticlockwise direction. Let at $t = 0$ it is at X and it is at P at any instant t making an angle θ with OX [Fig. 1.2]. Here $\theta = \omega t$, where ω is the uniform angular velocity of the particle. Now if a perpendicular PM is drawn from P to OX then it will give the displacement y of the particle at that instant t. This corresponds to a point P′ on the displacement-time curve on the right hand side of Fig. 1.2.

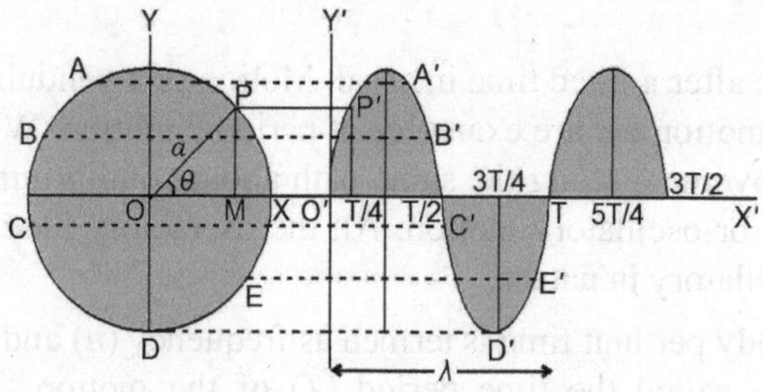

Fig. 1.2: SHM along with its reference circle

Similarly, the displacement of a particle, when it is at A, B, C, D and E in its circular motion, may be represented by A′, B′, C′, D′ and E′ on the displacement-time curve. Thus after completing a cycle in its circular path the graph on the right side indicates that the motion may be looked upon as simple harmonic. The circle is called the reference circle.

From Fig. 1.2 we have, $\sin\theta = \dfrac{PM}{OP} = \dfrac{y}{a}$

or, $$y = a\sin\theta = a\sin\omega t \tag{1.4}$$

If the particle is not at O initially, but at some other point then the equation for the displacement at any instant t is given by,

$$y = a\sin(\omega t + \varphi) \quad \text{[where } \varphi \text{ is the initial phase]} \tag{1.5}$$

EXAMPLE 1.1

Example 1.1 The displacement of a particle at any instant t is given by $x = 3\cos t + 4\sin t$. Show that the motion is simple harmonic. What is the amplitude of this oscillation?

Solution

Here $\quad x = 3\cos t + 4\sin t$

Therefore, $\quad \dfrac{dx}{dt} = -3\sin t + 4\cos t$ and $\dfrac{d^2x}{dt^2} = -(3\cos t + 4\sin t) = -x.$

So, the motion is simple harmonic.

Again, $\quad x = 5\left(\dfrac{3}{5}\cos t + \dfrac{4}{5}\sin t\right) = 5\sin(t+\theta)$

where $\sin\theta = \dfrac{3}{5}$ and $\cos\theta = \dfrac{4}{5}$.

Thus, the amplitude of the motion is 5 units.

1.3 VELOCITY OF A PARTICLE

We can define the velocity of an object as the rate of change of its position with respect to time. Velocity is a vector quantity having both magnitude and direction.

The displacement of a particle under SHM is, $x = a\sin(\omega t + \varphi)$.

Thus the velocity of the particle will be the rate of change of displacement of the particle and is given by,

$$v = \frac{dx}{dt} = a\omega\cos(\omega t + \varphi) = a\omega\sqrt{1 - \frac{x^2}{a^2}} = \omega\sqrt{a^2 - x^2} \tag{1.6}$$

When $x = 0$, then we have $v = v_{max} = \pm a\,\omega$ and when $x = \pm a$, then we get $v = v_{min} = 0$.

Example 1.2 The displacement of an oscillator is given by $x = a \sin(\omega t + \theta)$. If the oscillator started its motion at $t = 0$ from a position x_0 with velocity $x = v_0$ show that $\tan\theta = \omega x_0/v_0$ and $a = (x_0^2 + v_0^2/\omega^2)^{1/2}$.

Solution

The displacement of an oscillator is given by $x = a \sin(\omega t + \theta)$.

Therefore, velocity of the oscillator is $v = a\omega \cos(\omega t + \theta)$.

So, $\quad x_0 = a \sin\theta$ and $v_0 = a\omega\cos\theta$

$$\therefore \quad \omega \cot\theta = \frac{v_0}{x_0}$$

$$\text{or,} \quad \tan\theta = \frac{\omega x_0}{v_0}$$

$$\text{again,} \quad \sin^2\theta + \cos^2\theta = \frac{x_0^2}{a^2} + \frac{v_0^2}{a^2\omega^2}$$

$$\text{So,} \quad a = (x_0^2 + v_0^2/\omega^2)^{1/2}.$$

EXAMPLE 1.2

1.4 ACCELERATION OF A PARTICLE

The acceleration of an object is defined as the rate of change of its velocity with respect to time. Acceleration is a vector quantity and it has both magnitude and direction.

From Eq. (1.6) we have velocity of a particle executing SHM is

$$v = a\,\omega\cos(\omega t + \varphi)$$

Differentiating with respect to t we get the rate of change of velocity of the particle and it is called its acceleration as given by,

$$f = \frac{dv}{dt} = -a\omega^2 \sin(\omega t + \varphi) = -\omega^2 x \tag{1.7}$$

When $x = 0$, then we have $f = f_{min} = 0$ and when $x = \pm a$, then we have $f = f_{max} = \mp\omega^2 a$.

Since, at the extreme position of the motion x is maximum, thus the acceleration will be maximum and will be directed towards the mean position.

Example 1.3 The displacement of a particle of mass 0·2 kg executing SHM is given by $y = 10\sin\left(\frac{\pi}{3}t - \frac{\pi}{12}\right)$ m. Calculate the (i) amplitude, (ii) angular velocity, (iii) time period, (iv) maximum velocity, and (v) maximum acceleration.

Solution

The displacement is indicated by $y = 10\sin\left(\frac{\pi}{3}t - \frac{\pi}{12}\right)$ m.

EXAMPLE 1.3

EXAMPLE 1.3

Comparing this with the general equation $y = A \sin(\omega t - \theta)$ we get,

(i) amplitude $A = 10$ m,

(ii) the angular velocity $\omega = \frac{\pi}{3}$ rad/s,

(iii) the time period $T = \frac{2\pi}{\omega} = 6$ s,

Again the velocity and acceleration of the particle are given by,

$$v = \frac{dy}{dt} = \frac{10\pi}{3}\cos\left(\frac{\pi}{3}t - \frac{\pi}{12}\right)$$

and
$$f = \frac{dv}{dt} = \frac{10\pi^2}{9}\sin\left(\frac{\pi}{3}t - \frac{\pi}{12}\right).$$

So, (iv) the maximum velocity $= \frac{10\pi}{3}$ m/s and

(v) maximum acceleration $= \frac{10\pi^2}{9}$ m/s^2.

1.5 TOTAL ENERGY OF A PARTICLE

The total energy of any system represents the sum of its kinetic energy and potential energy, and generally this is a conserved quantity. Object having total energy less than zero is bound while those with zero or greater values are unbounded. Thus the total energy of a particle executing SHM represents the sum of its kinetic and potential energy.

Kinetic energy: The kinetic energy (T) of a particle of mass m moving with a velocity v is given by [from Eq. (1.6)], $T = \frac{1}{2}mv^2 = \frac{1}{2}m\omega^2\left(a^2 - x^2\right)$ (1.8)

Potential energy: Potential energy (U) stored in a particle of mass m performing SHM for a displacement x will be the total work done to overcome the restoring force acting on the particle for that displacement and is given by,

$$U = \int_0^x m\omega^2 x dx = \frac{1}{2}m\omega^2 x^2 \quad (1.9)$$

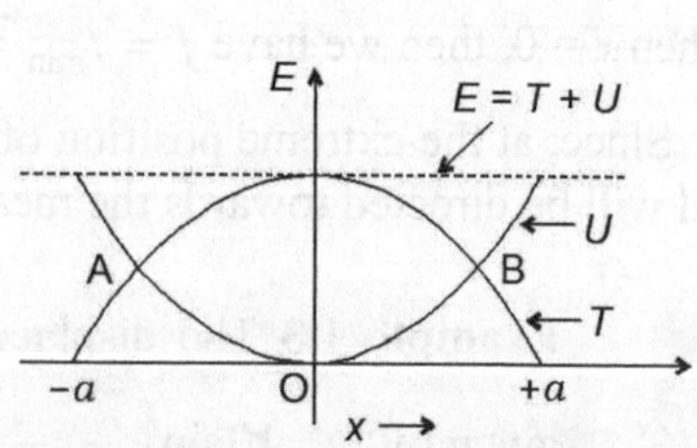

Fig. 1.3: Total energy distribution of a particle executing SHM

Hence the total energy of the particle is given by,

$$E = T + U = \frac{1}{2}m\omega^2\left(a^2 - x^2\right) + \frac{1}{2}m\omega^2 x^2$$

or,
$$E = \frac{1}{2}m\omega^2 a^2 \quad (1.10)$$

and it is a constant quantity. The variation of kinetic energy T, potential energy U and total energy E with displacement x is shown in Fig. 1.3. At points A and B the total energy is equally shared by the kinetic and potential energies.

Example 1.4 A particle of 0.5 kg mass is displaced during its SHM and this displacement is written as $y = 5\cos\left(\frac{\pi}{3}t - \frac{\pi}{6}\right)$ m. Calculate the

i) amplitude, angular velocity, time period of the particle;
ii) the displacement, velocity and acceleration of the particle after 1 s;
iii) the kinetic and potential energy of the particle when displacement is $x = 0.04$ m; and
iv) the total energy of the particle.

Solution

The equation of a SHM is given by, $y = a\cos(\omega t - \theta)$ (a)

Here $y = 5\cos\left(\frac{\pi}{3}t - \frac{\pi}{6}\right)$ (b)

(i) Comparing Eqs. (a) and (b) we get, the amplitude of SHM is a = 5m, the angular velocity $\omega = \frac{\pi}{3}$ rad/s and the time period $T = \frac{2\pi}{\omega} = 6$ s.

(ii) We have $y = 5\cos\left(\frac{\pi}{3}t - \frac{\pi}{6}\right)$

Thus at t = 1s the displacement is

$$y = 5\cos\left(\frac{\pi}{3} - \frac{\pi}{6}\right) = 5\cos\frac{\pi}{6} = 5\frac{\sqrt{3}}{2}m$$

$$= 4.33 \text{ m.}$$

The velocity of the particle is $v = \frac{dy}{dt} = -5\frac{\pi}{3}\sin\left(\frac{\pi}{3}t - \frac{\pi}{6}\right)$

Thus at t = 1s, $v = -5\frac{\pi}{3}\sin\left(\frac{\pi}{3} - \frac{\pi}{6}\right) - 5\frac{\pi}{3}\sin\frac{\pi}{6} = -2.62$ m/s.

And the acceleration of the particle is given by,

$$f = \frac{dv}{dt} = -5\left(\frac{\pi}{3}\right)^2\cos\left(\frac{\pi}{3}t - \frac{\pi}{6}\right)$$

Thus at t = 1s, $f = \frac{dv}{dt} = -5\left(\frac{\pi}{3}\right)^2\cos\left(\frac{\pi}{3} - \frac{\pi}{6}\right) - 5\left(\frac{\pi}{3}\right)^2\cos\frac{\pi}{6} = -4.74\text{ m/s}^2.$

(iii) Now the kinetic energy of the particle is given by,

$$\text{KE} = \frac{1}{2}m\omega^2\left(a^2 - x^2\right)$$

At $x = 0.04$m, $\text{KE} = \frac{1}{2} \times 0.5 \times \left(\frac{\pi}{3}\right)^2\left(5^2 - 0.04^2\right) = 6.845\text{ J}.$

EXAMPLE 1.4

EXAMPLE 1.4

and the potential energy of the particle is $PE = \frac{1}{2}m\omega^2 x^2$

At $x = 0.04$m, $PE = \frac{1}{2} \times 0.5 \times \left(\frac{\pi}{3}\right)^2 \times (0.04)^2 = 4.38 \times 10^{-4}$ J.

(iv) The total energy of the particle is $\frac{1}{2}m\omega^2 a^2 = \frac{1}{2} \times 0.5 \times \left(\frac{\pi}{3}\right)^2 \times (5)^2 = 6.846$ J.

1.6 DIFFERENTIAL EQUATION OF SHM

For a particle undergoing oscillatory motion if the acceleration is directed towards a fixed point along its path and is always proportional to the displacement from that point then the motion of the particle will be *simple harmonic*.

For a particle of mass *m* executing SHM; if *x* be the displacement of the particle from the mean position then the force acting on the particle is given by,

$$m\frac{d^2x}{dt^2} \propto -x$$

or, $$m\frac{d^2x}{dt^2} = -\mu x$$ where μ is the restoring force per unit displacement.

or, $$m\frac{d^2x}{dt^2} + \mu x = 0$$

or, $$\frac{d^2x}{dt^2} + \frac{\mu}{m}x = 0$$

or, $$\frac{d^2x}{dt^2} + \omega^2 x = 0 \tag{1.11}$$

where, $$\omega^2 = \frac{\mu}{m} \tag{1.12}$$

or, $$\mu = m\omega^2 \tag{1.13}$$

Eq. (1.11) is the basic differential equation of motion of a particle executing SHM.

Solution of the equation: We have, $\frac{d^2x}{dt^2} + \omega^2 x = 0$

Multiplying both sides by $2\frac{dx}{dt}$ we get,

$$2\frac{dx}{dt}\frac{d^2x}{dt^2} + 2\frac{dx}{dt}\omega^2 x = 0$$

or, $$\frac{d}{dt}\left\{\left(\frac{dx}{dt}\right)^2 + \omega^2 x^2\right\} = 0$$

or, $$\left(\frac{dx}{dt}\right)^2 + \omega^2 x^2 = K$$,where K is the integration constant.

Now at $x = a$, $$\frac{dx}{dt} = 0$$

$\therefore$ $$K = \omega^2 a^2$$

So, $$\left(\frac{dx}{dt}\right)^2 + \omega^2 x^2 = \omega^2 a^2$$

or, $$\left(\frac{dx}{dt}\right)^2 = \omega^2 (a^2 - x^2)$$

or, $$\frac{dx}{dt} = \omega\sqrt{a^2 - x^2}$$

or, $$\frac{dx}{\sqrt{a^2 - x^2}} = \omega dt$$

Integrating both sides we get, $$\int \frac{dx}{\sqrt{a^2 - x^2}} = \int \omega dt$$

or, $$\sin^{-1}\frac{x}{a} = \omega t + \varphi$$

or, $$\frac{x}{a} = \sin(\omega t + \varphi)$$

or, $x = a\sin(\omega t + \varphi)$; which is the required solution.

Alternative method for solution: We have, $\frac{d^2x}{dt^2} + \omega^2 x = 0$

Let the general solution of the equation is $x = ae^{\alpha t}$

Differentiating twice with respect to t, we get $\frac{dx}{dt} = a\alpha e^{\alpha t}$ and $\frac{d^2x}{dt^2} = a\alpha^2 e^{\alpha t}$

Thus from Eq. (1.8) we have, $a\alpha^2 e^{\alpha t} + a\omega^2 e^{\alpha t} = 0$

or, $$ae^{\alpha t}(\alpha^2 + \omega^2) = 0$$

or, $$(\alpha^2 + \omega^2) = 0 \qquad [\text{As, } ae^{\alpha t} \neq 0]$$

$\therefore$ $$\alpha = \pm i\omega$$

So the required solution is given by,

$$x = A_1 e^{i\omega t} + A_2 e^{-i\omega t} = A_1(\cos\omega t + i\sin\omega t) + A_2(\cos\omega t - i\sin\omega t)$$

or, $$x = \cos\omega t(A_1 + A_2) + \sin\omega t\,\{i(A_1 - A_2)\}$$

Let, $$i(A_1 - A_2) = a\cos\varphi \text{ and } (A_1 + A_2) = a\sin\varphi$$

$\therefore$ $$x = a\sin\omega t\cos\varphi + a\cos\omega t\sin\varphi$$

or, $$x = a\sin(\omega t + \varphi)$$; which is the required solution.

EXAMPLE 1.5

Example 1.5 Potential energy of a particle of mass 10 g is given by $V(x) = 32x^2 + 0.2$, where x is in m and V is in J. Write down the equation of motion and solve it. What is the frequency of oscillation?

Solution

Here the potential is $V(x) = 32x^2 + 0.2$

So the force associated is, $F = -\dfrac{dV}{dx} = -64x$

Thus the equation of motion of the particle will be,

$$m\frac{d^2x}{dt^2} = -64x$$

or, $$m\frac{d^2x}{dt^2} + 64x = 0$$

or, $$\frac{d^2x}{dt^2} + \frac{64}{m}x = 0 \qquad \text{(a)}$$

Also we have general equation of SHM as,

$$\frac{d^2x}{dt^2} + \omega^2 x = 0 \qquad \text{(b)}$$

Comparing Eq. (a) and (b) we get, $\omega^2 = \dfrac{64}{m}$

$\therefore$ $$\omega = \sqrt{\frac{64}{m}} = \sqrt{\frac{64}{10^{-2}}} = 80\,\text{rad/s}.$$

Thus the equation of motion of the particle is $\dfrac{d^2x}{dt^2} + 6400x = 0$ (as $\omega = 80$ rad/s)

General solution of this equation of motion is given by, $x = a\sin\omega t = a\sin 80t$

and the frequency of oscillation is, $n = \dfrac{\omega}{2\pi} = \dfrac{80}{2\times 3.14} = 12.74\,\text{Hz}.$

1.7 SUPERPOSITION OF TWO COLLINEAR SHM

The principle of superposition states that for all linear systems, the resultant response due to two or more stimuli becomes the sum of the responses by each stimulus. Let input A and input B exhibit responses X and Y respectively; then input $(A + B)$ will produce response $(X + Y)$.

We consider superposition of two collinear SHMs with same frequency and different amplitudes.

Let, the two collinear SHM be represented by,

$$x_1 = a_1 \sin \omega t \tag{1.14}$$

and
$$x_2 = a_2 \sin(\omega t + \theta) \tag{1.15}$$

where a_1 and a_2 are the respective amplitudes, and θ is the phase difference between the two given SHM.

Thus the resultant disturbance is given from the principle of superposition as,

$$x = x_1 + x_2 = a_1 \sin \omega t + a_2 \sin(\omega t + \theta)$$

or,
$$x = a_1 \sin \omega t + a_2 \sin \omega t \cos \theta + a_2 \cos \omega t \sin \theta$$
$$= (a_1 + a_2 \cos \theta) \sin \omega t + a_2 \cos \omega t \sin \theta$$

Let
$$(a_1 + a_2 \cos \theta) = A \cos \phi \qquad \text{(a)}$$

and
$$a_2 \sin \theta = A \sin \phi \qquad \text{(b)}$$

then
$$x = A \sin \omega t \cos \phi + A \cos \omega t \sin \phi$$

or,
$$x = A \sin(\omega t + \phi) \tag{1.16}$$

Hence the resultant displacement represents a SHM of amplitude A and it may be obtained from Eqs. (a) and (b) as,

$$A^2 = (a_1 + a_2 \cos \theta)^2 + (a_2 \sin \theta)^2 = a_1^2 + 2a_1 a_2 \cos \theta + a_2^2$$

$$\therefore \quad A = \left(a_1^2 + a_2^2 + 2a_1 a_2 cos\theta\right)^{1/2} \tag{1.17}$$

To evaluate ϕ, by dividing Eq. (b) by Eq. (a) we get, $\tan \phi = \dfrac{a_2 \sin \theta}{a_1 + a_2 \cos \theta}$

or,
$$\phi = \tan^{-1}\left(\frac{a_2 \sin \theta}{a_1 + a_2 \cos \theta}\right) \tag{1.18}$$

Hence, using Eq. (1.17) and Eq. (1.18) in Eq. (1.16) we get,

$$x = \left(a_1^2 + a_2^2 + 2a_1 a_2 \cos \theta\right)^{1/2} \sin\left\{\omega t + \tan^{-1}\left(\frac{a_2 \sin \theta}{a_1 + a_2 \cos \theta}\right)\right\} \tag{1.19}$$

Eq. (1.19) also represents a simple harmonic motion of same frequency as the components but of different amplitude A and phase difference ϕ, with the first component.

Special cases

Case I: If $\theta = 0$, $A = a_1 + a_2$ and $\phi = 0$

thus, $$x = (a_1 + a_2)\sin \omega t \tag{1.20}$$

So, the amplitude of the resultant wave is equal to the sum of the amplitudes of the components.

Case II: If $\theta = \pi$, $A = a_1 \sim a_2$ and $\phi = 0$

thus, $$x = (a_1 \sim a_2)\sin \omega t \tag{1.21}$$

So, the amplitude of the resultant wave is equal to the difference of the amplitudes of the components.

Case III: If $\theta = \frac{\pi}{2}$, $A = \left(a_1^2 + a_2^2\right)^{1/2}$ and $\phi = \tan^{-1}\left(\frac{a_2}{a_1}\right)$

thus, $$x = \left(a_1^2 + a_2^2\right)^{\frac{1}{2}} \sin\left\{\omega t + \tan^{-1}\left(\frac{a_2}{a_1}\right)\right\} \tag{1.22}$$

So, the amplitude of the resultant wave is same as the sum of square root of the amplitudes of the components.

1.8 COMPOSITION OF TWO MUTUALLY PERPENDICULAR SHM

We consider composition of two SHM having equal time period, different amplitude and phases along mutually perpendicular directions.

Let, two collinear SHM be represented by,

$$x = a_1 \sin \omega t \tag{1.23}$$

and $$y = a_2 \sin(\omega t + \theta) \tag{1.24}$$

where a_1 and a_2 are the respective amplitudes and θ is the phase difference between the two SHM.

From Eqs. (1.23) and (1.24) we have, $\frac{x}{a_1} = \sin \omega t$

and $$\frac{y}{a_2} = \sin(\omega t + \theta) = \sin \omega t \cos\theta + \cos \omega t \sin\theta \tag{1.25}$$

As, $$\sin \omega t = \frac{x}{a_1}$$

$$\therefore \quad \cos \omega t = \sqrt{1 - \sin^2 \omega t} = \pm\sqrt{1 - \frac{x^2}{a_1^2}}$$

So, from Eq. (1.25) we get,

$$\frac{y}{a_2} = \frac{x}{a_1}\cos\theta \pm \sqrt{1 - \frac{x^2}{a_1^2}}\sin\theta$$

or, $$\frac{y}{a_2}-\frac{x}{a_1}\cos\theta=\pm\sqrt{1-\frac{x^2}{a_1^2}}\sin\theta$$

Squaring both sides we get,

$$\left(\frac{y}{a_2}-\frac{x}{a_1}\cos\theta\right)^2=\left(1-\frac{x^2}{a_1^2}\right)\sin^2\theta$$

or, $$\frac{y^2}{a_2^2}+\frac{x^2}{a_1^2}\cos^2\theta-\frac{2xy}{a_1a_2}\cos\theta=\sin^2\theta-\frac{x^2}{a_1^2}\sin^2\theta$$

or, $$\frac{y^2}{a_2^2}+\frac{x^2}{a_1^2}\cos^2\theta-\frac{2xy}{a_1a_2}cos\theta+\frac{x^2}{a_1^2}\sin^2\theta=\sin^2\theta$$

or, $$\frac{y^2}{a_2^2}+\frac{x^2}{a_1^2}-\frac{2xy}{a_1a_2}\cos\theta=\sin^2\theta \tag{1.26}$$

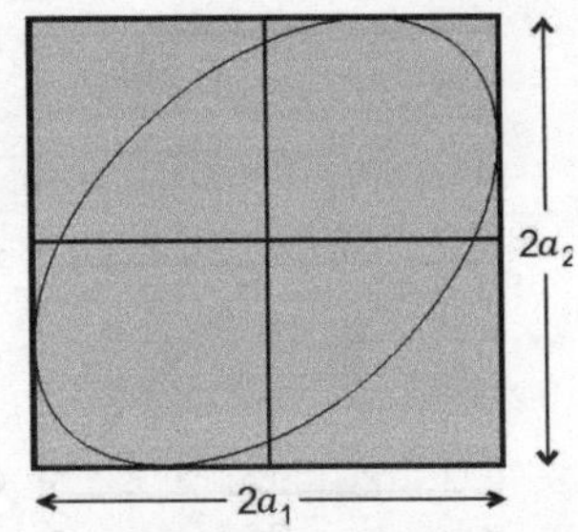

Fig. 1.4: Superposition of two SHM of equal time period but different amplitudes and phases

This gives the general equation of an ellipse inclined at an angle θ to the X-axis of the 2-dimensional Cartesian coordinate system and is contained within a rectangle of sides $2a_1$ and $2a_2$ (Fig. 1.4).

Special Cases: Depending on the different values of θ there may be different cases:

Case I: When $\theta = 0$ then $\sin\theta = 0$; $\cos\theta = 1$ and hence from Eq. (1.26) we get,

$$\frac{y^2}{a_2^2}+\frac{x^2}{a_1^2}-\frac{2xy}{a_1a_2}=0$$

or, $$\left(\frac{y}{a_2}-\frac{x}{a_1}\right)^2=0$$

or, $$y=\frac{a_2}{a_1}x \tag{1.27}$$

This gives a straight line which is passing through the origin and lying in the first and third quadrants, making an angle $\phi=\tan^{-1}\left(\frac{a_2}{a_1}\right)$ with the X-axis as shown in Fig. 1.5.

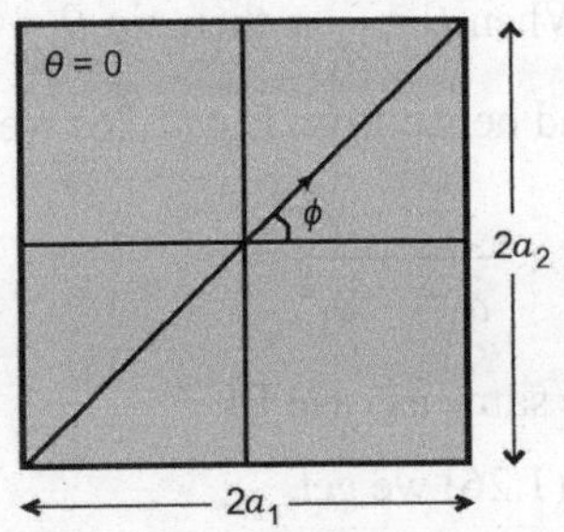

Fig. 1.5: Superposition of two SHM of equal time period but different amplitudes and with phase difference $\theta = 0$

Case II: When $\theta = \pi$, then $\sin\theta = 0$; $\cos\theta = -1$ and hence from Eq. (1.26) we get,

$$\frac{y^2}{a_2^2}+\frac{x^2}{a_1^2}+\frac{2xy}{a_1a_2}=0$$

or, $$\left(\frac{y}{a_2}+\frac{x}{a_1}\right)^2=0$$

or, $$y=-\frac{a_2}{a_1}x \tag{1.28}$$

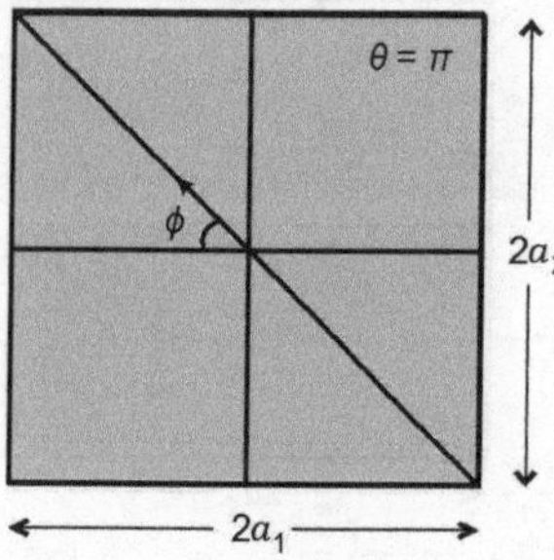

Fig. 1.6: Superposition of two SHM of equal time period but different amplitudes and with phase difference $\theta = \pi$

This represents a straight line passing through the origin and lying in the second and fourth quadrants, making an angle $\phi = \tan^{-1}\left(-\frac{a_2}{a_1}\right)$ with the X-axis as shown in Fig. 1.6.

Case III: When $\theta = \frac{\pi}{2}$, then $\sin\theta = 1$, $\cos\theta = 0$ and hence from Eq. (1.26) we get,

$$\frac{y^2}{a_2^2} + \frac{x^2}{a_1^2} = 1 \tag{1.29}$$

This represents an ellipse with semi major axis $2a_1$ and semi minor axis $2a_2$ and is shown in Fig. 1.7 (a).

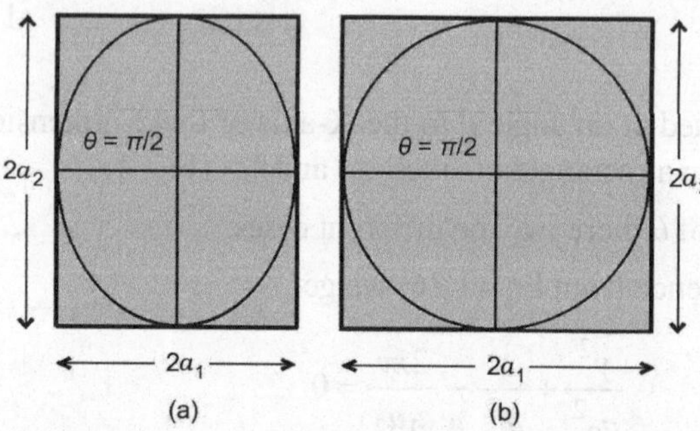

Fig. 1.7: (a) An ellipse and (b) a circle formed due to superposition of two SHM of equal time period but (a) different or (b) same amplitude and with phase difference $\theta = \pi/2$

Now if $a_1 = a_2 = a$ then Eq. (1.29) may be re-written as,

$$x^2 + y^2 = a^2 \tag{1.30}$$

and this represents a circle as shown in Fig. 1.7 (b).

Case IV: When $\theta = \frac{3}{2}\pi$ then $\sin\theta = -1$, $\cos\theta = 0$ and hence from Eq. (1.26) we get,

$$\frac{y^2}{a_2^2} + \frac{x^2}{a_1^2} = 1$$

which is the same as case III.

Case V: When $\theta = 2\pi$ then $\sin\theta = 0$ and $\cos\theta = 1$ and hence from Eq. (1.26) we get,

$$\frac{y^2}{a_2^2} + \frac{x^2}{a_1^2} - \frac{2xy}{a_1 a_2} = 0$$

or,
$$\left(\frac{y}{a_2} - \frac{x}{a_1}\right)^2 = 0$$

or,
$$y = \frac{a_2}{a_1}x$$, and it is the same case as in case I.

1.9 MUTUALLY PERPENDICULAR SHM WITH FREQUENCY RATIO 1:2

If the frequencies of two SHM be in the ratio 1:2 with initial phase difference θ, then we can represent the two SHM by the following two equations:

$$x = a_1 \cos\omega t \tag{1.31}$$

and $$y = a_2 \cos(2\omega t + \theta) \tag{1.32}$$

Now we can write,

$$\frac{y}{a_2} = \cos 2\omega t \cos\theta - \sin 2\omega t \sin\theta$$

or, $$\frac{y}{a_2} = (2\cos^2 \omega t - 1)\cos\theta - \sin 2\omega t \sin\theta$$

or, $$\frac{y}{a_2} = \left(2\frac{x^2}{a_1^2} - 1\right)\cos\theta - 2\sqrt{\left(1 - \frac{x^2}{a_1^2}\right)}\frac{x}{a_1}\sin\theta$$

or, $$\frac{y}{a_2} - \left(2\frac{x^2}{a_1^2} - 1\right)\cos\theta = -2\sqrt{\left(1 - \frac{x^2}{a_1^2}\right)}\frac{x}{a_1}\sin\theta$$

Squaring both sides of the above equation, we get

$$\frac{y^2}{a_2^2} - 2\frac{y}{a_2}\left(2\frac{x^2}{a_1^2} - 1\right)\cos\theta + \left(2\frac{x^2}{a_1^2} - 1\right)^2 \cos^2\theta = 4\left(1 - \frac{x^2}{a_1^2}\right)\frac{x^2}{a_1^2}\sin^2\theta$$

or, $$\frac{y^2}{a_2^2} - 4\frac{x^2 y}{a_1^2 a_2}\cos\theta + 2\frac{y}{a_2}\cos\theta + 4\frac{x^4}{a_1^4}\cos^2\theta - 4\frac{x^2}{a_1^2}\cos^2\theta + \cos^2\theta$$

$$= 4\frac{x^2}{a_1^2}\sin^2\theta - 4\frac{x^4}{a_1^4}\sin^2\theta$$

or, $$\frac{y^2}{a_2^2} - 4\frac{x^2 y}{a_1^2 a_2}\cos\theta + 2\frac{y}{a_2}\cos\theta + \cos^2\theta = 4\frac{x^2}{a_1^2}\left(1 - \frac{x^2}{a_1^2}\right)$$

or, $$\left(\frac{y}{a_2} + \cos\theta\right)^2 = 4\frac{x^2}{a_1^2}\left(1 - \frac{x^2}{a_1^2} + \frac{y}{a_2}\cos\theta\right) \tag{1.33}$$

The above equation gives the general equation of the resultant motion for any phase difference and amplitudes. For different values of θ different cases can be considered:

Case I: If $\theta = 0$, Eq. (1.33) gets reduced to

$$\left(2\frac{x^2}{a_1^2} - \frac{y}{a_2} - 1\right)^2 = 0$$

or, $$x^2 = \frac{a_1^2}{a_2}(y - a_2) \tag{1.34}$$

Eq. (1.34) represents two coincident parabolas as shown in Fig. 1.8.

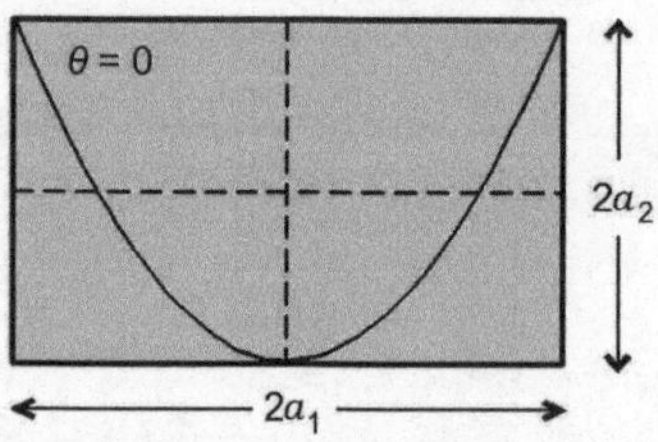

Fig. 1.8: Resultant of two mutually perpendicular SHM with frequency ratio 1:2 and phase difference $\theta = 0$

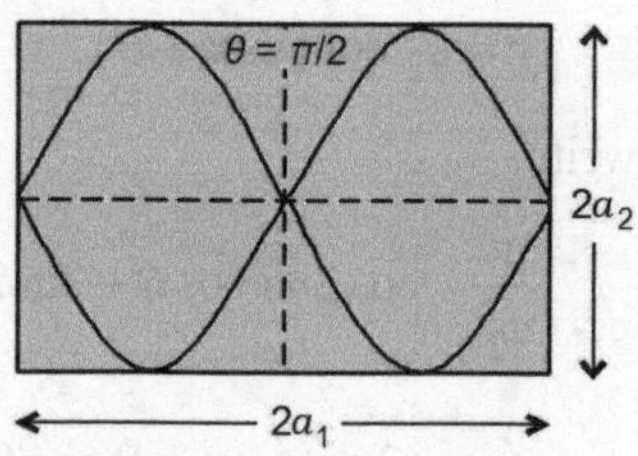

Fig. 1.9: Resultant of two mutually perpendicular SHM with frequency ratio 1 : 2 and phase difference $\theta = \pi/2$

Case II: If $\theta = \frac{\pi}{2}$, Eq. (1.33) gets reduced to

$$4\frac{x^2}{a_1^2}\left(\frac{x^2}{a_1^2}-1\right)+\frac{y^2}{a_2^2}=0 \tag{1.35}$$

Eq. (1.35) is a fourth degree equation and represents a curve which has two loops as has been shown in Fig. 1.9.

1.10 MUTUALLY PERPENDICULAR SHM WITH FREQUENCY RATIO 1:3

If the frequencies of two SHM be in the ratio 1:3 with initial phase difference θ then we can represent the two SHM by the following two equations:

$$x = a_1 \cos \omega t \tag{1.36}$$

and

$$y = a_2 \cos(3\omega t + \theta) \tag{1.37}$$

From the above equation we have, $\frac{y}{a_2} = \cos(3\omega t + \theta)$

or, $$\frac{y}{a_2} = \cos 3\omega t \cos\theta - \sin 3\omega t \sin\theta$$

or, $$\frac{y}{a_2} = \left(4\cos^3\omega t - 3\cos\omega t\right)\cos\theta - \left(3\sin\omega t - 4\sin^3\omega t\right)\sin\theta$$

or, $$\frac{y}{a_2} = \left(4\frac{x^3}{a_1^3} - 3\frac{x}{a_1}\right)\cos\theta - \left\{3\left(1-\frac{x^2}{a_1^2}\right)^{\frac{1}{2}} - 4\left(1-\frac{x^2}{a_1^2}\right)^{\frac{3}{2}}\right\}\sin\theta$$

or, $$\frac{y}{a_2} = \left(4\frac{x^3}{a_1^3} - 3\frac{x}{a_1}\right)\cos\theta - \left(1-\frac{x^2}{a_1^2}\right)^{\frac{1}{2}}\left(4\frac{x^2}{a_1^2}-1\right)\sin\theta$$

Now, transposing and squaring, we get

$$\left\{\frac{y}{a_2} - \left(4\frac{x^3}{a_1^3} - 3\frac{x}{a_1}\right)\cos\theta\right\}^2 = \left(1-\frac{x^2}{a_1^2}\right)\left(4\frac{x^2}{a_1^2}-1\right)^2\sin^2\theta \tag{1.38}$$

The above equation gives the general equation of the resultant motion for any phase difference and amplitudes. For different values of θ different cases can be considered.

Case I: If $\theta = 0$, then Eq. (1.38) may be written as,

$$\frac{y}{a_2} = \frac{4x^3}{a_1^3} - \frac{3x}{a_1}$$

or,
$$y = \frac{a_2}{a_1^3} x(4x^2 - 3a_1^2) \tag{1.39}$$

Eq. (1.39) represents two coincident cubic curves as shown in Fig. 1.10.

Case II: If $\theta = \frac{\pi}{2}$, Eq. (1.38) may be written as

$$\left(1 - \frac{x^2}{a_1^2}\right)\left(4\frac{x^2}{a_1^2} - 1\right)^2 - \frac{y^2}{a_2^2} = 0 \tag{1.40}$$

Eq. (1.40) represents an equation of sixth degree. It gives a curve having three loops as shown in Fig. 1.11. If anyone alters the phase difference slowly, then the shape of the loop will also gradually alter. For a ratio of frequencies 1: n, the curve will have n loops

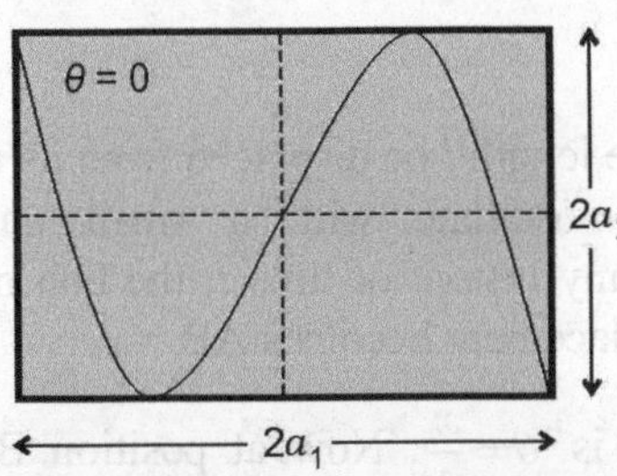

Fig. 1.10: Resultant of mutually perpendicular SHM with frequency ratio 1 : 3 and phase difference $\theta = 0$

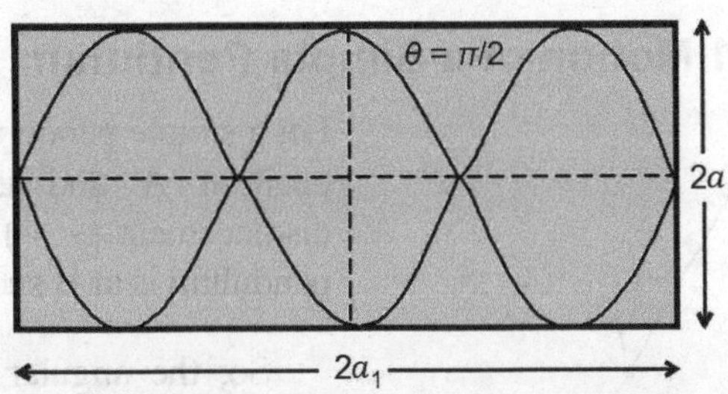

Fig. 1.11: Resultant of two mutually perpendicular SHM with frequency ratio 1 : 3 and phase difference $\theta = \pi/2$

1.11 LISSAJOUS FIGURES

These are the curves formed by the superposition of two simple harmonic motion acting at right angle to each other. The shape of the curves depends on frequencies, relative amplitudes and the phase difference between the two constituent waves. These curves are useful to determine the ratio of frequencies of two waves, or the unknown frequency of a tuning fork. The nature and shape of a wave or the amplitude can also be determined from the curves.

Production of Lissajous figures: Lissajous figures can be produced on the Cathode Ray Oscilloscope (CRO), which consists of a highly evacuated envelope called cathode ray tube (as shown in Fig. 1.12) containing

(i) An electron gun,

(ii) the horizontal (XX), and the vertical (YY) deflecting plates to deflect the narrow beam of high speed electrons produced by electron gun and

(iii) a fluorescent screen to focus the high speed electrons.

When at the vertical deflecting plates YY a voltage is applied, it deflects the spot vertically on the screen. Similarly the spot moves horizontally on the screen due to the application of external voltage across the horizontal deflecting plates XX.

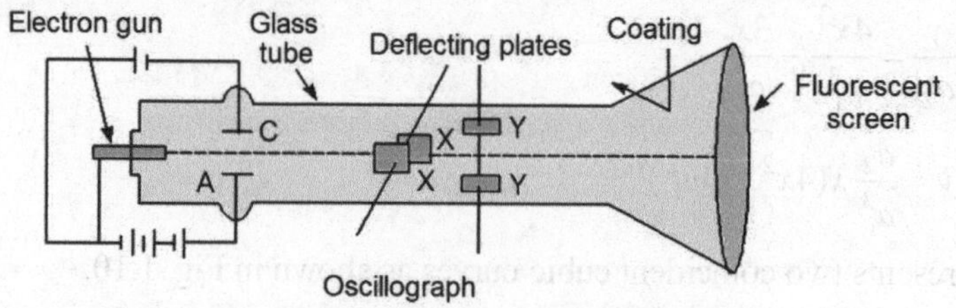

Fig. 1.12: Cathode Ray Oscilloscope

To obtain Lissajous figures on the CRO screen, the two shapes of the SHM are first converted into sinusoidal voltages with the help of microphones and then they are simultaneously applied to the horizontal and vertical deflecting plates of the CRO. Due to this, the spot of light on the screen moves under the simultaneous action of the sinusoidal voltages and hence traces out distinct visible pattern, known as Lissajous figures.

1.12 APPLICATIONS OF SHM

1.12.1 Motion of a Simple Pendulum

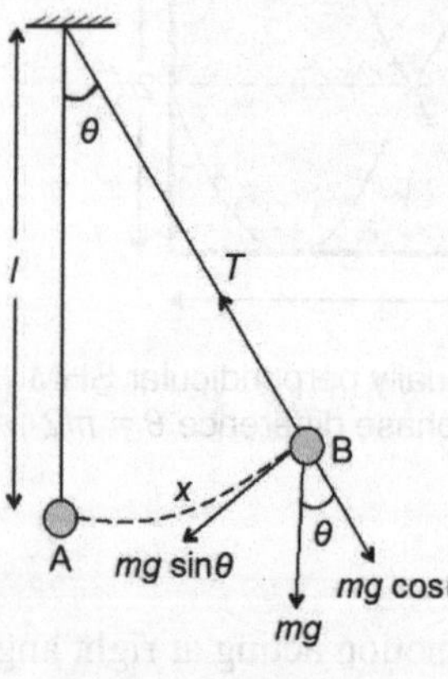

Fig. 1.13: Motion of a simple pendulum

Let a simple pendulum of effective length l be displaced from its mean position A and is allowed to oscillate with a small angular displacement ($\sim 4^0$). Also let, at any instant of time t, the bob of the pendulum is at B such that its displacement becomes AB = x.

So, the angular displacement is $\theta = \frac{x}{l}$. Now at position B, the weight mg of the bob, acting vertically downwards may be resolved into its vertical and horizontal components as,

(i) $mg\cos\theta$, which is acting along the length of the pendulum and it balances the tension T of the string and

(ii) $mg \sin\theta$, which is acting perpendicular to the length of the pendulum and is responsible for the motion of the pendulum towards its mean position.

So, the tension on the string is

$$T = mg \cos\theta \tag{1.41}$$

and the restoring force is

$$F = -mg \sin\theta \tag{1.42}$$

Again, $\sin\theta = \theta - \frac{\theta^3}{3!} + \frac{\theta^5}{5!} - \frac{\theta^7}{7!} + ...$ (using Taylor's series expansion)

Now, for small value of θ, $\sin\theta \rightarrow \theta$.

SCAN ME

for more on simple pendulum

So, Eq. (1.42) may be written as, $F = -mg\theta$.

or, $$\frac{F}{m} = -g\theta$$

or, $$f = -g\theta$$

or, $$\frac{d^2x}{dt^2} = -\frac{g}{l}x \tag{1.43}$$

Since the acceleration of the bob is proportional to its displacement and is directed opposite to that of the displacement, the motion of the simple pendulum is simple harmonic.

Now as, $$\frac{d^2x}{dt^2} + \frac{g}{l}x = 0 \tag{1.44}$$

Comparing Eq. (1.44) with the general differential equation for SHM given by $\frac{d^2x}{dt^2} + \omega^2 x = 0$ we get,

$$\omega^2 = \frac{g}{l}$$

or, $$\left(\frac{2\pi}{T}\right)^2 = \frac{g}{l}$$

$\therefore$ $$T = 2\pi\sqrt{\frac{l}{g}} \tag{1.45}$$

The time period of the simple pendulum can be represented by this equation.

1.12.2 Motion of a Horizontal Spring

It is a vibrating system with its vibration in a simple harmonic manner consisting a mass m attached to a spring of spring constant k as shown in Fig. 1.14.

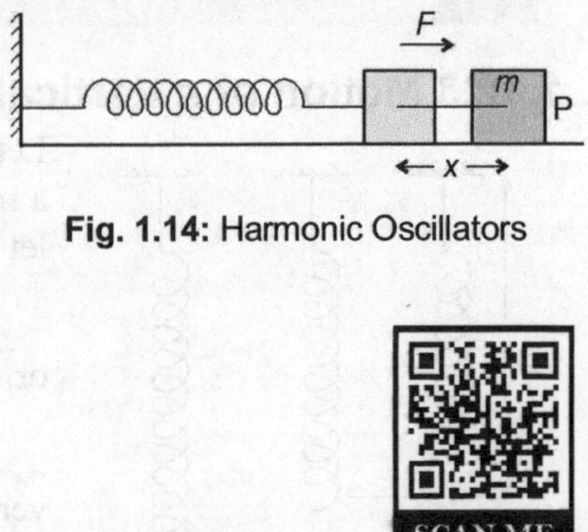

Fig. 1.14: Harmonic Oscillators

Let the mass m is displaced to point P by a distance x and then released. The restoring force developed will act in the negative direction of x and is given by

$$F \propto -x$$

or, $$F = -kx$$

where the force constant is represented by k.

SCAN ME for more on harmonic oscillator

So, $$m\frac{d^2x}{dt^2} = -kx \tag{1.46}$$

or, $$\frac{d^2x}{dt^2} = -\frac{k}{m}x \tag{1.47}$$

So, the motion of the body is simple harmonic. Comparing Eq. (1.47) with the general differential equation of SHM we get,

$$\omega^2 = \frac{k}{m}$$

or, $$\omega = \sqrt{\frac{k}{m}}$$

So, $$T = 2\pi\sqrt{\frac{k}{m}} \tag{1.48}$$

This gives the time period of the harmonic oscillator.

EXAMPLE 1.6

Example 1.6 If a spring is cut into two equal parts; find the spring constants of the individual parts.

Solution

Let F be the amount of force applied on a spring of length L and spring constant K. Now if x be the elongation of the spring under the application of the force F then, $F = Kx$.

$$\therefore \quad \frac{F}{x} = K$$

Now if the spring is divided into two equal parts then owing to the same force applied each part will be elongated by a distance $\frac{x}{2}$. Then for identical value of the applied force F, the force constants of the individual spring will be,

$$K_1 = \frac{\text{Force}}{\text{Elongation}} = \frac{F}{x/2} = \frac{2F}{x} = 2K$$

Thus the new the force constant of the individual spring is $2K$.

1.12.3 Motion of a Vertical Spring

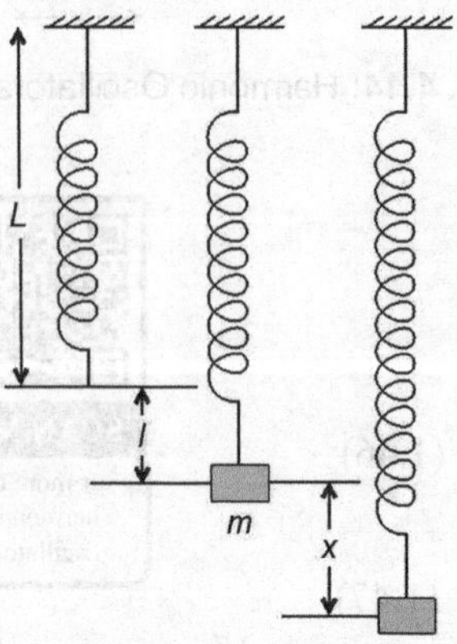

Fig. 1.15: Motion of a vertical spring

Let a spring of length L is suspended freely from a rigid support and a mass m is attached at its free end. Due to the weight of the mass m, let the length of the spring is increased by l.

$$\therefore \quad mg = kl$$

or, $$k = \frac{mg}{l} \tag{1.49}$$

Now, let the mass m is further stretched a little by applying a vertical force such that the displacement of the mass is x. The restoring force acting on the spring after releasing the mass will be,

$$F = -kx.$$

So, $$F = -\frac{mg}{l}x$$

or, $$f = -\frac{g}{l}x \qquad [\because F = mf]$$

or, $$\frac{d^2x}{dt^2} = -\frac{g}{l}x = -\frac{k}{m}x \qquad (1.50)$$

So, the motion of the vertical spring is simple harmonic. Comparing Eq. (1.50) with the general differential equation of SHM we get,

$$\omega^2 = \frac{k}{m}$$

or, $$\omega = \sqrt{\frac{k}{m}} \qquad (1.51)$$

∴ $$T = 2\pi\sqrt{\frac{k}{m}}$$

which gives the time period of the vertical spring.

Example 1.7 A body of mass 5 kg hangs from a spring. A 500 g body added to the mass stretches the spring 2 cm further. If now the mass is removed, then find the period of oscillation of the body.

Solution

We have, $$mg = kx$$

or, $$500 \times 980 = k \times 2$$

or, $$k = 24.5\,\text{N/m}.$$

Thus the period of oscillation of the body is,

$$T = 2\pi\sqrt{\frac{m}{k}} = 2\pi\sqrt{\frac{5}{24.5}} = 8.971\,\text{s}.$$

EXAMPLE 1.7

1.12.4 Motion of a Liquid in a U tube

Let us consider a U tube of uniform cross sectional area a which is containing a liquid of density ρ. If pressure is applied on the liquid column in one arm of the tube and then released, the liquid column inside the tube will oscillate in a simple harmonic manner. Let initial level of liquid column inside the tube is ab and at the arm A, the liquid column is depressed by x due to the applied force F. So, if the liquid is incompressible, the rise of liquid column at the other arm will be x and then the difference in liquid levels in the two arms will be $2x$. Due to the weight of this extra liquid column of height $2x$ the motion of the liquid inside the U tube takes place and under equilibrium the force equation is,

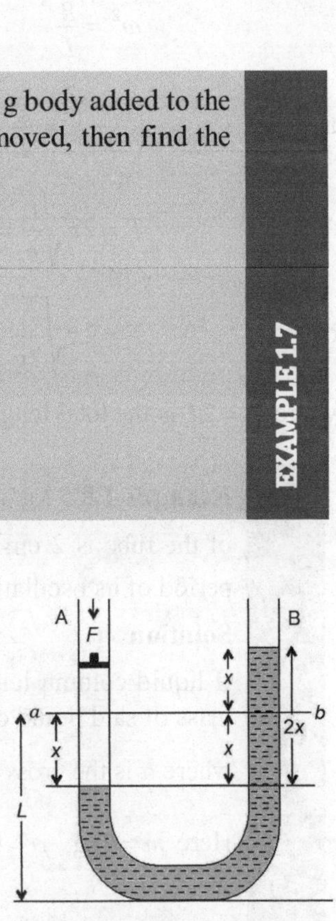

Fig. 1.16: Motion of a liquid in a U tube

$$F = -2xa\rho g$$

The negative sign indicates that for a rise of liquid column, the force tends to depress it. Now, if L be the height of the liquid columns at initial condition then, $F = 2La\rho\frac{d^2x}{dt^2}$.

or, $$2L\alpha\rho\frac{d^2x}{dt^2}=-2x\alpha\rho g$$

or, $$\frac{d^2x}{dt^2}=-\frac{g}{L}x \tag{1.52}$$

or, $$\frac{d^2x}{dt^2}+\frac{g}{L}x=0 \tag{1.53}$$

Eq. (1.52) shows that the motion of the liquid column inside the U tube is simple harmonic.

Comparing Eq. (1.53) with the general differential equation for SHM we get,

$$\omega^2=\frac{g}{L}$$

or, $$\left(\frac{2\pi}{T}\right)^2=\frac{g}{L}$$

$\therefore$ $$T=2\pi\sqrt{\frac{L}{g}}=2\pi\sqrt{\frac{2L}{2g}}$$

or, $$T=2\pi\sqrt{\frac{\Lambda}{2g}} \tag{1.54}$$

where $\Lambda=2L$ is the total length of liquid column inside the tube.

EXAMPLE 1.8

Example 1.8 5 kg of glycerin is poured into a U tube made by glass. The diameter of the tube is 2 cm and density of glycerin is $1.26\times10^3\,\text{kg/m}^3$. Calculate the time period of its oscillations in the tube.

Solution

If liquid column length is taken as L inside the tube and if ρ be the density then mass of said liquid column will be, $m=L\alpha\rho$

where α is the cross sectional area of the tube.

Here $m=5\text{kg}$, $\rho=1.26\times10^3\,\text{kg/m}^3$ and $\alpha=\pi\left(\frac{d}{2}\right)^2=\pi\left(\frac{2}{2}\right)^2\text{cm}^2=3.14\times10^{-4}\,\text{m}^2$.

$\therefore$ $$L=\frac{m}{\alpha\rho}=\frac{5}{1.26\times10^3\times3.14\times10^{-4}}\text{m.}$$

Thus the time period of oscillation is

$$T=2\pi\sqrt{\frac{L}{2g}}=2\pi\sqrt{\frac{\frac{5}{1.26\times10^3\times3.14\times10^{-4}}}{2\times9.8}}=5.04\text{s.}$$

1.12.5 Motion of a Floating Cylinder

Let us consider a cylindrical block of mass m, length L and radius r with density ρ. It is floating in a liquid of density σ with h depth under the liquid. Now, if this cylindrical block is further depressed by x then it will displace a mass $\pi r^2 \sigma x$ of the liquid and thus will experience an upward thrust of magnitude $\pi r^2 \sigma x g$ and the equation of motion of the cylinder will be,

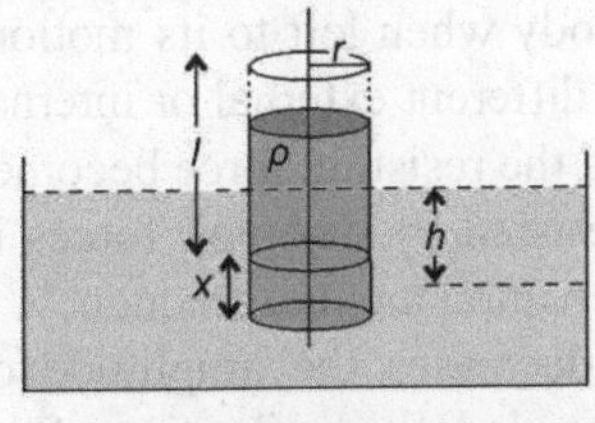

Fig. 1.17: Motion of a floating cylinder

$$m\frac{d^2x}{dt^2} = -\pi r^2 x\sigma g$$

or, $$\pi r^2 L\rho\frac{d^2x}{dt^2} = -\pi r^2 x\sigma g$$

or, $$\frac{d^2x}{dt^2} = -\frac{\sigma g}{L\rho}x \tag{1.55}$$

Eq. (1.55) shows that the motion of the floating cylinder is simple harmonic.

Comparing Eq. (1.55) with the general differential equation for SHM we get, $\omega^2 = \frac{\sigma g}{L\rho}$. Thus the time period of the motion is,

$$T = 2\pi\sqrt{\frac{\rho L}{\sigma g}} \tag{1.56}$$

1.12.6 LC Oscillation

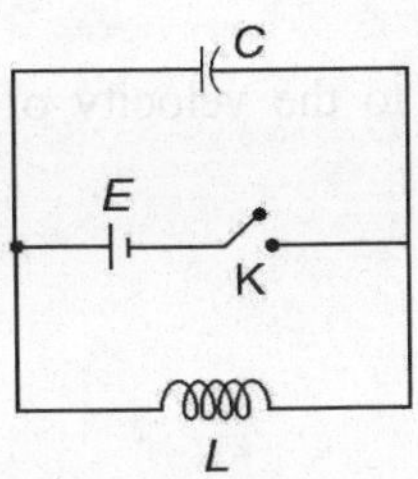

Fig. 1.18: LC oscillator

Consider LC oscillation in an LC circuit as shown in Fig. 1.18 consisting of an inductor L and a capacitor C in parallel. At equilibrium the capacitor is uncharged and no current will flow but when the key K is pressed, the capacitor is charged and the voltage drop across the capacitor and the inductor will be respectively $V_C = \frac{q}{C}$ and $V_L = -L\frac{di}{dt}$.

From KVL the net voltage in the circuit is zero and thus, $-L\frac{di}{dt} = \frac{q}{C}$.

or, $$L\frac{d^2q}{dt^2} + \frac{q}{C} = 0$$ [where $i = \frac{dq}{dt}$ is the current flowing in the circuit]

or, $$\frac{d^2q}{dt^2} + \frac{q}{LC} = 0$$

or, $$\frac{d^2q}{dt^2} + \omega^2 q = 0 \tag{1.57}$$

which shows that the circuit will act as a simple harmonic oscillator. Comparing Eq. (1.57) with the general differential equation for SHM we get $\omega^2 = \frac{1}{LC}$.

Thus the time period of motion is

$$T = 2\pi\sqrt{\frac{1}{LC}} \tag{1.58}$$

1.13 DAMPED VIBRATION

In practice it is found that the amplitude of vibration of any vibrating body when left to its motion, gradually diminishes and ultimately becomes zero. Since it is resisted by different external or internal forces or both, the medium withdraws energy from the vibrating body and the resisting force becomes proportional to the velocity of the particle. Such vibration of particle resisted by damping forces is called damped vibration. When a body performs steady vibration with its natural angular frequency, it experiences a number of dissipative forces (frictional forces) which decreases the amplitude of vibration of the body with time and ultimately the motion of the body is stopped. Such vibration of the body with decaying amplitude is referred as the damped vibration.

1.13.1 Differential Equation of Damped Vibration

Let x be the displacement of a particle of mass m executing damped SHM from its initial position of rest at any instant t.

The forces acting on the particle are:

(i) the force of inertia on the particle by virtue of its motion and is given by $m\frac{d^2x}{dt^2}$,

(ii) the force of restitution tending to bring the particle to its initial position of rest and is given by μx, μ being the restoring force per unit displacement, and

(iii) the resisting or dissipative force due the damping and is proportional to the velocity of the particle given by $b\frac{dx}{dt}$.

Thus at equilibrium the force equation of the particle will be,

$$m\frac{d^2x}{dt^2} = -\mu x - b\frac{dx}{dt}$$

or, $$m\frac{d^2x}{dt^2} + b\frac{dx}{dt} + \mu x = 0$$

or, $$\frac{d^2x}{dt^2} + \frac{b}{m}\frac{dx}{dt} + \frac{\mu}{m}x = 0$$

or, $$\frac{d^2x}{dt^2} + 2k\frac{dx}{dt} + \omega^2 x = 0 \tag{1.59}$$

where, $2k = \frac{b}{m}$ and $\omega^2 = \frac{\mu}{m}$.

Eq. (1.59) is the general differential equation of motion of a particle executing damped vibration.

Let the general solution of Eq. (1.59) be, $x = ae^{\alpha t}$.

Differentiating twice with respect to t, we get $\dfrac{dx}{dt} = a\alpha e^{\alpha t}$ and $\dfrac{d^2x}{dt^2} = a\alpha^2 e^{\alpha t}$.

Thus from Eq. (1.59) we have,

$$a\alpha^2 e^{\alpha t} + 2ka\alpha e^{\alpha t} + a\omega^2 e^{\alpha t} = 0$$

or, $$ae^{\alpha t}\left(\alpha^2 + 2k\alpha + \omega^2\right) = 0$$

But $$ae^{\alpha t} \neq 0$$

Thus, $$\alpha^2 + 2k\alpha + \omega^2 = 0$$

or, $$\alpha = \frac{-2k \pm \sqrt{4k^2 - 4\omega^2}}{2}$$

$$= -k \pm \sqrt{k^2 - \omega^2}$$

Thus, the general solution of Eq. (1.59) may be written as,

$$x = A_1 e^{\left(-k+\sqrt{k^2-\omega^2 t}\right)} + A_2 e^{\left(-k-\sqrt{k^2-\omega^2)t}\right)}$$

$$= e^{-kt}\left[A_1 e^{\sqrt{k^2-\omega^2} t} + A_2 e^{-\sqrt{k^2-\omega^2} t}\right] \tag{1.60}$$

where A_1 and A_2 are two constants to be evaluated later. Depending on the relative values of k and ω there may be three cases –

Case 1 - Under-damped motion with light damping: When $k < \omega$ *i.e.,* the damping force is small $\sqrt{k^2 - \omega^2}$ is imaginary and is given by, $\sqrt{k^2 - \omega^2} = i\sqrt{\omega^2 - k^2}$.

Hence Eq. (1.60) reduces to –

$$x = e^{-kt}\left[A_1 e^{i\sqrt{\omega^2-k^2} t} + A_2 e^{-i\sqrt{\omega^2-k^2} t}\right]$$

or, $$x = e^{-kt}\left[A_1\left(\cos\sqrt{\omega^2 - k^2}t + i\sin\sqrt{\omega^2 - k^2}t\right) + A_2\left(\cos\sqrt{\omega^2 - k^2}t - i\sin\sqrt{\omega^2 - k^2}t\right)\right]$$

or, $$x = e^{-kt}\left[(A_1 + A_2)\cos\sqrt{\omega^2 - k^2}t + i(A_1 - A_2)\sin\sqrt{\omega^2 - k^2}t\right]$$

Let, $(A_1 + A_2) = P\sin\theta$ and $i(A_1 - A_2) = P\cos\theta$

$\therefore$ $$x = e^{-kt}\left[P\cos\sqrt{\omega^2 - k^2}t\sin\theta + P\sin\sqrt{\omega^2 - k^2}t\cos\theta\right]$$

$$= Pe^{-kt}\sin\left(\sqrt{\omega^2 - k^2}t + \theta\right) \tag{1.61}$$

This represents a *damped oscillatory* motion.

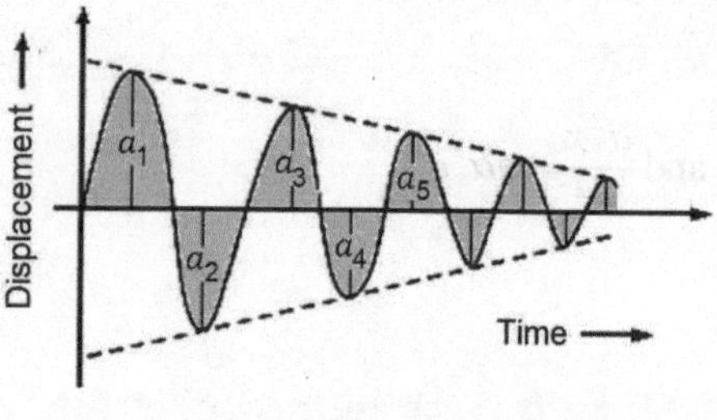

Fig. 1.19: Graphical illustration of damped oscillatory motion

The graphical representation is shown in Fig. 1.19. The amplitude of the motion is Pe^{-kt} which decreases exponentially at a rate determined by the decay constant k and the time period of the motion is,

$$T = \frac{2\pi}{\sqrt{\omega^2 - k^2}} \tag{1.62}$$

From Eq. (1.61) it is clear that the presence of the damping force has the following two effects—

(a) The amplitude is no longer constant but dies away according to the factor e^{-kt}.

(b) The frequency reduces slightly below the free natural frequency and is given by $\frac{\sqrt{\omega^2 - 2k^2}}{2\pi}$, instead of $\frac{\omega}{2\pi}$.

e.g., Motion of a pendulum in air, electric oscillations of LCR circuits *i.e.*, discharge of a charged capacitor through an inductive coil of low inductance, oscillations of coils in ballistic galvanometer *etc.*

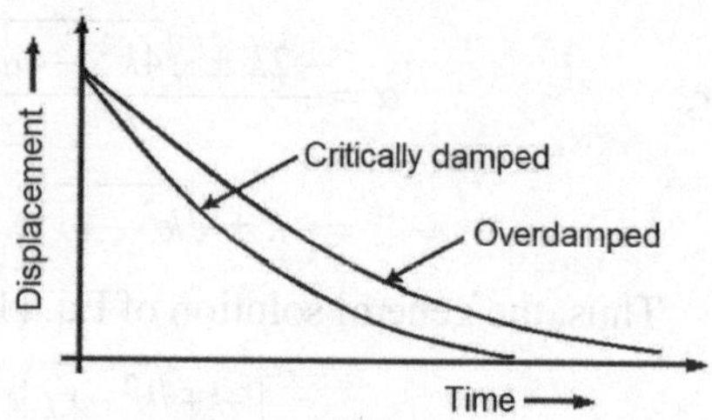

Fig. 1.20: Illustration of critically damped and overdamped motion

EXAMPLE 1.9

Example 1.9 The equation for displacement of a point of a damped harmonic oscillator is, $x = 5e^{-0.25t}\sin\left(\frac{\pi}{2}\right)t$ m. Find the velocity of the oscillating point at $t = T/4$, where T is the time period of the oscillator.

Solution

Displacement of the damped harmonic oscillator is, $x = 5e^{-0.25t}\sin\left(\frac{\pi}{2}\right)t$ m (a)

Comparing this with, $x = Pe^{-kt}\sin\omega t$, we get $\omega = \pi/2$.

Thus, the period is $T = \frac{2\pi}{\omega} = \frac{2\pi}{\pi/2} = 4$ s.

Now differentiating Eq. (a) with respect to t, we get,

$$\frac{dx}{dt} = 5\times(-0.25)e^{-0.25t}\sin\left(\frac{\pi}{2}\right) + 5\frac{\pi}{2}e^{-0.25t}\cos\left(\frac{\pi}{2}\right)$$

At, $t = \frac{T}{4} = \frac{4}{4} = 1$s,

velocity, $v = -1.25e^{-0.25t}\sin\frac{\pi}{2} + 5\left(\frac{\pi}{2}\right)e^{-0.25t}\cos\frac{\pi}{2} = -0.974$ m/s.

So, the velocity is in the opposite direction.

Case 2 - Over-damped motion with heavy damping (dead beat motion): When $k > \omega$ *i.e.*, the damping force is large, then the motion will be non-oscillatory. In this case $\sqrt{k^2 - \omega^2} < k$ and so from Eq. (1.60) we get the displacement x of the body which gradually decreases exponentially with time and it comes to its equilibrium position at $t \to \infty$ without performing any oscillation. This type of motion is called *over-damped* or *aperiodic* or *dead-beat motion*.

The motion is illustrated in Fig. 1.20.

e.g., The motion of pendulum in a viscous liquid or discharges of capacitor through a resistor.

Case 3 - Critical damping: When $k = \omega$ then we get from Eq. (1.60),

$$x = e^{-kt}\left[A_1 e^{\sqrt{k^2-\omega^2}t} + A_2 e^{-\sqrt{k^2-\omega^2}t}\right] = e^{-kt}\left[A_1\left(1+\sqrt{k^2-\omega^2}\,t\right)+A_2\left(1-\sqrt{k^2-\omega^2}\,t\right)\right]$$

$$= e^{-\omega t}\left[\left(A_1 + A_2\right)+\left(A_1 - A_2\right)\sqrt{k^2-\omega^2}\,t\right]$$

$$\therefore \qquad x = \left(A + Bt\right)e^{-\omega t} \qquad (1.63)$$

where $A = A_1 + A_2$ and $B = \left(A_1 - A_2\right)\sqrt{k^2 - \omega^2}$.

This type of motion is known as *critically damped motion*. It is also illustrated in Fig. 1.20. In this case the damped oscillatory motion suddenly changes into a dead beat motion but the particle tends to move to equilibrium much more rapidly than the over-damped motion. It is also a non-oscillatory motion.

EXAMPLE 1.10

Example 1.10 A body of mass 10 g is acting due to a restoring force (per unit displacement) of 10 dyne cm^{-1} and resisting force (per unit velocity) of 2 dyne cm^{-1} s.

(i) Find whether the motion is aperiodic or oscillatory.

(ii) Determine the resisting force for which the motion will be critically damped.

(iii) Determine the mass for which the applied force will produce the motion critically damped.

Solution

Here, $\mu = \dfrac{\text{Resisting force}}{\text{Displacement}} = 10$ dyne cm^{-1}, $b = \dfrac{\text{Frictional force}}{\text{Velocity}} = 2$ dyne cm^{-1} s and $m = 10$ g.

(i) So, $\quad k = \dfrac{b}{2m} = \dfrac{2}{20} = 0.1\ \text{s}^{-1}$ and $\omega = \sqrt{\dfrac{\mu}{m}} = \sqrt{\dfrac{10}{10}} = 1\ \text{s}^{-1}$

Since, $\omega > k$ therefore the motion is oscillatory.

(ii) For critically damped motion $\sqrt{\dfrac{\mu}{m}} = \dfrac{b}{2m}$

EXAMPLE 1.10

or, $b=\sqrt{4\mu m}$

or, $b=\sqrt{4\times 10\times 10}=20$ dyne cm^{-1}s.

(iii) We have $b=\sqrt{4\mu m}$

Thus, $m=\dfrac{b^2}{4\mu}=\dfrac{20^2}{4\times 10}=10$ g.

EXAMPLE 1.11

Example 1.11 A particle of mass 0.2 kg is acted on by a restoring force per unit displacement 10×10^{-3} N/m and a frictional force per unit velocity 2×10^{-3} N m^{-1}s. Find

(i) whether the motion is oscillatory or not,

(ii) the value of resistive force for which the damping will be critical, and

(iii) if the particle is displaced through 2 cm then find its period.

Solution

Here, $\mu=\dfrac{\text{Resisting force}}{\text{Displacement}}=10\times 10^{-3}$ N/m,

$b=\dfrac{\text{Frictional force}}{\text{Velocity}}=2\times 10^{-3}$ N m^{-1}s,

and $m=0.2$ kg.

(i) So, $k=\dfrac{b}{2m}=\dfrac{2\times 10^{-3}}{2\times 0.2}=0.005\ \text{s}^{-1}$,

and $\omega=\sqrt{\dfrac{\mu}{m}}=\sqrt{\dfrac{10\times 10^{-3}}{0.2}}=0.224\ \text{s}^{-1}$.

As, $k<\omega$ the motion is oscillatory.

(ii) For the motion to be critically damped,

$$\sqrt{\frac{\mu}{m}}=\frac{b}{2m}$$

or, $b=\sqrt{4\mu m}$

or, $b=\sqrt{4\times 0.2\times 10\times 10^{-3}}=8.94\times 10^{-2}$ N m^{-1}s.

(iii) Time period is given by

$$T=\frac{2\pi}{\sqrt{(0.22)^2-(0.005)^2}}=28.1\text{s}.$$

1.14 ATTENUATION COEFFICIENTS

In damped oscillatory motion the displacement of the body is given by Eq. (1.61) as,

$$x = Pe^{-kt} \sin\left(\sqrt{\omega^2 - k^2}\, t + \theta\right)$$

The amplitude of vibration in time t is Pe^{-kt} and thus the energy will decay as the square of amplitude *i.e.*, $\left(e^{-kt}\right)^2 = e^{-2kt}$ with time.

The parameters giving the attenuation of energy of the vibrating system are –

(i) *Logarithmic decrement,*

(ii) *Relaxation time* and

(iii) *Quality factor.*

1.14.1 Logarithmic Decrement

It measures the rate when the amplitude of the damped oscillatory system falls off or dies away.

In Fig. 1.19; let $a_1, a_2, a_3, ...$ be the amplitudes of the vibrating system at $\frac{T}{4}, \frac{3T}{4}, \frac{5T}{4}$ respectively, where T is the period of the motion.

Now, the general equation of the damped oscillatory motion of a body is,

$$x = Pe^{-kt} \sin\left(\sqrt{\omega^2 - k^2}\, t + \theta\right)$$

Thus, $a_1 = Pe^{-kT/4}$, $a_2 = Pe^{-3kT/4}$ and $a_3 = Pe^{-5kT/4}$.

So, $$\frac{a_1}{a_2} = \frac{a_2}{a_3} = ... = \frac{a_{n-1}}{a_n} = e^{\frac{kT}{2}} = d = \text{constant}$$

where, d denotes the decrement in amplitude. The *log* of this decrement is termed as logarithmic decrement λ and is given by,

$$\lambda = \ln d = \ln e^{\frac{kT}{2}} = \frac{kT}{2} \tag{1.64}$$

Logarithmic decrement basically determines the natural logarithm of the ratio of amplitudes of two consecutive oscillations.

1.14.2 Relaxation Time

Let us now determine the rapidity with which the motion is damped by the resistive forces.

The energy of a particle of mass m for a damped oscillatory vibration is given by,

$$E = \text{Kinetic energy} + \text{Potential energy}$$

$$= E_K + E_p = \frac{1}{2}mv^2 + \frac{1}{2}\mu x^2$$

Now, $x = Pe^{-kt} \sin\left(\sqrt{\omega^2 - k^2}\,t + \theta\right)$

Thus velocity of the body,

$$v = \frac{dx}{dt} = Pe^{-kt}\sqrt{\omega^2 - k^2}\cos\left(\sqrt{\omega^2 - k^2}\,t + \theta\right) - kPe^{-kt}\sin\left(\sqrt{\omega^2 - k^2}\,t + \theta\right) \quad (1.65)$$

Now if the body is excited by giving an initial velocity v_0 at $t = 0$ and if we choose $\theta = 0$ then from Eq. (1.65) we get,

$$v_0 = P\sqrt{\omega^2 - k^2}$$

or, $$P = \frac{v_0}{\sqrt{\omega^2 - k^2}} = \frac{v_0}{\omega'} \quad \text{(where, } \omega' = \sqrt{\omega^2 - k^2}\text{)} \quad (1.66)$$

Thus, $$x = \frac{v_0}{\omega'} e^{-kt} \sin\omega' t \quad (1.67)$$

and $$v = \frac{v_0}{\omega'} e^{-kt}\omega' \cos\omega' t - \frac{v_0}{\omega'} e^{-kt} k \sin\omega' t$$

If the damping is small *i.e.*, $k \to 0$, then

$$v = v_0 e^{-kt} \cos\omega' t \quad (1.68)$$

Thus the energy of the vibrating body will be,

$$E = \frac{1}{2}mv^2 + \frac{1}{2}\mu x^2 = \frac{1}{2}m\left(v_0 e^{-kt}\cos\omega' t\right)^2 + \frac{1}{2}\mu\left(\frac{v_0}{\omega'}e^{-kt}\sin\omega' t\right)^2 \quad (1.69)$$

Now, for low damping, $k \to 0$ and $\sqrt{\omega^2 - k^2} = \omega' \approx \omega$.

Thus, $$E = \frac{1}{2}m\left(v_0 e^{-kt}\cos\omega t\right)^2 + \frac{1}{2}\mu\omega^2\left(\frac{v_0}{\omega}e^{-kt}\cos\omega t\right)^2 = \frac{1}{2}mv_0^2 e^{-2kt} \quad [\text{where } \mu = m\omega^2]$$

or, $$E = E_0 e^{-2kt} \quad (1.70)$$

where $E_0 = \frac{1}{2}mv_0^2$ is the initial energy of the system.

At $t = \frac{1}{2k} = \tau$, $E = \frac{E_0}{e} = 0.37E_0$

where τ is called the relaxation time and is defined as the time taken by the system to fall the energy of the damped oscillatory motion to $\frac{1}{e}$ (about 37%) of its initial energy E_0.

Hence energy of the vibrating body at any instant is given by

$$E = E_0 e^{-\frac{t}{\tau}} \quad (1.71)$$

Example 1.12 If the time of relaxation of a damped harmonic oscillator be 50s, obtain the time in which (i) amplitude falls to $\frac{1}{e}$ times of initial value, and (ii) energy falls to $\frac{1}{e^4}$ times of initial value.

Solution

We have $x = Pe^{-kt}\sin\left(\sqrt{\omega^2 - k^2}\,t + \theta\right)$

So, amplitude $A = Pe^{-kt}$. At $t = 0$, $A = A_0$.

Let at $t = t_1$, $A = \frac{A_0}{e} = Pe^{-kt_1}$

Hence, $t_1 = \frac{1}{k} = 2\tau = 2 \times 50 = 100\,\text{s}.$

Again, the energy falls to $\frac{1}{e}$ times its initial value in the relaxation time.

Energy falls off as, $E = E_0 e^{-2kt}$.

Let at $t = t_2$, $E = \frac{E_0}{e^4} = e^{-2kt_2}$

Hence, $t_2 = \frac{2}{k} = 4\tau = 4 \times 50 = 200\,\text{s}.$

Thus the energy falls to $\frac{1}{e^4}$ times its initial value in four relaxation times.

EXAMPLE 1.12

1.14.3 Quality Factor

It is used to differentiate between light and heavy damping. The ratio of the maximum value of restoring force to that of the damping force is called quality factor. So, for less value of damping, quality factor will be higher and thus the quality will be better. Now the damping force is $b\frac{dx}{dt}$ and its maximum value is,

$$b\left(\frac{dx}{dt}\right)_{max} = b\frac{d}{dt}\left(\frac{v_0}{\omega}e^{-kt}\sin\omega' t\right)_{max} \quad \text{[using Eq. (1.67)]}$$

$$= b\frac{v_0}{\omega'}\omega' e^{-kt} = bP\omega' e^{-kt} \quad \text{[using Eq. (1.68)]}$$

$$= bA\omega, \text{ (where } A = Pe^{-kt} \text{ and for low damping } \omega' \approx \omega) \tag{1.72}$$

and the maximum value of restoring force is $= \mu A = m\omega^2 A$ (1.73)

Thus, quality factor is given by,

$$Q = \frac{\text{maximum value of restoring force}}{\text{maximum value of damping force}} = \frac{m\omega^2 A}{bA\omega} = \frac{m\omega^2 A}{2kmA\omega} = \frac{\omega}{2k}$$

or, $$Q = \omega\tau \quad [\text{where } \tau = \frac{1}{2k} \text{ is the relaxation time}] \qquad (1.74)$$

So, quality factor = natural angular frequency × relaxation time.

1.15 DISSIPATION OF POWER IN DAMPED VIBRATION

The energy of a particle of mass m for damped oscillatory motion is given from Eq. (1.70) as,

$$E = E_o e^{-2kt}$$

So, after one full time period T, the energy is,

$$E_1 = E_o e^{-2k(t+T)} = E_o e^{-2kt} e^{-2kT} = E e^{-2kT}$$

So, loss of energy in one time period T is,

$$E - E_1 = E - E e^{-2kT} = E(1 - e^{-2kT})$$

or, $$E - E_1 = E[1 - (1 - 2kT + 4k^2T^2/2 + \ldots\ldots)] = E - E(1 - 2kT) = 2EkT \qquad (1.75)$$

[As k is small, its higher order terms are neglected]

So the rate of loss of energy is $= \frac{2EkT}{T} = 2Ek = \frac{bE}{m}$ for a complete period.

1.16 EXAMPLE OF DAMPED VIBRATION

1.16.1 LCR Circuit

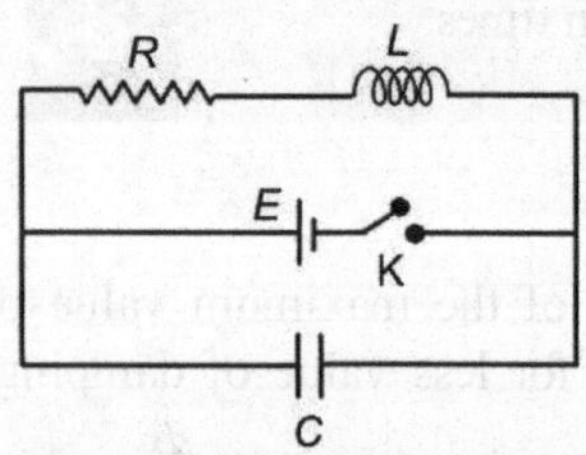

Fig. 1.21: Damped vibration in LCR circuit

In Fig. 1.21, let the current flowing through the circuit is i. At any instant of time t, the voltage noted across the capacitor is $V_C = \frac{q}{C}$ and the *emf* induced across the inductance is $\frac{di}{dt}$.

Now applying KVL to the circuit we get, $\frac{q}{C} = -L\frac{di}{dt} - Ri$

or, $$\frac{q}{C} + L\frac{di}{dt} + Ri = 0$$

or, $$\frac{di}{dt} + \frac{R}{L}i + \frac{q}{LC} = 0$$

or, $$\frac{d^2q}{dt^2} + \frac{R}{L}\frac{dq}{dt} + \frac{q}{LC} = 0 \qquad [\because i = \frac{dq}{dt}]$$

or, $$\frac{d^2q}{dt^2} + 2k\frac{dq}{dt} + \omega^2 q = 0 \qquad [\text{where, } 2k = \frac{R}{L} \text{ and } \omega^2 = \frac{1}{LC}]$$

and thus the above equation defines a damped oscillatory motion. The angular frequency of the circuit is given by,

$$\varpi = \sqrt{\omega^2 - k^2} = \sqrt{\frac{1}{LC} - \frac{R^2}{4L^2}} \tag{1.76}$$

and hence the time period is given by,

$$T = \frac{2\pi}{\varpi} = \frac{2\pi}{\sqrt{\frac{1}{LC} - \frac{R^2}{4L^2}}} \tag{1.77}$$

1.17 CONSERVATION OF ENERGY IN DAMPED VIBRATION

We know the equation of motion of a particle executing damped vibration is,

$$m\frac{d^2x}{dt^2} + b\frac{dx}{dt} + \mu x = 0$$

or, $$m\frac{d^2x}{dt^2} + \mu x = -b\frac{dx}{dt} = F_{res} = \text{resistive force}$$

Now, the total energy of the particle is

$$E = E_K + E_p = \frac{1}{2}mv^2 + \frac{1}{2}\mu x^2$$

$\therefore$ $$E = \frac{1}{2}m\left(\frac{dx}{dt}\right)^2 + \frac{1}{2}\mu x^2$$

Now $$\frac{dE}{dt} = \frac{m}{2}2\left(\frac{dx}{dt}\right)\frac{d^2x}{dt^2} + \frac{\mu}{2}2x\frac{dx}{dt} = \left(m\frac{d^2x}{dt^2} + \mu x\right)\frac{dx}{dt} = \left(-b\frac{dx}{dt}\right)\frac{dx}{dt}$$

$\therefore$ $$\frac{dE}{dt} = F_{res}\frac{dx}{dt} = \frac{d}{dt}(F_{res}\,.\,x) = \frac{dW}{dt}$$

where $W = F_{res}\,x$ is the work done against resistive force.

So, $$\frac{dE}{dt} = \frac{dW}{dt} \tag{1.78}$$

i.e., the rate of loss of energy of the damped vibration of the body is equal to the rate of work done against the resisting force leading to the conservation of energy.

1.18 FORCED VIBRATION

When one object forces another interconnected object producing vibrational motion, it is termed as forced vibration. Musical instruments and other objects are set into vibration at their natural frequency by hitting, striking, strumming, plucking or somehow disturbing the object. For instance, a guitar string is strummed or plucked; the arms of a tuning fork are knocked with a rubber mallet; whatever the case, a person or a thing puts energy into the instrument by direct contact with it. This energy input disturbs the particles and as a consequence it forces the object to vibrational motion at natural frequency. If a guitar string is stretched to a given length and plucked, one can hear a noise; but the noise would not even be close in comparison to the loudness produced by an acoustic guitar. On the contrary, when the string is connected to sound box of a guitar, the vibrating string forces the sound box into a vibration with the same natural frequency. As a result, the sound box forces air particles within the box into vibrational motion, same as the natural frequency of the string. The whole system which includes string, guitar and enclosed air starts to vibrate. This forces surrounding air particles into the vibrational motion. For the guitar, its string is mounted with the sound box. The surface area of a sound box is more than the surface area of the string which means that greater amount of surrounding air particles will be forced into the vibration. This causes an increase in the amplitude and thus loudness of the sound. In forced vibration there is an external force which acts on the system. This external force which acts on the system executes the vibration of the system.

The force applied externally may be non-harmonic and periodic, harmonic and periodic or non-periodic. The amplitude of oscillation of a body undergoing damped vibration goes on decreasing with time due to loss of energy to overcome the resistive or dissipative forces. When an external periodic force is applied to the system to make up for the losses, the amplitude of vibration does not decay with time and the body vibrates regularly under the action of the periodic force. This vibration of the body under the action of the periodic forces is termed as the forced vibration.

1.18.1 Differential Equation of Forced Vibration

Let us apply an external periodic force $F\cos pt$, where F is the amplitude and p is the angular frequency, to a body of mass m. We assume that the displacement x of a particle which is executing forced vibration from its initial position of rest at any instant t. The forces acting on the particle are,

(i) Force of inertia $m\dfrac{d^2x}{dt^2}$ due to its motion.

(ii) Force of restitution tending to bring the particle to its initial position of rest and is given by μx, μ being the restoring force per unit displacement.

(iii) Damping force which is directly proportional to the velocity of the particle and is given by $b\dfrac{dx}{dt}$, where b is the resisting force per unit velocity.

(iv) The external periodic force $F\cos pt$.

Hence, the equation of motion of the particle is given by, $m\dfrac{d^2x}{dt^2} = -\mu x - b\dfrac{dx}{dt} + F\cos pt$

or, $$m\frac{d^2x}{dt^2} + \mu x + b\frac{dx}{dt} = F\cos pt$$

or, $$\frac{d^2x}{dt^2} + \frac{b}{m}\frac{dx}{dt} + \frac{\mu}{m}x = \frac{F}{m}\cos pt$$

or, $$\frac{d^2x}{dt^2} + 2k\frac{dx}{dt} + \omega^2 x = f\cos pt \tag{1.79}$$

where, $2k = \frac{b}{m}$, $\omega^2 = \frac{\mu}{m}$ and $f = \frac{F}{m}$.

Eq. (1.79) represents the differential equation of motion of a particle of mass m undergoing forced vibration under the action of a periodic force $F\cos pt$.

1.18.2 Solution of Forced Vibration Equation

In case of forced vibration, initially the system will try to vibrate with its natural period but ultimately it will vibrate with the period and frequency of the external periodic force and hence the solution of Eq. (1.79) may be written as,

$$x = A\cos(pt - \alpha) \tag{1.80}$$

If we differentiate both sides of Eq. (1.80) twice with respect to t we get,

$$\frac{dx}{dt} = -Ap\sin(pt - \alpha)$$

and

$$\frac{d^2x}{dt^2} = -Ap^2\cos(pt - \alpha)$$

Using this in Eq. (1.79) we get,

$$-Ap^2\cos(pt-\alpha) - 2kAp\sin(pt-\alpha) + \omega^2 A\cos(pt-\alpha) = f\cos pt$$

$$= f\cos(pt - \alpha + \alpha) = f\cos(pt-\alpha)\cos\alpha - f\sin(pt-\alpha)\sin\alpha$$

or,

$$A(\omega^2 - p^2)\cos(pt-\alpha) - 2kAp\sin(pt-\alpha)$$

$$= f\cos(pt-\alpha)\cos\alpha - f\sin(pt-\alpha)\sin\alpha$$

As this is true for all values of t, the coefficients of $\cos(pt-\alpha)$ and $\sin(pt-\alpha)$ on both sides must be separately equal. Hence,

$$A(\omega^2 - p^2) = f\cos\alpha \tag{1.81}$$

and

$$2kAp = f\sin\alpha \tag{1.82}$$

Squaring and adding Eq. (1.81) and Eq. (1.82) we get, $f^2 = A^2(\omega^2 - p^2)^2 + 4A^2k^2p^2$

or,

$$A^2 = \frac{f^2}{(\omega^2 - p^2)^2 + 4k^2p^2}$$

or,

$$A = \frac{f}{\sqrt{(\omega^2 - p^2)^2 + 4k^2p^2}} \tag{1.83}$$

Again dividing Eq. (1.82) by Eq. (1.81) we get, $\tan\alpha = \frac{2kp}{(\omega^2 - p^2)}$.

or, $$\alpha = \tan^{-1} \frac{2kp}{\left(\omega^2 - p^2\right)} \tag{1.84}$$

Thus solution for Eq. (1.79) may be written as,

$$x = \frac{f}{\sqrt{\left(\omega^2 - p^2\right)^2 + 4k^2 p^2}} \cos\left(pt - \tan^{-1} \frac{2kp}{\left(\omega^2 - p^2\right)}\right) \tag{1.85}$$

This is the particular solution for SHM having the same frequency as that of the driving force, but lagging behind by an angle α. This is true only for steady state. But, at the beginning as the force sets the particle into vibration, the particle will vibrate with its natural frequency. The solution for the displacement is complementary solution obtained by solving equation for damped oscillation,

$$\frac{d^2 x}{dt^2} + 2k\frac{dx}{dt} + \omega^2 x = 0$$

which have the solution, $x = Pe^{-kt} \sin(\sqrt{\omega^2 - k^2}\, t + \theta)$.

Hence the general solution is given by adding both the early stage and the steady stage solutions as,

$$x = \frac{f}{\sqrt{\left(\omega^2 - p^2\right)^2 + 4k^2 p^2}} \cos\left(pt - \tan^{-1} \frac{2kp}{\left(\omega^2 - p^2\right)}\right) + Pe^{-kt} \sin(\sqrt{\omega^2 - k^2}\, t + \theta) \tag{1.86}$$

1.19 DISCUSSION FOR THE STEADY STATE

In a damped system, some energy is dissipated at each cycle of vibration and must be replaced from an external source if a steady vibration is to be maintained and such vibration is commonly referred as the forced vibratory motion. The displacement amplitude and velocity amplitude of a body undergoing forced vibration depend on the frequency of the applied external periodic force. When the displacement amplitude of a particle undergoing forced vibration is highest for a particular frequency (resonance frequency) of the applied periodic force, the phenomena is called *amplitude resonance.*

1.19.1 Amplitude Resonance

In steady state the amplitude of vibration is obtained from Eq. (1.83) as,

$$A = \frac{f}{\sqrt{\left(\omega^2 - p^2\right)^2 + 4k^2 p^2}}$$

When the amplitude becomes maximum, the amplitude resonance takes place. But A will be greatest when the denominator $\sqrt{\left(\omega^2 - p^2\right)^2 + 4k^2 p^2}$ will be lowest, because f is fixed. In this case, ω and k are constants and the only variable is p.

So, if the denominator is minimum, its differential coefficient with respect to p must be zero.

Hence, $$\frac{d}{dp}\left[\left(\omega^2-p^2\right)^2+4k^2p^2\right]=0$$

or, $$-4p\left(\omega^2-p^2\right)+8k^2p=0$$

or, $$p=\sqrt{\omega^2-2k^2} \qquad (1.87)$$

Thus, when the frequency of the driving force is $\frac{1}{2\pi}\sqrt{\omega^2-2k^2}$, then amplitude resonance takes place and the maximum amplitude is given by,

$$A_{max}=\frac{f}{2k\sqrt{\omega^2-k^2}} \qquad (1.88)$$

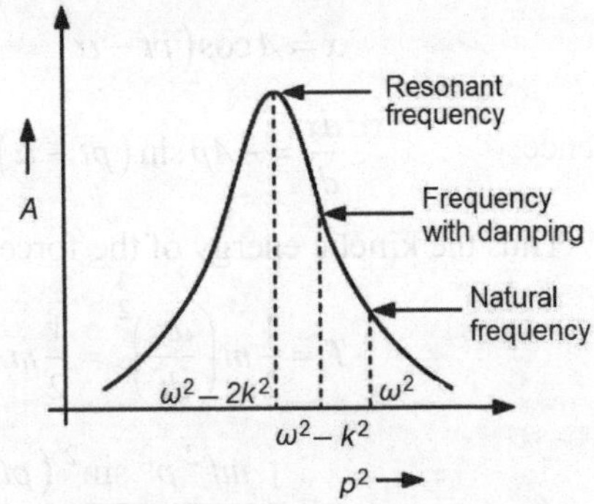

Fig. 1.22: Occurrence of amplitude resonance at resonant frequency

Hence, the amplitude is greater when the value of the damping factor k is lower. This is illustrated in Fig. 1.22.

Example 1.13 A body of mass 10 g is acted upon by a restoring force per unit displacement of 10^7 dyne cm^{-1}, a frictional force per unit velocity 4×10^3 dyne cm^{-1} s and a driving force of $10^5 \cos pt$ dyne. Find the value of maximum amplitude.

Solution

The amplitude of a particle executing forced vibration is given by,

$$A=\frac{f}{\sqrt{\left(\omega^2-p^2\right)^2+(2kp)^2}}=\frac{F/m}{\sqrt{\left(\omega^2-p^2\right)^2+(2kp)^2}}$$

When, $$p=\omega, A=A_{max}=\frac{F/m}{\left(\sqrt{2kp}\right)^2}=\frac{F}{2kmp}$$

Here, $$k=\frac{b}{2m}=\frac{4\times10^3}{2\times10}=2\times10^2\ \text{s}^{-1}$$

and $$p=\omega=\sqrt{\frac{\mu}{m}}=\sqrt{\frac{10^7}{10}}=10^3\ \text{s}^{-1}.$$

Thus, $$A_{max}=\frac{10^5/10}{2\times2\times10^3\times10^3}=\frac{1}{40}=0.025\text{cm}.$$

EXAMPLE 1.13

1.19.2 Velocity and Energy Resonance

When the velocity amplitude of a body undergoing forced vibration attains a maximum value for a certain frequency of the applied periodic force, the phenomena is known as velocity resonance.

The motion of a particle in the steady state under the action of a periodic force is given from Eq. (1.80) as,

$$x = A\cos(pt - \alpha)$$

Hence, $$\frac{dx}{dt} = -Ap\sin(pt - \alpha)$$

Thus the kinetic energy of the forced system at any instant is given by,

$$T = \frac{1}{2}m\left(\frac{dx}{dt}\right)^2 = \frac{1}{2}mA^2p^2\sin^2(pt - \alpha)$$

$$= \frac{1}{2}\frac{mf^2p^2\sin^2(pt-\alpha)}{\left(\omega^2 - p^2\right)^2 + 4k^2p^2} \quad (1.89)$$

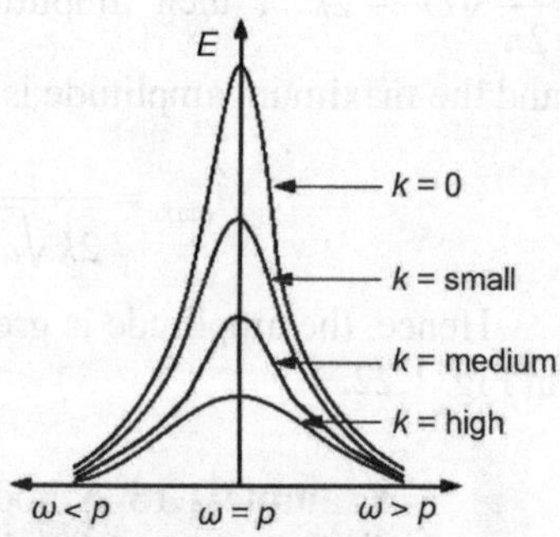

Fig. 1.23: Variation of energy with frequency of the external periodic force

Since the motion is steady SHM, the whole energy of the system at any instant is equal to the maximum kinetic energy. Thus for the system the total energy is,

$$E = T_{\max} = \frac{1}{2}\frac{mf^2p^2}{\left(\omega^2 - p^2\right)^2 + 4k^2p^2}$$

or, $$E = \frac{1}{2}\frac{mf^2}{\omega^2\left(\frac{\omega}{p} - \frac{p}{\omega}\right)^2 + 4k^2}$$

$$= \frac{1}{2}\frac{mf^2}{\omega^2\Delta^2 + 4k^2} \quad (1.90)$$

where, $$\Delta = \frac{\omega}{p} - \frac{p}{\omega} \quad (1.91)$$

Energy resonance is said to be occurr when the energy of the vibrating system is maximum. Again the total energy of the vibrating system is equal to the maximum kinetic energy, which is the case when velocity is maximum. So, this type of resonance is known as velocity resonance. The variation of energy with the frequency of the external periodic force is illustrated in Fig. 1.23. When $\omega = p$, $\Delta = 0$ then the energy of the system becomes maximum for any given value of k. Thus when frequency of the forcing system coincides with that of the natural angular frequency f of the forced system, the energy of the forced system is maximum. This phenomena is called energy resonance and the energy at resonance is given from Eq. (1.90) as,

$$E_{\max} = \frac{mf^2}{8k^2} \quad (1.92)$$

From Eq. (1.92) it is clear that for smaller values of k, greater is the value of E.

1.19.3 Sharpness of Resonance

At resonant frequency the response of the driving body *i.e.*, the energy taken by it from the drived force is maximum. But this will diminish if the frequency differs from the resonant frequency. The rapidity with which the response falls is a measure of *sharpness of resonance*. The sharpness of resonance is quantitatively defined as the reciprocal of Δ for which the energy is reduced to half the resonance value.

Let for $\Delta = \Delta_1$, $E = \frac{1}{2} E_{\max} = \frac{1}{2} \frac{mf^2}{8k^2}$

Thus, using Eq. (1.90), putting $\Delta = \Delta_1$, we have,

$$\frac{1}{2} \frac{mf^2}{\Delta_1^2 \omega^2 + 4k^2} = \frac{1}{2} \frac{mf^2}{8k^2}$$

or,
$$\Delta_1^2 \omega^2 + 4k^2 = 8k^2$$

or,
$$\Delta_1 = \pm \frac{2k}{\omega} \tag{1.93}$$

Hence sharpness of resonance

$$S = \frac{1}{\Delta_1} = \pm \frac{\omega}{2k} \tag{1.94}$$

From Eq. (1.94) it is clear that sharpness of resonance increases as the damping factor decreases and for a particular value of k it is greater as higher the value of natural frequency ω. The sharpness of resonance can be more rapidly represented by plotting $\frac{E}{E_{\max}}$ against ω. When, $\omega = p$, $\Delta = 0$ we have, $\frac{E}{E_{\max}} = 1$.

It is clear from the curves that smaller the value of damping factor k, the sharper is the resonance.

1.20 POWER SUPPLIED IN FORCED VIBRATION

During each cycle of oscillation of the oscillating system, loss of energy takes place due to the damping forces. So, to maintain a steady forced vibration the loss of energy must be supplied by the driving force. Let us first calculate the energy loss by the forced system against the damping force $b\frac{dx}{dt}$. Let in time dt the displacement is dx. Hence the work done by the forced system in time dt is,

$$dW_1 = b\frac{dx}{dt}dx$$

Thus the rate of work done in time dt is,

$$\frac{dW_1}{dt} = b\left(\frac{dx}{dt}\right)^2$$

So, rate of average work done is,

$$\left\langle\frac{dW_1}{dt}\right\rangle=\frac{1}{T}\int_0^T b\left(\frac{dx}{dt}\right)^2 dt \tag{1.95}$$

Now from Eq. (1.80), $x=A\cos(pt-\alpha)$

Hence, $\quad \dfrac{dx}{dt}=-Ap\sin(pt-\alpha)$

So using Eq. (1.95) we have,

$$\left\langle\frac{dW_1}{dt}\right\rangle=\frac{1}{T}\int_0^T b\{\text{Ap}\sin(\text{pt}-\alpha)\}^2\,dt=\frac{b\text{A}^2\text{p}^2}{\text{T}}\int_0^T\sin^2(pt-\alpha)\,dt$$

$$=\frac{b\text{A}^2\text{p}^2}{2\text{T}}\int_0^T 2\sin^2(pt-\alpha)\,dt=\frac{b\text{A}^2\text{p}^2}{2\text{T}}\int_0^T[1-\cos 2(pt-\alpha)]\,dt$$

$$=\frac{b\text{A}^2\text{p}^2}{2\cancel{T}}\cdot\cancel{T}=\frac{1}{2}b\text{A}^2\text{p}^2 \tag{1.96}$$

To calculate the rate of supply of average energy by the driving force let the applied periodic force be, $F\cos pt$. If due to this the system moves through a distance dx in time dt then the work done by the system under the applied force will be, $dW_2=F\cos pt\,dx$.

Thus the rate of work done in time dt is,

$$\frac{dW_2}{dt}=F\cos pt\frac{dx}{dt}$$

So, rate of average work done is,

$$\left\langle\frac{dW_2}{dt}\right\rangle=\frac{1}{T}\int_0^T F\cos pt\frac{dx}{dt}\,dt=-\frac{FAp}{T}\int_0^T\cos pt\sin(pt-\alpha)\,dt$$

or, $$\left\langle\frac{dW_2}{dt}\right\rangle=-\frac{FAp}{T}\int_0^T\cos pt(\sin pt\cos\alpha-\cos pt\sin\alpha)\,dt$$

$$=-\frac{FAp\cos\alpha}{2T}\int_0^T\sin 2pt\,dt+\frac{FAp\sin\alpha}{2T}\int_0^T 2\cos^2 pt\,dt$$

$$=\frac{1}{2}FAP\sin\alpha=\frac{1}{2}FAP\frac{2kAp}{f} \quad \text{[using Eq. (1.82)]}$$

$$=\frac{Fk\text{A}^2\text{p}^2}{f}=\frac{mfk\text{A}^2\text{p}^2}{f}=\frac{mb\text{A}^2\text{p}^2}{2m}=\frac{1}{2}b\text{A}^2\text{p}^2 \tag{1.97}$$

Hence, from Eqs. (1.96) and (1.97) it is clear that the rate of supply of energy by the driving force is equal to the rate of loss of energy by the forced system.

1.21 EXAMPLE OF FORCED VIBRATION

1.21.1 LCR Circuit

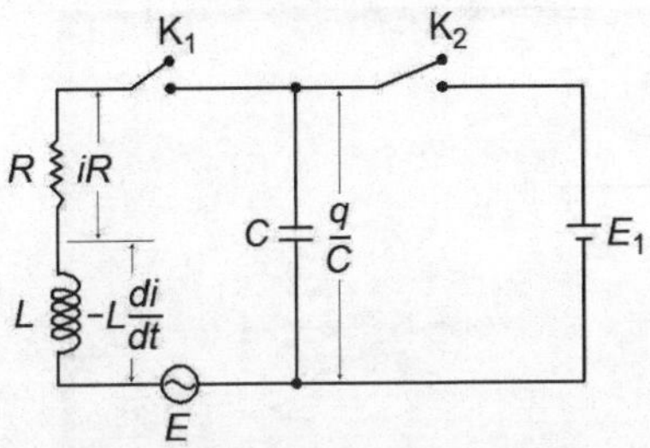

Fig. 1.24: LCR Circuit

Let us consider an *LCR* circuit as shown in Fig. 1.24 which can exhibit forced vibration if it is set to do so. Initially the key K_2 is closed keeping the key K_1 open. The *d.c.* source E_1 charges the capacitor *C*. When the capacitor *C* gets fully charged, the key K_1 is closed and the key K_2 is opened. At this stage the circuit consisting of the component K_1 (key), *R* (resistance), *L* (inductance), *E* (*a.c.* source) and *C* (capacitor) becomes active and through the process of discharging of the capacitor, the *LCR* circuit starts oscillate. Due to the resistance *R* in the circuit its electrical energy is reduced.

The energy loss is balanced by the supplied energy owing to the *a.c.* source E [$= E_0 \cos \omega_0 t$]. Hence the concerned *LCR* circuit becomes a forced oscillator. At an instant of time *t*, $\frac{q}{C}$ is the potential across the capacitor *C*, $L\frac{di}{dt} = L\frac{d^2q}{dt^2}$ is the induced *emf* across the inductor *L*, $Ri = R\frac{dq}{dt}$ is the potential across the resistance *R* and $E_0 \cos(\omega_0 t)$ is the *emf* of the *a.c.* source. So, by analyzing the *LCR* circuit, we can write the following differential equation:

$$L\frac{d^2q}{dt^2} + R\frac{dq}{dt} + \frac{q}{C} = E_0 \cos \omega_0 t \tag{1.98}$$

where ω_0 is the angular frequency of the *a.c.* source.

Dividing throughout by *L*, we get

$$\frac{d^2q}{dt^2} + 2k\frac{dq}{dt} + \omega^2 q = f \cos \omega_0 t \tag{1.99}$$

where $\frac{R}{L} = 2k, \frac{1}{LC} = \omega^2$ and $\frac{E_0}{L} = f$

Now, we can compare this equation [(*i.e.* Eq. 1.99)] with the Eq. (1.79) which is rewritten below:

$$\frac{d^2x}{dt^2} + 2k\frac{dx}{dt} + \omega^2 x = f \cos pt$$

(1.100)

Both the equations, *i.e.*, Eq. (1.79) and Eq. (1.99) are exactly similar. We have already solved Eq. (1.79). Using the same method, we can solve the differential Eq. (1.99) of the *LCR* circuit for forced vibration.

UNIT SUMMARY

- ***Velocity of a particle executing SHM***

$$v=\frac{dx}{dt}=a\omega\cos(\omega t+\varphi)=a\omega\sqrt{1-\frac{x^2}{a^2}}=\omega\sqrt{a^2-x^2}$$

- ***Acceleration of a particle executing SHM***

$$f=\frac{dv}{dt}=-a\omega^2\sin(\omega t+\varphi)=-\omega^2 x$$

- ***Total energy of a particle executing SHM***

$$T=\frac{1}{2}mv^2=\frac{1}{2}m\omega^2\left(a^2-x^2\right) \qquad U=\int_0^x m\omega^2 x\,dx=\frac{1}{2}m\omega^2x^2$$

$$E=\frac{1}{2}m\omega^2a^2$$

- ***Differential equation of SHM and its solution***

$$\frac{d^2x}{dt^2}+\omega^2x=0 \qquad x=a\sin(\omega t+\varphi)$$

- ***Superposition of two collinear SHM with same frequency and different amplitudes***

$$x=\left(a_1^2+a_2^2+2a_1a_2\cos\theta\right)^{1/2}\sin\left\{\omega t+\tan^{-1}\left(\frac{a_2\sin\theta}{a_1+a_2\cos\theta}\right)\right\}$$

- ***Composition of two SHM having equal time period, different amplitude and phases along mutually perpendicular directions***

$$\frac{y^2}{a_2^2}+\frac{x^2}{a_1^2}-\frac{2xy}{a_1a_2}\cos\theta=\sin^2\theta$$

- ***Superposition of two mutually perpendicular SHM having frequencies in the ratio 1:2***

$$\left(\frac{y}{a_2}+\cos\theta\right)^2=4\frac{x^2}{a_1^2}\left(1-\frac{x^2}{a_1^2}+\frac{y}{a_2}\cos\theta\right)$$

- ***Superposition of mutually perpendicular SHM with angular frequencies in the ratio 1:3***

$$\left\{\frac{y}{a_2}-\left(4\frac{x^3}{a_1^3}-3\frac{x}{a_1}\right)\cos\theta\right\}^2=\left(1-\frac{x^2}{a_1^2}\right)\left(4\frac{x^2}{a_1^2}-1\right)^2\sin^2\theta$$

- ***Lissajous figures***

  Curves formed by the superposition of SHM acting at right angle to each other.

- ***Applications of SHM***

Motion of a simple pendulum

$$\frac{d^2x}{dt^2}+\frac{g}{l}x=0 \qquad T=2\pi\sqrt{\frac{l}{g}}$$

Harmonic oscillator: motion of a horizontal spring

$$\frac{d^2x}{dt^2}=-\frac{k}{m}x \qquad T=2\pi\sqrt{\frac{k}{m}}$$

Motion of a vertical spring

$$\frac{d^2x}{dt^2}=-\frac{g}{l}x=-\frac{k}{m}x \qquad T=2\pi\sqrt{\frac{k}{m}}$$

Motion of a liquid in a U tube

$$\frac{d^2x}{dt^2}=-\frac{g}{L}x \qquad T=2\pi\sqrt{\frac{\Lambda}{2g}}$$

$\Lambda=2L$ is the total length of liquid column inside the tube.

Motion of a floating cylinder

$$\frac{d^2x}{dt^2}=-\frac{\sigma g}{L\rho}x \qquad T=2\pi\sqrt{\frac{\rho L}{\sigma g}}$$

LC oscillation

$$\frac{d^2q}{dt^2}+\frac{q}{LC}=0 \qquad T=2\pi\sqrt{\frac{1}{LC}}$$

- ***Damped Vibration***

Differential equation of damped vibration

$$\frac{d^2x}{dt^2}+2k\frac{dx}{dt}+\omega^2x=0 \quad \text{where, } 2k=\frac{b}{m} \text{ and } \omega^2=\frac{\mu}{m}$$

$$x=e^{-kt}\left[A_1e^{\sqrt{k^2-\omega^2}t}+A_2e^{-\sqrt{k^2-\omega^2}t}\right]$$

When $k<\omega$; $x=Pe^{-kt}\sin\left(\sqrt{\omega^2-k^2}t+\theta\right)$

When $k=\omega$; $x=(A+Bt)e^{-\omega t}$ where $A=A_1+A_2, B=(A_1-A_2)\sqrt{k^2-\omega^2}$

- ***Attenuation coefficients***

Logarithmic decrement

$$\lambda=\ln d=\ln e^{\frac{kT}{2}}=\frac{kT}{2}$$

Relaxation time

$$E=E_0e^{-\frac{t}{\tau}}$$

At $t = \frac{1}{2k} = \tau$, $E = \frac{E_0}{e} = 0.37E_0$

Quality factor

$$Q = \frac{\text{maximum value of restoring force}}{\text{maximum value of damping force}} = \frac{m\omega^2 A}{bA\omega} = \frac{m\omega^2 A}{2\text{k}mA\omega} = \frac{\omega}{2k}$$

- ***Dissipation (loss) of power in damped vibration***

Rate of loss of energy is $= \frac{2EkT}{T} = 2Ek = \frac{bE}{m}$ for a complete period

- ***Example of damped vibration***

LCR Circuit

$$\frac{d^2 q}{dt^2} + 2k\frac{dq}{dt} + \omega^2 q = 0$$

$$\varpi = \sqrt{\omega^2 - k^2} = \sqrt{\frac{1}{LC} - \frac{R^2}{4L^2}} \qquad T = \frac{2\pi}{\varpi} = \frac{2\pi}{\sqrt{\frac{1}{LC} - \frac{R^2}{4L^2}}}$$

- ***Conservation of energy in damped vibration***

$$\frac{dE}{dt} = \frac{dW}{dt}$$

- ***Forced vibration***

Differential equation of forced vibration

$$\frac{d^2 x}{dt^2} + 2k\frac{dx}{dt} + \omega^2 x = f\cos pt \quad \text{where } 2k = \frac{b}{m},\ \omega^2 = \frac{\mu}{m} \text{ and } f = \frac{F}{m}$$

Solution of the differential equation

$$x = \frac{f}{\sqrt{\left(\omega^2 - p^2\right)^2 + 4k^2 p^2}}\cos\left(pt - \tan^{-1}\frac{2kp}{\left(\omega^2 - p^2\right)}\right)$$

- ***Discussion for the steady state***

Amplitude resonance

$$A = \frac{f}{\sqrt{\left(\omega^2 - p^2\right)^2 + 4k^2 p^2}} \qquad A_{\max} = \frac{f}{2k\sqrt{\omega^2 - k^2}}$$

Velocity and energy resonance

$$E = T_{\max} = \frac{1}{2}\frac{mf^2 p^2}{\left(\omega^2 - p^2\right)^2 + 4k^2 p^2} \qquad E_{\max} = \frac{mf^2}{8k^2}$$

Sharpness of resonance

$$S = \frac{1}{\Delta_1} = \pm\frac{\omega}{2k}$$

- ***Power supplied by the driving force in forced vibration***

$$\left\langle \frac{dW_1}{dt} \right\rangle = \left\langle \frac{dW_2}{dt} \right\rangle$$

- ***Example of forced vibration***

LCR Circuit

$$\frac{d^2 q}{dt^2} + 2k\frac{dq}{dt} + \omega^2 q = f \cos \omega_0 t \quad \text{where } \frac{R}{L} = 2k, \ \frac{1}{LC} = \omega^2 \text{ and } \frac{E_0}{L} = f$$

EXERCISES

Multiple Choice Questions

1.1 Two mutually perpendicular oscillations with same frequency, amplitude but phase difference δ will produce closed curve with non-zero area enclosed

(a) for all values of δ except $\delta = 0$ (b) only for $\delta = \frac{\pi}{2}$

(c) for all values of δ except $\delta = 0$ and π (d) for all values of $\delta > \frac{\pi}{2}$

1.2 Two mutually perpendicular SHM with equal time periods but different amplitudes are superposed. If the phase difference between these oscillations is 45°, then they form a

(a) circle (b) parabola (c) ellipse (d) straight line

1.3 A plane wave travels in a medium in the positive X-direction. The displacements of the particles are given by, $y(x, t) = 10 \sin 2\pi(t - 0.2x)$; where x and y are measured in metre and t is measured in seconds. The wavelength of the wave is

(a) 0.2 m (b) 5 m (c) 2π m (d) 0.4 m

1.4 Superposition of two simple harmonic waves of equal time period and equal amplitude with phase difference $\phi = \frac{\pi}{2}$ forms a

(a) circle (b) ellipse (c) parabola (d) oblique ellipse

1.5 If a spring with spring constant k is cut into three equal parts, the force constant of each part would be,

(a) $3k$ (b) k (c) $k/3$ (d) none of these

1.6 If a particle is executing simple harmonic motion with frequency v then its potential energy

(a) remains constant over time (b) is oscillating with frequency v

(c) is oscillating with a frequency $v/2$ (d) is oscillating with a frequency $2v$

1.7 To get a circular Lissajous figure, the phase difference (ϕ) and the amplitudes (a and b) of two superimposing, mutually perpendicular SHM are respectively

(a) $\phi = 0, a = b$ (b) $\phi = \pi/2, a = b$ (c) $\phi = \pi/2, a \neq b$ (d) $\phi = \pi/4, a \neq b$

1.8 Resultant of two perpendicular SHM of equal frequency and equal amplitude but a phase difference of π radian is

(a) a straight line (b) an ellipse
(c) a circle (d) a spiral collapsing inward

1.9 The time period of a simple pendulum of infinite length is

(a) finite and non-zero (b) zero (c) infinite (d) none of these

1.10 The potential energy of a particle executing SHM of amplitude a is equal to its kinetic energy when displacement of the particle is

(a) $\pm a$ (b) $\pm\frac{a}{2}$ (c) $\pm\frac{a}{4}$ (d) $\pm\frac{a}{\sqrt{2}}$

1.11 The acceleration of a particle executing SHM is

(a) inversely proportional to the displacement from the mean position
(b) directly proportional to the displacement from the mean position
(c) constant (d) none of these

1.12 Lissajous figures can be viewed through an instrument called –

(a) disdrometer (b) spectrometer (c) sonometer (d) CRO

1.13 Potential energy and kinetic energy of a particle executing SHM is equally shared when the displacement is

(a) $\pm a$ (b) $\pm\frac{a}{2}$ (c) $\pm\frac{a}{\sqrt{2}}$ (d) none of these

1.14 Total mechanical energy of a particle executing SHM is –

(a) $\frac{1}{2}m\omega^2a^2$ (b) $\frac{1}{2}m\omega^2$ (c) $m\omega^2a^2$ (d) none of these

1.15 The maximum velocity of a particle executing SHM is represented by $x = a \sin \omega t$ at time t occurs at

(a) $x = 0$ (b) $x = a$ (c) $x = -a$ (d) $x = a/2$

1.16 If the amplitude of a SHM is 2 cm, then its energy will be equally shared by its kinetic and potential energy for a displacement of

(a) 1 cm (b) 2 cm (c) $\sqrt{2}$ cm (d) none of these

1.17 Due to superposition of two mutually perpendicular SHM of same frequency but different amplitudes, the resultant wave will be straight line when the phase difference between them is

(a) 0 and π (b) 0 and $\frac{\pi}{2}$ (c) π and $\frac{\pi}{2}$ (d) $\frac{\pi}{2}$ only

1.18 If two SHM are of the same amplitude and time period and their phase act at right angle to each other, then the resultant vibration is

(a) parabolic (b) linear (c) elliptical (d) circular

1.19 If acceleration of a particle executing SHM is maximum, displacement will be

(a) maximum (b) minimum (c) zero (d) none of these

1.20 Simple harmonic motion (SHM) is a

(a) periodic motion (b) circular motion (c) both (a) and (b) (d) none of these

1.21 Natural angular frequency of a body executing SHM is

(a) ω (b) $\omega/2\pi$ (c) $2\pi/\omega$ (d) none of these

1.22 Time period of a horizontal spring (force constant k) can be expressed as,

(a) $T = 2\pi\sqrt{k/m}$ (b) $T = 2\pi\sqrt{m/k}$ (c) $T = \pi\sqrt{k/m}$ (d) none of these

1.23 Time period (T) of a particle executing SHM with displacement x and acceleration f is

(a) $2\pi\sqrt{\dfrac{f}{x}}$ (b) $2\pi\sqrt{xf}$ (c) $2\pi\sqrt{\dfrac{x}{f}}$ (d) none of these

1.24 If ω_0 is the natural frequency of a body and k is the damping constant, then its quality is

(a) ω_0 / k (b) $\omega_0 / 2k$ (c) $2\omega_0 / k$ (d) none of these

1.25 The effect of the damping force on a harmonic oscillator is to

(a) reduce angular frequency and amplitude of vibration

(b) increase amplitude and angular frequency of vibration

(c) reduce angular frequency and increase amplitude of vibration

(d) increase angular frequency and reduce amplitude of vibration

1.26 Relaxation time is the time in which the amplitude A_0 of a damped oscillator falls to

(a) $\dfrac{A_0}{3}$ (b) $\dfrac{A_0}{e}$ (c) $\dfrac{A_0}{\gamma}$ (d) $A_0 e$

1.27 If γ be the damping constant and T the periodic time of a damped oscillator, then logarithmic decrement of motion is

(a) $\dfrac{\gamma T}{4}$ (b) γT (c) $\dfrac{T}{\gamma}$ (d) $\dfrac{\gamma}{4T}$

1.28 The quality factor of a series LCR circuit is

(a) $\dfrac{1}{R}\sqrt{\dfrac{1}{LC}}$ (b) $\dfrac{1}{R}\sqrt{\dfrac{L}{C}}$ (c) $\sqrt{\dfrac{RL}{C}}$ (d) $\sqrt{RLC}$

1.29 If the damping constant is small, the resonance curve is

(a) flat (b) sharp (c) unchanged (d) none of these

1.30 The amplitude of a damped oscillator with damping factor γ varies as

(a) $e^{\gamma t}$ (b) $e^{\frac{\gamma t}{2}}$ (c) $e^{-\frac{\gamma t}{2}}$ (d) $e^{\frac{1}{\gamma t}}$

1.31 In terms of Q, the logarithmic decrement λ is given by

(a) $\dfrac{\pi}{2Q}$ (b) $\dfrac{2Q}{\pi}$ (c) $2\pi Q$ (d) $\dfrac{\omega_0}{Q}$

1.32 The relaxation time (τ) of a damped oscillator with damping coefficient (K) is

(a) $\tau = 1/K$ (b) $\tau = 1/2K$ (c) $\tau = K$ (d) $\tau = 2K$

1.33 The energy decay time of a damped oscillator is

(a) $T = \frac{2}{\tau}$ (b) $T = 2\tau$ (c) $T = \frac{\tau}{2}$ (d) none of these

1.34 In terms of Q, the energy decay time is given by

(a) $T = \frac{T_0 Q}{2\pi}$ (b) $T = \frac{2\pi}{T_0 Q}$ (c) $T = \frac{2\pi T_0}{Q}$ (d) $T = 2\pi T_0 Q$

1.35 For oscillatory motion of damped oscillator (damping constant γ and natural frequency ω_0)

(a) $\gamma > 2\omega_0$ (b) $\gamma = 2\omega_0$ (c) $\gamma < 2\omega_0$ (d) $\gamma = \omega_0$

1.36 The time period of a damped vibration is

(a) $2\pi\sqrt{\omega^2 - k^2}$ (b) $2\pi\sqrt{k^2 - 2\omega^2}$ (c) $\frac{2\pi}{\sqrt{\omega^2 - k^2}}$ (d) $\frac{2\pi}{\sqrt{2k^2 - \omega^2}}$

1.37 For critical damping, the motion of a damped vibration will be non-oscillatory when

(a) $\omega^2 > k^2$ (b) $\omega^2 < k^2$ (c) $\omega = k$ (d) none of these

1.38 Motion of a system in critical damped condition is

(a) damped oscillatory (b) oscillatory (c) harmonic (d) non-oscillatory

1.39 The relaxation time of a damped vibration is defined as the time taken for its initial energy to fall to

(a) 50% (b) 70% (c) 37% (d) 63%

1.40 The quality factor of an electrical oscillator is

(a) $\omega R/L$ (b) LR/ω (c) $L\omega/R$ (d) $L/(\omega R)$

1.41 The quality factor Q for an LCR circuit is

(a) $\frac{\omega R}{L}$ (b) $\frac{\omega L}{R}$ (c) $\frac{\omega}{LR}$ (d) $\frac{R}{\omega L}$

1.42 For a series LC circuit, time period of oscillation is given by

(a) $2\pi\sqrt{LC}$ (b) $2\pi\sqrt{\frac{L}{C}}$ (c) $2\pi\sqrt{\frac{1}{LC}}$ (d) $\frac{1}{2\pi}\sqrt{\frac{L}{C}}$

1.43 In case of damped vibration the damping force is proportional to its –

(a) velocity (b) displacement (c) acceleration (d) none of these

1.44 In case of critical damping the system is –

(a) non-oscillatory (b) oscillatory (c) vibratory (d) none of these

1.45 Relation between damping constant and relaxation time is –

(a) $\tau = \frac{1}{2k}$ (b) $\tau = \frac{1}{k}$ (c) $\tau = k$ (d) none of these

1.46 For small value of damping constant, Q value is –

(a) $Q = \omega\tau$ (b) $Q = \frac{\omega}{\tau}$ (c) $Q = \frac{\tau}{\omega}$ (d) none of these

1.47 Example of weakly damped harmonic oscillator is

(a) dead-beat galvanometer (b) tangent galvanometer
(c) ballistic galvanometer (d) discharge of a capacitor through a resistance

1.48 If a is the force constant of an oscillating body of mass m, the Q factor is

(a) $Q = C\sqrt{am}$ (b) $Q = C\sqrt{\frac{m}{a}}$ (c) $Q = C\sqrt{\frac{a}{m}}$ (d) none of these

1.49 If the damping force on an one dimensional harmonic oscillator of natural frequency ω_0, is $20mv$, where m is the mass and v is the instantaneous velocity of the oscillator then the frequency of oscillation (when $b < \omega_0$) is

(a) ω_0 (b) b (c) $\omega_0\left(1-\frac{b^2}{2\omega_0^2}\right)$ (d) $\omega_0\left(1+\frac{b^2}{2\omega_0^2}\right)$

1.50 The resonant frequency of a forced oscillator of natural frequency ω_0 in a medium of damping coefficient γ is

(a) $\omega_r = \omega_0$ (b) $\omega_r = \left(\omega_0^2 - \frac{\gamma^2}{2}\right)^{\frac{1}{2}}$ (c) $\omega_r = \left(\omega_0^2 - \frac{\gamma^2}{4}\right)^{\frac{1}{2}}$ (d) $\omega_r = \sqrt{\gamma^2 - \omega_0^2}$

1.51 If F_0 be the amplitude of the applied force and k is the restoring force per unit displacement, then the amplitude of the forced oscillator for $\omega = 0$ is

(a) $\frac{k}{F_0}$ (b) kF_0 (c) $\frac{F_0}{k}$ (d) $\frac{1}{kF_0}$

1.52 If ω and ω_0 be the applied frequency and natural frequency of a driven oscillator, then the velocity resonance occurs at

(a) $\omega > \omega_0$ (b) $\omega = \omega_0$ (c) $\omega = \frac{\omega_0}{2}$ (d) $\omega = 2\omega_0$

1.53 The phase difference between external force and velocity of the forced oscillator is

(a) $\frac{\pi}{2}$ (b) 0 (c) π (d) none of these

1.54 For higher sharpness of the resonance curve, the Q value must be

(a) small (b) large (c) medium (d) none of these

1.55 In case of forced oscillator, transient beats are observed in

(a) steady state (b) early state (c) middle stage (d) none of these

1.56 In a series LCR circuit, the discharge of the capacitor is oscillatory, if

(a) $R > 2\sqrt{\frac{L}{C}}$ (b) $R = \sqrt{\frac{L}{C}}$ (c) $R < 2\sqrt{\frac{L}{C}}$ (d) none of these

1.57 For the motion of the forced oscillator to be resistance controlled, near $\omega = \omega_0$

(a) $\gamma < \omega_0$ (b) $\gamma = \omega_0$ (c) $\gamma > \omega_0$ (d) $\gamma\omega_0 =$ constant

1.58 For the motion of the forced oscillator to be stiffness controlled under the condition, $\gamma < \omega_0$ is

(a) $\omega = \omega_0$ (b) $\omega << \omega_0$ (c) $\omega >> \omega_0$ (d) none of these

1.59 Mass controlled motion of the forced oscillator occurs for $\gamma < \omega_0$, if

(a) $\omega = \omega_0$ (b) $\omega < \omega_0$ (c) $\omega > \omega_0$ (d) none of these

1.60 Velocity resonance of the forced oscillator occurs at

(a) $\omega = 2\omega_0$ (b) $\omega = \frac{\omega_0}{2}$ (c) $\omega = \omega_0$ (d) none of these

1.61 At resonance, the maximum amplitude of the forced oscillator is

(a) $\frac{F_0}{m\gamma}$ (b) $\frac{mF_0}{\gamma}$ (c) $\frac{\gamma}{mF_0}$ (d) $mF_0\gamma$

1.62 The dissipated power varies as

(a) ω (b) ω_0 (c) ω^* (d) none of these

1.63 If the restoring force constant of a body is 98 Nm^{-1}, then the restoring force of the body for a displacement of 10 cm is

(a) 98 N (b) 9.8 N (c) 0.98 N (d) none of these

Answers of Multiple Choice Questions

1.1 (c), 1.2 (c), 1.3 (b), 1.4 (a), 1.5 (a), 1.6 (d), 1.7 (b), 1.8 (a), 1.9 (c), 1.10 (d), 1.11(b), 1.12 (d), 1.13 (c), 1.14 (a), 1.15 (a), 1.16 (c), 1.17 (a), 1.18 (d), 1.19 (a), 1.20 (c), 1.21 (b), 1.22 (b), 1.23 (c), 1.24 (a), 1.25 (a), 1.26 (b), 1.27 (a), 1.28 (b), 1.29 (a), 1.30 (b), 1.31 (d), 1.32 (b), 1.33 (c), 1.34(a), 1.35 (c), 1.36 (c), 1.37 (c), 1.38 (d), 1.39 (c), 1.40 (c), 1.41 (b), 1.42 (c), 1.43 (a), 1.44 (a), 1.45 (a), 1.46 (a), 1.47(c), 1.48 (c), 1.49 (c), 1.50 (b), 1.51 (c), 1.52 (b), 1.53 (b), 1.54 (b), 1.55 (b), 1.56 (c), 1.57 (a), 1.58 (b), 1.59 (c), 1.60 (c), 1.61 (a), 1.62(b), 1.63 (b)

Short and Long Answer Type Questions

Category I

1.1 State the characteristics of SHM. Define phase. Prove the principle of conservation of energy.

1.2 Establish the differential equation of a simple harmonic motion and solve the equation.

1.3 Derive the expression for the total energy of a simple harmonic oscillator and show that it is constant and proportional to the square of the amplitude.

1.4 Show that for a particle executing SHM, the average kinetic energy is half of the corresponding maximum energy.

1.5 Calculate the time period of the liquid column of length l in a U-tube if it is depressed in one arm by x. Take d as the density of liquid and A as the cross-sectional area of each arm of the tube.

1.6 Mention the characteristics of SHM.

1.7 Find the difference between phase and epoch in relation to SHM.

1.8 Prove that total mechanical energy of a particle executing SHM is constant.

1.9 Explain Lissajous figures. Also explain their formation.

1.10 Show that the motion of a simple pendulum is simple harmonic and hence, find the expression for the period of oscillation.

1.11 Show that the vertical oscillation of a loaded spring is simple harmonic and hence find an expression for the time period of its oscillation.

1.12 Two SHM having the same period but differing in phase and amplitude are acting in the same direction on a particle. Deduce the expression of amplitude and phase for the resulting motion.

1.13 Two simple harmonic vibrations of same frequency at right angles to each other superpose at a point. Discuss the nature of resultant motion when the phase difference between them is (i) $\varphi = 0$ and (ii) $\varphi = \pi$.

1.14 Prove that the resultant of two SHM of same period but different amplitudes and phases, gives an elliptical motion.

1.15 Explain the terms logarithmic decrement, relaxation time and quality factor of a damped oscillatory system. How are they related?

1.16 Explain damped harmonic motion. Establish the differential equation of damped harmonic oscillator explaining each term in the equation. What is Lisasajous figure?

1.17 Show that, the amplitude of a weakly damped harmonic oscillator reduces to half its initial value in time $t = \tau \ln 2$, where τ is the relaxation time.

1.18 Show that in case of damped vibratory motion, the rate of loss of energy of the vibrator is equal to the rate of work done by the vibrator against the resisting force.

1.19 In damped vibratory motion find out the time in which the energy of the system falls to e^{-1} times of the initial value. Establish a relation between logarithmic decrement and relaxation time for the motion.

1.20 Prove the principle of conservation of energy in damped vibration.

1.21 Establish the differential equation of damped harmonic motion and explain the different terms in the expression.

1.22 Explain the term critical damping.

1.23 Explain the terms free and damped vibration. Does the principle of conservation of energy applicable in case of damped vibration? Explain.

1.24 (a) Establish the differential equation of a damped oscillator and solve it.
(b) Show graphically the variation of amplitude of damped oscillation with time.

1.25 Explain the early state and steady state of forced vibration. Discuss the factors characterizing them.

1.26 Explain the term impedance of an oscillator. Obtain an expression for it and discuss its variation with the frequency of the applied periodic force.

1.27 Find the difference between resonance and resonant frequency of a forced oscillator. Discuss amplitude resonance.

1.28 Discuss the velocity resonance in forced vibration. Distinguish between amplitude resonance and velocity resonance.

1.29 Explain the resistance controlled, stiffness controlled and mass controlled motion of a forced oscillator.

1.30 Show that in the steady state, the time-averaged input power equals the time averaged power dissipated through friction.

1.31 Show that in forced vibration the rate of supply of energy by the forcing system is equal to the rate of dissipation of energy by the forced system.

Category II

1.32 (a) A cubical block of side L cm and density d is floating in water of density ρ ($\rho > d$). The block is slightly depressed and released. Show that it will execute simple harmonic motion. Determine the frequency and time period of oscillation.

(b) Write down the difference between standing and progressive wave and establish the differential equation of a progressive wave.

(c) Establish the differential equation of SHM from energy conservation principle. An oscillator executing SHM has zero displacement at time $t = 0$. If the displacements are 1 mm and 1.5 mm at instants 0.1 s and 0.2 s, calculate the frequency and amplitude of oscillation.

1.33 (a) Two simple harmonic oscillators of different masses oscillate separately under the action of same restoring force at frequencies 3 Hz and 5 Hz. Calculate the ratio of their masses.

(b) Find the amplitude, phase and instantaneous velocity of the vibrational motion represented by

$$x = a\cos\omega t + \frac{a}{2}\cos\left(\omega t + \frac{\pi}{2}\right) + \frac{a}{4}\cos\left(\omega t + \pi\right) + \frac{a}{8}\cos\left(\omega t + \frac{3\pi}{2}\right)$$

1.35 (a) Show that the frequency of oscillation of a body of mass M suspended from a uniform spring of force constant k and mass m is given by $\frac{1}{2\pi}\sqrt{\frac{k}{M + m/3}}$. It is given that the velocity of any element of the spring is proportional to its distance from the fixed end of the spring.

(b) A particle of mass 0.1 kg is placed in a field of potential energy $U = 2x^2 + 4x + 4$ Jkg^{-1}. Find the frequency, force constant and equilibrium positions.

1.36 Write down the differential equation for damped vibratory motion explaining the physical significance of each term in the equation. What is the difference between damped and forced vibration?

1.37 (a) Under the condition of underdamping ($b < \omega$), find the expression for displacement as a function of time. Draw the displacement-time graph.

(b) Distinguish between free and damped oscillation.

(c) Write down the differential equation of a series LCR circuit driven by sinusoidal voltage. Identify the natural frequency of the circuit and also find the relaxation time.

1.38 (a) If a harmonic oscillator of mass m and natural frequency ω_0 is driven by a force $F_0 \sin\omega t$ and the damping is proportional to $2p$ times the velocity of the oscillator, then show that the displacement is given by $x = \frac{F_0/m}{\sqrt{\left(\omega_0^2 - \omega^2\right)^2 + 4p^2\omega^2}}\sin\left(\omega t - \theta\right)$; where $\tan\theta = \frac{2\omega p}{\left(\omega_0^2 - \omega^2\right)}$.

Using the above relation, show that at velocity resonance, the velocity is in phase with the driving force.

(b) Find the energy of an oscillator at velocity resonance. Define sharpness of resonance.

1.39 (a) Write down the differential equation of forced vibration explaining each term in the equation.
(b) Starting from the steady state solution of forced vibration explain the phenomenon of amplitude and velocity resonance deriving the value of the driving frequency in each case.

1.40 (a) Set up the differential equation of motion of a simple harmonic oscillator subjected to a damping force and an external simple harmonic force.
(b) Obtain expression for the amplitude and the phase angle of the displacement in the steady state.

1.41 For the forced vibration of a mechanical damped harmonic oscillator driven by a sinusoidal force of angular frequency ω, obtain the value of ω for amplitude resonance. What is the amplitude of the oscillator at resonance? What are the time-averaged kinetic and potential energies of the oscillator? Determine the value of ω for which they are equal.

1.42 Draw the plots of the displacement amplitude and velocity amplitude against the frequency of the sinusoidal force driving a mechanical oscillator for the different values of damping. What is the power factor at velocity resonance? What is sharpness of resonance? Give a measure for it.

1.43 (a) Obtain an expression for the average power supplied by the sinusoidal force driving a mechanical oscillator over a complete cycle in steady state. What is power factor and what is its value at resonance?
(b) Show that in the steady state the time-averaged input power equals the time-averaged power dissipated through friction.

1.44 (a) Distinguish between amplitude resonance and velocity resonance. Show that at velocity resonance the maximum velocity is inversely proportional to damping constant and the velocity of the oscillator is in phase with the driving force.
(b) Explain half power frequencies. How are they related to sharpness of resonance? Show that the resonant frequency is the geometric mean between the half power frequencies.

Numerical Problems

1.1 Calculate the time period of the liquid column of length l in a U-tube, if it depressed in one arm by x. Take d as the density of the liquid and A as the cross sectional area of each arm of U-tube.

1.2 The potential energy of a particle with mass 10 g is given by $V(x) = 32x^2 + 0.2$, where x is in m and V is in J. Write down the equation of motion and solve it.

1.3 A particle of mass 2 kg is acted on by a restoring force/unit displacement 10×10^{-3} N/m, and a frictional force/unit velocity 2×10^{-3} N/ms^{-1}. Find whether the motion is oscillatory or non-oscillatory. For what value of frictional (resistive) force, the damping will be critical?

1.4 The displacement of any particle at any instant t is given by $x = 3\cos t + 4\sin t$. Show that the motion is simple harmonic. What is the amplitude of this oscillation? Show that its kinetic energy oscillates with angular frequency 2ω.

1.5 The displacement of a simple harmonic oscillator is given by $x = a\sin(\omega t + \theta)$. If the oscillations started at time $t = 0$ from a position x_0 with velocity v_0 show that,

$$\tan\theta = \omega x_0/v_0 \text{ and } a = (x_0^2 + v_0^2/\omega^2)^{1/2}.$$

1.6 The displacement of a particle of mass 0·2 kg executing SHM is indicated by

$$y = 10\sin\left(\frac{\pi}{3}t - \frac{\pi}{12}\right) \text{ m.}$$

Calculate the (i) amplitude, (ii) angular velocity, (iii) time period, (iv) maximum velocity and (v) maximum acceleration.

1.7 A solid cylinder of mass M is attached to a massless horizontal spring of constant k and it executes SHM, when rolls without slipping. Show that, its period of oscillation is $T = 2\pi\sqrt{\dfrac{3M}{2k}}$.

1.8 Calculate the displacement to amplitude ratio for a SHM when its kinetic energy is 90% of total energy.

1.9 The potential energy of a particle of mass m is given by $\frac{1}{2}m\omega^2x^2$, where ω is a constant. Show that the particle is executing SHM.

1.10 The displacement of a particle executing SHM is given by y = sinkt + coskt. Find its amplitude of vibration. [*Ans*: $\sqrt{2}$ unit]

1.11 Two vibrations at right angles to each other are described by the following equations

$$x = a\sin\omega t,\ y = a\sin(\omega t + \theta).$$

Construct the Lissajous' figure of the combined motion.

1.12 Two vibrations at right angles to each other are described by the equations,

$$x = 4\sin\pi t, y = 4\cos\left(\pi t + \frac{\pi}{3}\right),$$ where x and y are in cm and t is in s.

Construct the Lissajous' figure of the combined motion.

1.13 The potential energy of a particle is given by $V = ax^2$. Show that the motion is simple harmonic.

1.14 9 kg mercury is poured into a glass U tube of uniform internal diameter of 1.2 cm. It oscillates freely about the equilibrium position. Calculate the period of oscillation. [*Ans*: 3.43 s]

1.15 The displacements of a particle of mass 0.2 kg executing SHM is given by $y = 10\sin\left(\frac{\pi}{3}t - \frac{\pi}{12}\right)$ m

Calculate the kinetic and potential energy of the particle at x = 0.04 m. Also calculate the total energy of the particle.

1.16 A particle executes SHM of period 8 s, amplitude 4 cm. Find the velocity and acceleration when the particle is 2 cm from the central position and also find their maximum values.

1.17 The equation of motion of a particle executing SHM is given by, $y = A\sin(\omega t + \varphi)$. Find the ratio of KE and PE of the particle when the amplitude is half the maximum.

1.18 A mass of 2 kg hangs from a spring. A 50 gm body hung below the mass stretches the spring 1 cm further. If now the mass is removed then find the period of oscillation of the body.

1.19 The amplitude of oscillation of a body performing SHM is 10 cm and the frequency of oscillation is 4 Hz. Find the equation of motion of the body and also find the velocity and acceleration of the body at t = 5 s from the starting of the motion.

1.20 In one dimensional motion of a mass of 10 g, it is acted upon by a restoring force per unit displacement of 10 dynecm^{-1} and a resisting force per unit velocity of 2 dynecm^{-1}s.

(i) Find whether the motion is aperiodic or oscillatory.

(ii) Find the value of resisting force which will make the motion critically damped.

(iii) Find the value mass for which the given force will make the motion critically damped.

1.21 For the motion of a damped vibratory body, the restoring force per unit displacement is 10 dyne/cm and a resisting force per unit velocity is 2 dyne cm^{-1} s. Find whether the motion is oscillatory or aperiodic and also find the resisting force that makes the motion critically damped.

1.22 The displacement of a damped harmonic oscillator is given by, $x = 5e^{-5t}\sin\frac{\pi}{2}t$. Find the velocity of the oscillating point at $t = T/4$, T being the time period of the motion.

1.23 A spring is suspended vertically and loaded with mass of 0.3 kg, so that it is stretched by 0.05 m. For a displacement of 0.01 m, it is given a downward velocity 0.3 m/s. Find the time period and amplitude of vibration of the spring.

1.24 The amplitude of oscillation of a body of frequency 300 Hz decays to 1/10 of its initial value after 1800 cycles. Find its damping constant, Q factor, relaxation time and the time in which the energy is reduced to 1/2 of its initial value. Establish the differential equation of the damped harmonic motion.

1.25 A mass of 0.1 kg is attached at the free end of a massless spring hanging from a rigid support. The system oscillates in a liquid that exerts a viscous force $-pv$, where p is a constant and v is the velocity. It is observed that the energy decays to half of its original value in 5 s.

(i) What is the value of p?

(ii) What is the Q-value of the oscillator?

(iii) What is the change in frequency due to damping? [Assume spring constant = 9 N/m]

PRACTICALS

1. To find the value of acceleration due to gravity using angular oscillations of a long bar

Aim

To find the acceleration due to gravity of a rigid body in the form of a long bar with the help of the angular oscillations of the body

Apparatus

A long bar, knife-edge, stop watch and a meter scale

Theory

The long bar, shown in the Fig. (*i*), is hung from a knife-edge through one of the holes along the length. The bar is free to oscillate, taking the knife-edge as axis. With reference to the vertical position of equilibrium for any displacement θ, it would give rise to an oscillatory motion. For this being a rotating rigid body, the torque of the gravitational force produces the angular acceleration. For an angular displacement θ we can write the restoring torque τ as

$$\tau = -Mgd\sin\theta \qquad (i)$$

where,

M represents the mass of the composite system *i.e.*, the compound pendulum and d is the distance from the point of suspension to the centre of mass of the bar.

Now, the time period for the simple pendulum is equal to the time period for any compound pendulum. From the theoretical analysis, we can write,

$$L = \frac{d^2 + k^2}{d} \quad (ii)$$

where, L is the equivalent length. Rearranging we have,

$$d^2 - Ld + k^2 = 0$$

which is a quadratic equation in d. Let d_1 and d_2 are the two solutions of the equation. Solving the equation we obtain

$$d_1 + d_2 = L \quad (iii)$$

and

$$d_1 d_2 = k \quad (iv)$$

Since both d_1 and d_2 are positive we can infer that for the two points of suspensions on one side of the centre of gravity (CG) of the pendulum the time periods are identical. Similarly, there are two points of suspension on the opposite side of the CG having the identical time period. Hence, for the compound pendulum, there are altogether four points of suspension; on either side of the CG two points each where the time periods are identical. Equivalent length L represents the sum of two points of suspension situated on two sides of the CG asymmetrically.

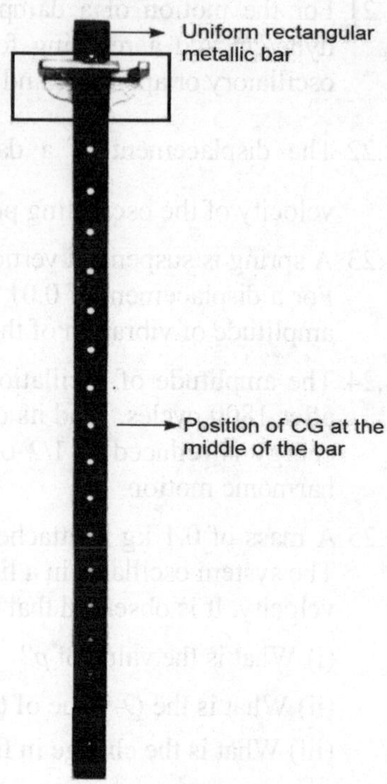

Fig. (*i*)

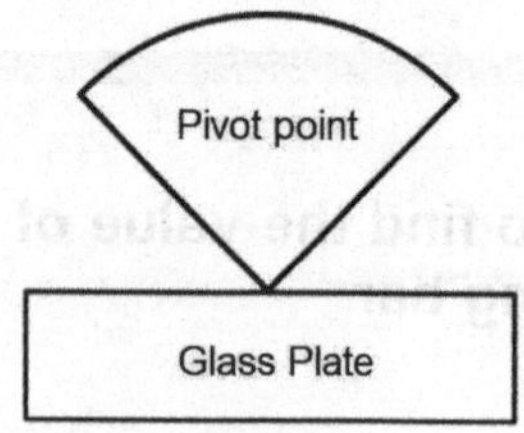

Fig. (*ii*)

Description of the apparatus

A bar pendulum is shown in Fig. (*i*). It is taken as a uniform rectangular metallic bar of length of the order of 1 m, with holes drilled along its length at 5 cm apart (say).

In the figure the CG is marked in the middle of the bar. Two knife edges are there, which are symmetrically placed on either side of CG so that it can be suspended at various distances from CG. In Fig. (*ii*), we have shown separately the pivot or the knife edge of the bar pendulum.

Procedure

1. Balance the given bar on sharp wedge and carefully mark the position of its CG.
2. Ensure that the frame on which movable pivot or the knife edge is to be rested is properly horizontal.
3. Now, suspend the pendulum in the first hole. The pivot or the knife-edge is to be kept on the glass plate as shown in Fig. (*ii*).
4. Measure the distance d between the point of suspension and the CG.

5. Now start to oscillate the pendulum and find the time for 20 oscillations. The recording of time, is to be started after a few oscillations of the pendulum, when it becomes regular.
6. Repeat the processes 3-5 by suspending the pendulum in all the consecutive holes.

Experimental data

Table 1: To measure acceleration due to gravity using a compound pendulum

Serial No.	d (cm)	No. of oscillations			Time for 20 oscillations T_{20} (s)	Mean T_{20} (s)
		1	2	3		

Drawing the graph

By taking distance along the *X*-axis and time period along the *Y*-axis, a graph is to be drawn as shown in Fig. (*iii*). This is a plot of time *versus* distance from CG (*C*) of the bar. Shifting the axes, a full page graph is to be drawn.

Calculation

From the graph, distances d_1 and d_2 are to be measured, wherefrom the value of g can be calculated.

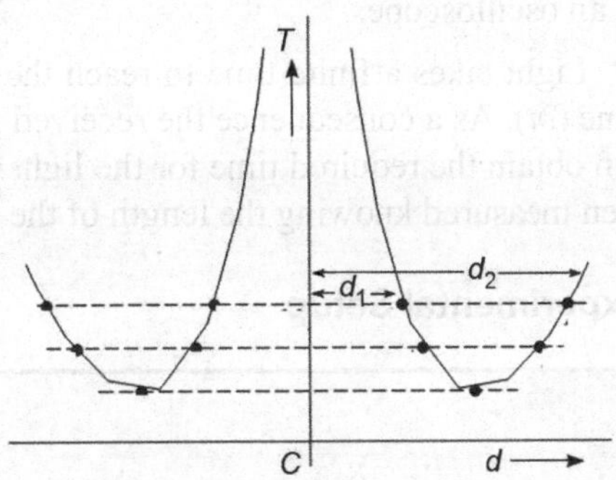

Fig. (*iii*)

Result

The value of g = ... m/s^2.

Precautions

1. The knife-edge is to be made perfectly horizontal. If it is not so then the bar will be twisted during the time of swinging.
2. The motion of the bar pendulum should be maintained strictly in a vertical plane.
3. Small amplitude of the swing is to be maintained.
4. The time period of oscillations should be measured by noting the time for a large number of oscillations of the bar with a good quality stop watch.
5. All the distances should be correctly measured and are to be plotted from one end of the rod.

Viva-Voce Questions

1. What are the amplitude range for the oscillation of the bar in the vertical plane?
2. Why the period of the pendulum increases when it is close to the CG?

2. Measuring the speed of light using PC-based oscilloscope

Aim

To determine the speed of light using a PC-based oscilloscope through an optical cable of polymer fibre

Apparatus

1. An optical transmitter and a receiver with high frequency square modulation with a range 250 kHz to 1 MHz
2. A PC with PicoScope installed
3. A PicoScope PC based oscilloscope with 50 MHz or more bandwidth
4. Suitable leads and connectors

Theory

A square wave oscillator at a frequency of 1 MHz produces a signal with a time period of 1 μs having a square pulse of 0.5 μs. This signal can be utilized for modulating a light beam, allowed to pass along a fibre cable to get a square pulse of light that lasts for 0.5 μs [Fig. (*i*)]. At other side of the optical cable, a phototransistor is used for detecting the pulse. This converts it back into a pulse of electricity. The original and the received pulses are displayed on the screen of an oscilloscope.

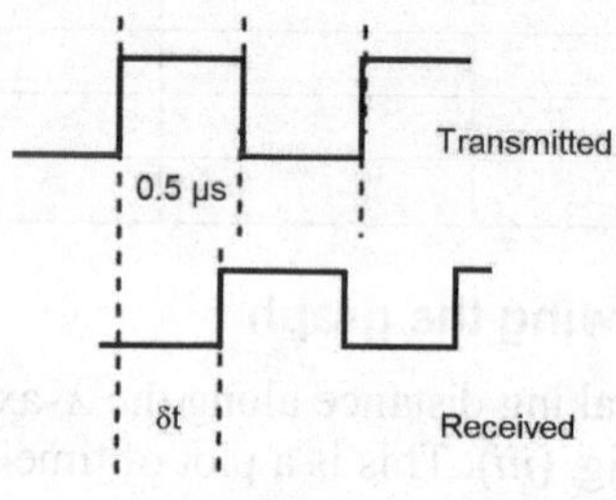

Fig. (*i*)

Light takes a finite time to reach the receiver and as a result the second signal will be delayed in time (δt). As a consequence the received pulse shifts along the time base axis. Measuring this shift one can obtain the required time for the light pulse to travel the length of the fibre cable. The light speed is then measured knowing the length of the cable.

Experimental Setup

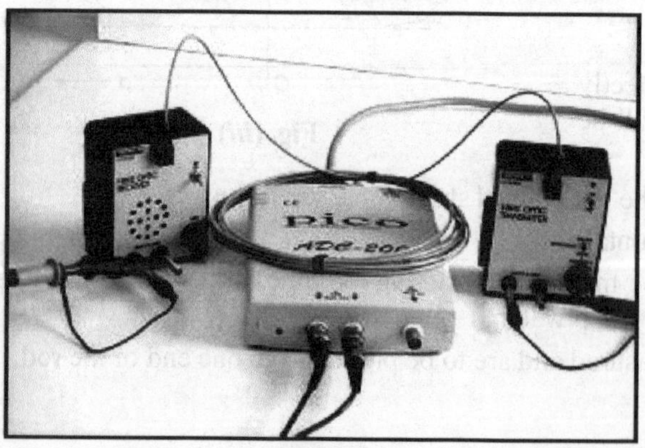

Fig. (*ii*)

Make the required arrangement of the fibre optic system to transmit HF modulated light beam along the coaxial cable which is nearly 5 m in length. Next join the receiver unit to the cable and thereby monitor the transmitted pulse at Channel A on the Pico ADC while the pulse is received at Channel B. The photograph shown in Fig. (*ii*) exhibits the experimental equipment connected in a manner as described.

Some typical values of the PC oscilloscope settings are: (a) Time base 200 ns/div, (b) Channel A and B input + 5 V DC, (c) Trigger single shot, rising to capture a pulse when GO is selected, (d) Use a sensitivity of 400 mV or so at different pre-trigger values.

It may be noted here that DrDAQ is a versatile instrument, as shown in Fig. (*iii*) that connects to the USB port of any PC. Using the installed PicoScope software it can be conveniently used as an oscilloscope, spectrum analyzer and signal generator.

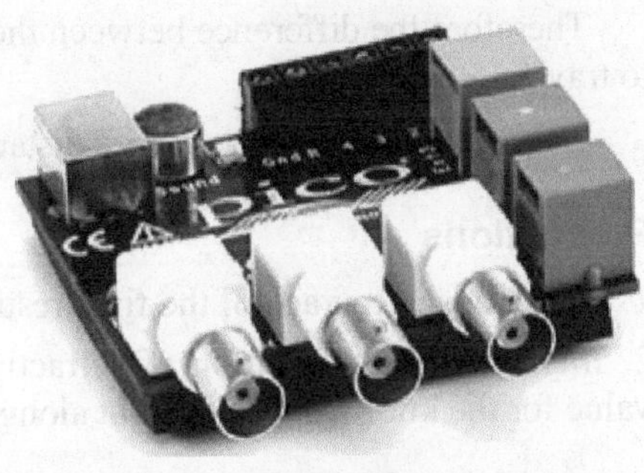

Fig. (*iii*)

Procedure

1. Switch on both the transmitter and the receiver and select GO on the oscilloscope screen. Adjust the pre-trigger and then reselect GO to get a trace showing the "square" shaped transmitted pulse. Then the received pulse will be shifted to the right.

   After a familiarity with the shape of the pulses, reset the time base for a possible short time to display the two leading edges of the pulses. Then save it and take a print out of the screen image. A typical pattern is shown in Fig. (*iv*) (a).
2. The units are then to be switched off and the 5 m cable is to be replaced with a 20 m cable. By switching on the units again and with a careful adjustment of the time base, the whole experiment then can be repeated. Again save the display for the lading edges of the pulse and take a print out of it for the cable of larger length. This time there will be more delay between the two pulses. A typical pattern is shown in Fig. (*iv*) (b).

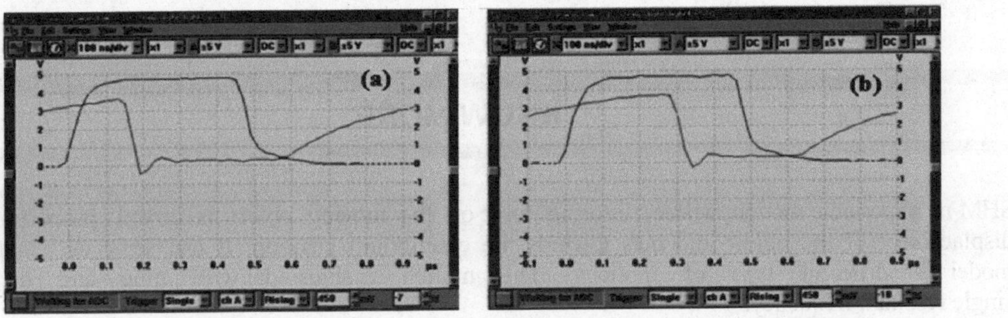

Fig. (*iv*) [https://www.picotech.com/library/experiment/speed-of-light]

3. As the light travels with 3×10^8 m/s in air, so if the fibre optic cable is 30 m long there will be a time delay of 0.1 μs. A 50 MHz oscilloscope is required for the purpose of doing this experiment. When we monitor the electrical pulses, we can measure the delay between transmitted and received electrical signals. At high frequencies "square waves" become rounded owing to stray capacitances and non-linear responses of the converters.

Calculation

If there is any time delay then that delay should be same for both the parts of the experiments with two cables of different lengths, *viz.*, 5 m and 20 m cables. The primary difference between two parts is that the light will travel a further 15 m in the second part. If we determine δt from each print-out graph then subtracting one from the other we can get the amount of time taken to travel the extra 15 m.

From the two print-outs using leading edges we have, for the 5 m cable δt, 0.55 (– 0.02) μs and for the 20 m cable $\delta t = 0.64$ (– 0.01) μs *i.e.*, 0.57 μs and 0.65 μs.

Therefore the difference between these two values is 0.08 μs and this is the time taken by the light to travel 15 m.

Hence, the speed of the light = distance/time = $15/(0.08 \times 10^{-6}) = 1.9 \times 10^{8}$ m/s.

Precautions

1. To establish accuracy of the final result, the errors in all the measurements are to be estimated.
2. Implement the definition of refractive index to find whether you can determine an approximate value for the known speed of light along the fibre optic cable.

Reference

https://www.picotech.com/library/experiment/speed-of-light

Viva-Voce Questions

1. What is the velocity of light?
2. How does the present experimental result can be compared with the speed of light in air?
3. Using the definition of refractive index, suggest how you can calculate an approximate value for the expected speed of light along the fibre optic cable.
4. How do the results compare when trailing edges are utilized in the calculation?
5. Estimate the errors in all the measurements and establish the estimated accuracy of the final result.

KNOW MORE

SHM is a periodic motion where restoring force on the moving object is directly proportional to the displacement of the object and acts towards the equilibrium position. It can serve as a mathematical model for different type of motions. The motion is sinusoidal with time and demonstrates single resonant frequency.

Activity

Many phenomena can be modeled by using SHM. SHM can also be used to model molecular vibration. Some popular examples are, (i) A mass attached to a spring (ii) Uniform circular motion and (iii) Motion of a simple pendulum in the small-angle approximation. Some computer model can also be developed to understand the inter-relations among the displacement, velocity and acceleration of a particle executing SHM.

Interesting facts

A Scotch yoke mechanism can be utilized for converting between rotational motion and linear reciprocating motion. Depending on the shape of the slot, the linear motion can take different forms of simple harmonic motion. The Scotch yoke is a reciprocating motion mechanism that converts the linear motion of a slider into rotational motion or *vice versa.*

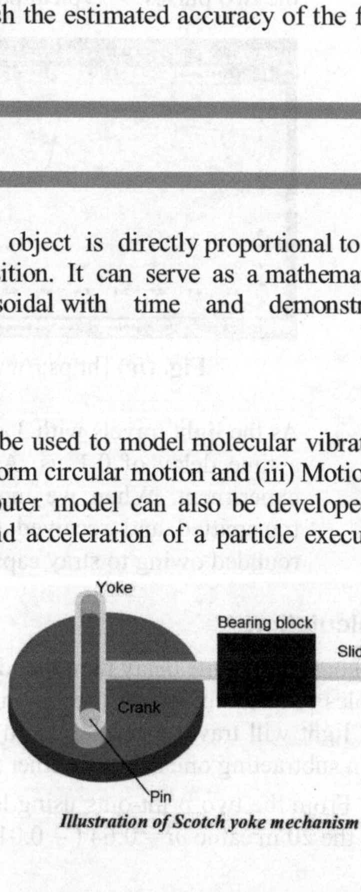

Illustration of Scotch yoke mechanism

The reciprocating part (piston) is directly coupled to a sliding yoke with a slot that engages a pin on the rotating part. Location of the piston vs. time is simple harmonic motion, *i.e.*, having constant amplitude and constant frequency and hence exhibits a constant rotational speed.

Analogy

All SHM are oscillatory and periodic but not all oscillatory motions are simple harmonic. In real world, oscillations hardly follow true SHM. Friction acts to dampen the motion and so needs more force to continue. A guitar string stops to oscillate in a few seconds after being plucked. We can sometimes consider friction and other non-conservative forces small or negligible as completely undamped motion is rare. Under some situations we may even want to damp oscillations, as with car shock absorbers.

History

The origin of this branch of science is generally attributed to the Greek philosopher Pythagoras (6th century BC). His experiment on the properties of vibrating strings produced pleasing musical intervals. These were of such merit that they led to a tuning system to bear his name. Leonardo DaVinci, the well known Italian thinker and artist, was credited with discovering that sound moves in the form of waves. He made this discovery around the year 1500. However, some accounts pointed out that the Roman philosopher Seneca actually discovered sound waves in the first century AD.

Timelines

1564–1642: Modern study of waves is said to have originated with Galileo Galilei.

1636: French mathematician Marin Mersenne investigated the vibration of stretched strings; the results were summarized in the three Mersenne's laws. Later in the century Robert Hooke, an English physicist, first produced a sound wave of known frequency, by making use of a rotating cog wheel as a measuring device. Further, in the 19th century French physicist Felix Savart discovered a disk called Savart's disk, often used today for demonstrations during physics lectures.

1650: The concept of sound as a wave was challenged but it was not established experimentally until the first bell-in-vacuum experiment was performed by Athanasius Kircher, a German scholar.

1660: The Anglo-Irish scientist Robert Boyle improved vacuum technology where he could observe sound intensity decreasing virtually to zero as air was pumped out. Boyle then concluded correctly that a medium like air is essential for transmission of sound waves.

1740: G. L. Bianconi demonstrated that the speed of sound in air increases with temperature.

Late 17th and early 18th centuries: Studies of the relationship between frequency and pitch and of waves in stretched strings were carried out by the French physicist Joseph Sauveur.

Applications (Real Life / Industrial)

Essentially a Scotch yoke is used to generate a sinusoidal motion. The Scotch yoke is used when the slot in the yoke is shorter than the diameter of the circle made by the crank pin. For example, the side rods of a locomotive may have Scotch yokes to permit vertical motion of intermediate driving axles. Though it is not a very common metal working machine today, crude shapers can use Scotch yokes. Those use a whit worth linkage that gives a slow speed forward cutting stroke and a faster return. It has been used in different internal combustion engines, like the Bourke engine, SyTech engine and many hot air engines and steam engines.

Case Study (Environmental / Sustainability / Social / Ethical Issues)

Observing the nature, we find that many physical processes (*e.g.*, the Earth rotation around its polar axis) are repetitive, happening the facts cyclically. In these cases we speak about periodic motion and characterize it by its period, which is the required time to complete a cycle or by its frequency that represents the number of complete cycles in the unit of time. A very interesting case of periodic motion appears when a physical system oscillates around an equilibrium position. The system describes the same path, first in a sense and then in the opposite one, inverting the motion sense in both ends of the path. A complete cycle covers the crossing of the equilibrium position twice. The mass hanging from a pendulum or a spring, the electrical charge of a condenser, the strings of a musical instrument and the molecules of a crystalline net are examples of physical systems that often realize oscillatory motion.

The simplest case of oscillating motion is known as simple harmonic motion and takes place when the total force on the system is a restoring linear force. Fourier's theorem provides the reason of its importance. In fact, any periodic function may be built from a set of simple harmonic functions.

Inquisitiveness and Curiosity Topics

Under ideal engineering conditions, any force can be applied directly in the line of travel of the assembly. The sinusoidal motion and sinusoidal acceleration result in smoother operation. The higher percentage of time spent at top dead centre improves theoretical engine efficiency of constant volume combustion cycles. It permits the elimination of joints typically served by a wrist pin and near elimination of piston skirts and cylinder scuffing, as side loading of piston due to sine of connecting rod angle is mitigated.

REFERENCES AND SUGGESTED READINGS

1. J. Walker, *Principles of Physics*, 9th Edition, Hoboken, New Jersey: Wiley, 2011.
2. S. T. Thornton and J. B. Marion, *Classical Dynamics of Particles and Systems*, 5th Edition, Brooks Cole, 2003.
3. R. Eshleman, *Basic Machinery Vibrations*: *An Introduction to Machine Testing, Analysis, and Monitoring*, 1999.
4. B. Tongue, *Principles of Vibration*, Oxford University Press, 2001.
5. D. J. Inman, *Engineering Vibration*, Prentice Hall, 2001.
6. W. T. Thompson, *Theory of Vibrations*, Nelson Thornes Ltd, 1996.
7. D. Hartog, *Mechanical Vibrations*, Dover Publications, 1985.
8. D. D. Reynolds, *Engineering Principles of Mechanical Vibration*, 4th Edition, Bloomington, Indiana, USA: Trafford on Demand Publishing, 2016.
9. P. A. Simionescu, *Computer Aided Graphing and Simulation Tools for AutoCAD Users*, 1st Edition, Boca Raton, FL: CRC Press, 2014.
10. Mobius Institute; *Vibration Analyst Category* 2 - *Course Notes* 2013.
11. https://www.austincc.edu/mmcgraw/files_2425/Chap_15Ha-Oscillations.pdf
12. https://www.deepspace.ucsb.edu/wp-content/uploads/2015/01/Chap-12-Oscillations1.pdf
13. https://ww2.odu.edu/~jdudek/Phys111N_materials/8_oscillations_waves.pdf
14. https://nptel.ac.in/courses/115/105/115105083/
15. https://nptel.ac.in/courses/115/106/115106119/

2 Wave Propagation and Dispersion

UNIT SPECIFICS

We have emphasized the following aspects in this unit:

- Transverse waves on a string and their wave equation
- Harmonic waves, reflection and transmission of waves at a boundary
- Impedance matching, standing waves and their eigen frequencies
- Longitudinal waves and their wave equation
- Acoustics waves, speed of sound and standing waves
- Waves with dispersion: water waves
- Superposition of waves: Fourier method
- Wave groups: group velocity

The applications of the topics are discussed for generating further curiosity and creativity as well as improving problem solving capacity.

Besides giving a large number of multiple choice questions as well as questions of short and long answer types marked in two categories following lower and higher order of Bloom's taxonomy, assignments through a number of numerical problems, a list of references and suggested readings are given in the unit so that one can go through them for practice. It is important to note that for getting more information on various topics of interest some QR codes have been provided in different sections which can be scanned for relevant supportive knowledge.

After the end of the unit, based on the content, there is a "Know More" section. This section has been carefully designed so that the supplementary information provided in this part becomes beneficial for the users of the book. This section mainly highlights the initial activity, examples of some interesting facts, analogy, history of the development of the subject focusing the salient observations and finding, timelines starting from the development of the concerned topics up to the recent time, applications of the subject matter for our day-to-day real life or/and industrial applications on variety of aspects, case study related to environmental, sustainability, social and ethical issues whichever applicable, and finally inquisitiveness and curiosity topics of the unit.

RATIONALE

This unit on Wave propagation and Dispersion will help students to get a clear idea about the transverse and longitudinal waves in one dimension and some preliminaries on dispersion. It establishes a clear idea on the wave equation on a string and the characteristics of harmonic waves including reflection and transmission of waves at a boundary, impedance matching and the standing waves. Superposition of waves and Fourier method as well as wave groups and group velocity are discussed at length.

Dispersion is caused either by geometric boundary conditions or due to interaction of the waves in the transmitting medium. Elementary particles have a nontrivial dispersion relation even in the absence of other media. A dispersive medium is a medium where waves of different frequencies propagate with different velocities. In radio waves dispersion corresponds to a frequency-dependent variation with the refractive index of the medium.

PRE-REQUISITES

Mathematics: Differential equations (Class XII)
Physics: Wave motion (Class XII)

UNIT OUTCOMES

List of outcomes of this unit is as follows:

U2-O1: Apply boundary conditions for longitudinal and transverse wave motion

U2-O2: Demonstrate wave dispersion phenomenon from the consideration of group velocity and phase velocity

U2-O3: Interpret impedance matching concept from the idea of reflection and transmission of waves at a boundary

U2-O4: Apply superposition principle in different types of wave motion

U2-O5: Analyze the nature of harmonics in case of longitudinal and transverse wave motion

U2-O6: Develop the wave equation for the water waves

Unit-2 Outcome	EXPECTED MAPPING WITH COURSE OUTCOMES *(1- Weak Correlation; 2- Medium Correlation; 3- Strong Correlation)*					
	CO-1	CO-2	CO-3	CO-4	CO-5	CO-6
U2-O1	-	-	3	-	-	2
U2-O2	-	-	2	-	-	1
U2-O3	-	-	2	-	-	2
U2-O4	-	-	3	-	-	3
U2-O5	-	-	3	-	-	2
U2-O6	-	-	2	-	-	3

2.1 INTRODUCTION

The transfer of energy from any particular point to another through a medium may be referred as a wave. Some examples are: radio waves, water waves etc. We can classify waves as transverse and longitudinal waves. The first one acts perpendicular to the direction of propagation while the other one acts along the line of wave propagation. Example of transverse waves would be that generated from a vibrating guitar string, while that of a longitudinal wave would be sound wave or ultrasound wave.

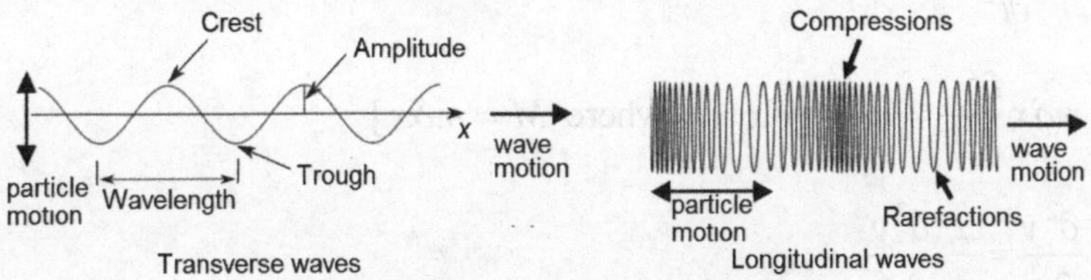

Fig. 2.1: Illustrative properties of waves

In a transverse wave, the crests and troughs are characterized by the locations of maximum up or down displacements. By amplitude we mean the maximum displacement. Wavelength means the distance of one complete cycle. It may also be regarded as the separation from crest to crest or trough to trough.

For longitudinal wave, maximum displacement is measured considering compressions and rarefactions. If the wave is stronger, the medium becomes more compressed and spread out.

2.2 TRANSVERSE WAVE ON A STRING

The wave that occurs in a string is a type of transverse wave. In this type of wave, the wave direction is at right angle to the direction along which the string oscillates. For all practical vibrators, as for example, in the case of a string the masses are distributed along the length of the string.

For this assembly of mass points we have to formulate the equation of motion and solve them thereafter. Instead of looking at individual masses we can find out the displacement of a segment of such a vibrator from its undisturbed position with respect to position and time. The factors causing departure from the ideal condition may be due to the influences of different factors like:

i) non-uniform density; ii) yielding of the support and iii) stiffness.

2.2.1 Wave Equation of Transverse Vibration

When a string is fixed at its ends and drawn perpendicular to its length such that the vibrational amplitude is small enough so as to have constant tension in the string with respect to time, transverse waves travel along the string. Consider a small section of a string between x and $x+\delta x$ as shown in Fig. 2.2.

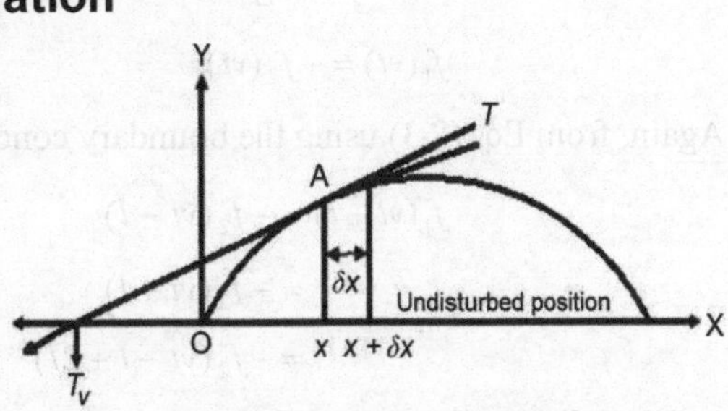

Fig. 2.2: Transverse vibration of a string

Let m be the mass per unit length of the string. The vertical component of tension exerted at any point A is

$$T_V = T\sin\theta \cong T\tan\theta = T\frac{\partial y}{\partial x}$$

where, $$T\left(\frac{\partial y}{\partial x}\right)_{x+\delta x} - T\left(\frac{\partial y}{\partial x}\right)_x \cong T\frac{\partial^2 y}{\partial x^2}\delta x$$ [expanding by Taylor's series expansion]

Now from Newton's law, neglecting the effect of gravity, the equation of motion of the vibrating string may be written as

$$M\frac{\partial^2 y}{\partial t^2} = T\frac{\partial^2 y}{\partial x^2}\delta x$$

or, $$m\delta x\frac{\partial^2 y}{\partial t^2} = T\frac{\partial^2 y}{\partial x^2}\delta x \quad [\text{where } M = m\delta x]$$

or, $$\frac{\partial^2 y}{\partial t^2} = \frac{T}{m}\frac{\partial^2 y}{\partial x^2} \tag{2.1}$$

Comparing this with the general differential wave equation

$$\frac{\partial^2 y}{\partial t^2} = v^2\frac{\partial^2 y}{\partial x^2}$$

we get $$v^2 = \frac{T}{m}$$

or, $$v = \sqrt{\frac{T}{m}} \tag{2.2}$$

The general solution of Eq. (2.1) may be written as,

$$y = f_1(vt + x) + f_2(vt - x) \tag{2.3}$$

Here $f_1(vt + x)$ represents waves travelling towards the negative X-axis with velocity v and $f_2(vt - x)$ represents waves travelling towards the positive X-axis with the same velocity.

When the ends of the string are fixed rigidly at its end, the train of waves gets reflected from the fixed ends and after superposition with the original one, it results a stationary wave.

Now, from Eq. (2.3) using the boundary condition that $y = 0$ at $x = 0$ for all values of t we have,

$$f_1(vt) = -f_2(vt)$$

Again, from Eq. (2.3) using the boundary condition that $y = 0$ at $x = l$ for all values of t we have,

$$f_1(vt + l) = -f_2(vt - l)$$
$$= -f_2(vt + l)$$
$$= -f_2(vt - l + 2l)$$

which indicates that the motion repeats exactly after a spatial distance $2l$ or after a time interval taken by the disturbance to travel the same distance.

So, we can conclude that the vibration of a string is a periodic function of period $\frac{2l}{v}$.

Example 2.1 With what tension must a rope of length 5 m and mass 0.16 kg should be stretched to generate transverse waves of frequency 60 Hz and wavelength 0.8 m?

Solution

We know, $v = \sqrt{\dfrac{T}{m}} = \sqrt{\dfrac{T}{M/L}}$

Here, $M = 0.16$ kg, $L = 5$m and $v = 60 \times 0.8 = 48$ m/s.

$$\therefore \quad 48 = \sqrt{\frac{T}{0.16/5}}$$

or, $\quad T = 73.7$ N.

EXAMPLE 2.1

2.2.2 Solution of the Differential Equation

Considering both the spatial and temporal dependence, let the solution of Eq. (2.1) is,

$$y = X(x)F(t) \tag{2.4}$$

where $X(x)$ is a function of x and $F(t)$ is a function of t only.

Differentiating Eq. (2.4) twice with respect to x we get,

$$\frac{\partial^2 y}{\partial x^2} = F\frac{\partial^2 X}{\partial x^2}$$

Again differentiating Eq. (2.4) twice with respect to t we get,

$$\frac{\partial^2 y}{\partial t^2} = X\frac{\partial^2 F}{\partial t^2}$$

So, from Eq (2.1) we have

$$X\frac{\partial^2 F}{\partial t^2} = \frac{T}{m}F\frac{\partial^2 X}{\partial x^2}$$

or,
$$\frac{1}{F}\frac{\partial^2 F}{\partial t^2} = v^2\frac{1}{X}\frac{\partial^2 X}{\partial x^2} = -\omega^2 \text{ (say)}$$

So,
$$\frac{\partial^2 F}{\partial t^2} = -\omega^2 F \tag{2.5}$$

and
$$\frac{\partial^2 X}{\partial x^2} = -\frac{\omega^2}{v^2}X \tag{2.6}$$

Solutions of Eqs. (2.5) and (2.6) may be written as,

$$F = A\cos\omega t + B\sin\omega t \tag{2.7}$$

and
$$X = C\cos\frac{\omega x}{v} + D\sin\frac{\omega x}{v} \tag{2.8}$$

Using Eqs. (2.7) and (2.8); Eq (2.4) may be written as,

$$y(x,t)=(A\cos\omega t+B\sin\omega t)\left(C\cos\frac{\omega x}{v}+D\sin\frac{\omega x}{v}\right) \tag{2.9}$$

The constants *A*, *B*, *C* and *D* can be calculated from the following boundary conditions:

a) $y=0$ at $x=0$ for all values of *t* and

b) $y=0$ at $x=l$ for all values of *t*.

Putting condition a) in Eq. (2.9) we get $C=0$ and putting condition b) in Eq. (2.9) we get,

$$D\sin\frac{\omega l}{v}=0 \tag{2.10}$$

So, $D\neq 0$ as in that case $X=0$ and hence $y(x,t)=0$ for all values of *t*, and this contradicts the motion of the vibrating string.

Thus $$y(x,t)=D\sin\frac{\omega x}{v}(A\cos\omega t+B\sin\omega t) \tag{2.11}$$

2.2.3 Eigen Frequencies and Eigen Functions

Eigen values and eigen functions help to simplify the problems with linear operation. From Eq. (2.10) with $D\neq 0$ we must have,

$$\sin\frac{\omega l}{v}=0=\sin n\pi$$

or, $$\frac{\omega l}{v}=n\pi \tag{2.12}$$

So, $$y(x,t)=D\sin\frac{n\pi x}{l}\left(A\cos\frac{n\pi vt}{l}+B\sin\frac{n\pi vt}{l}\right) \tag{2.13}$$

The values of n in Eq (2.12) are known as the *eigen values* and functions corresponding to them are the *eigen functions*.

The frequencies $f=\frac{\omega}{2\pi}=\frac{nv}{2l}$ are termed as the normal frequencies of the vibrating string. For a given *n*, the solution of Eq. (2.1) may be expressed by using Eq. (2.13) as,

$$y_n(x,t)=\sin\frac{n\pi x}{l}\left(A_n\cos\frac{n\pi vt}{l}+B_n\sin\frac{n\pi vt}{l}\right) \tag{2.14}$$

Eq. (2.14) represents the normal mode of vibration of the string. Each point of the string vibrates with the same frequency $\omega_n=\frac{n\pi v}{l}$ with amplitude, varying sinusoidally along the length of the string. So, Eq. (2.1) represents a linear and homogeneous equation and by superposition principle the sum of the solutions (2.14) for different values of *n* will be a solution as given below,

$$y(x,t)=\sum_{n=1}^{\infty}y_n(x,t)=\sum_{n=1}^{\infty}\sin\frac{n\pi x}{l}\left(A_n\cos\frac{n\pi vt}{l}+B_n\sin\frac{n\pi vt}{l}\right) \tag{2.15}$$

Eq. (2.15) represents the general equation of the displacement for the transverse vibration of a string.

Example 2.2: A transverse wave is described by the equation $y = y_0 \sin 2\pi\left(\upsilon t - \frac{x}{\lambda}\right)$.

If the maximum particle velocity becomes equal to four times the wave velocity, show that $\lambda = \frac{\pi y_0}{2}$.

Solution

Given wave is represented as, $y = y_0 \sin 2\pi\left(\upsilon t - \frac{x}{\lambda}\right)$

where, υ is the frequency and λ is the wavelength.

Thus, the wave velocity is $v = \upsilon\lambda$

The particle velocity for the given wave is, $v_p = \frac{\partial y}{\partial t} = 2\pi\upsilon y_0 \cos 2\pi\left(\upsilon t - \frac{x}{\lambda}\right)$

Thus, the maximum particle velocity is $v_{p_{max}} = 2\pi\upsilon y_0$

According to the condition of the problem, $v_{p_{max}} = 4v$

$\therefore \quad 2\pi\upsilon y_0 = 4\upsilon\lambda$

or, $\quad \lambda = \frac{\pi y_0}{2}$.

EXAMPLE 2.2

2.2.4 Reflection and Transmission of Waves: Impedance Matching

Consider two ropes, one thick and one thin, are connected together (Fig. 2.3). If a sinusoidal wave moves from left side to right side along the ropes, the speed (v) of the wave along a rope under tension T will be $v = \sqrt{T/m}$, where m is the mass per unit length. So, the speed and the wavelength will be less in the thicker rope. Let, the speed in the left hand rope is v_1 and that in the right hand rope is v_2.

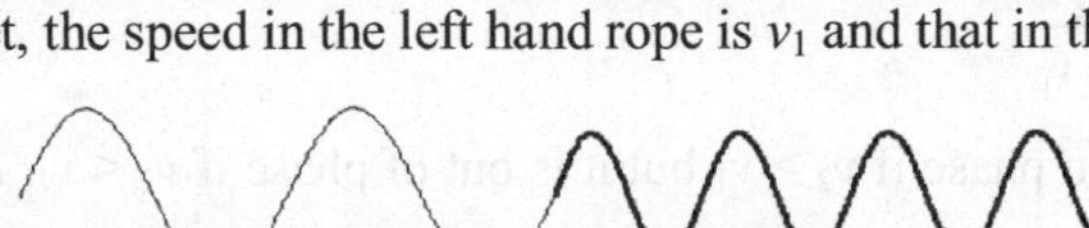

Fig. 2.3: Impedance matching at a boundary

At the boundary ($x = 0$), some of the wave is transmitted and some is reflected. When a wave meets a boundary it gets reflected. A hard boundary will invert the reflection while a soft boundary keeps the original sense. At any discontinuity in the medium (*e.g.*, passing from higher to lower density), partial transmission and partial reflection will be observed. From minimum value to the maximum value of the density there will be an inversion at the point of reflection.

We wish to find how much is transmitted and how much is reflected at the boundary. For the incident, transmitted and reflected waves we can express the displacement equations as follows:

$$y_i = \cos\omega\left(t - \frac{x}{v_1}\right) \tag{2.16 (a)}$$

$$y_t = \tau \cos\omega\left(t - \frac{x}{v_2}\right) \tag{2.16 (b)}$$

and $$y_r = R\cos\omega\left(t + \frac{x}{v_1}\right) \tag{2.16 (c)}$$

If we consider the right side and the left side of the boundary, we can respectively express the displacement as,

$$y = \tau\cos\omega\left(t - \frac{x}{v_2}\right) \text{ and } y = \cos\omega\left(t - \frac{x}{v_1}\right) + R\cos\omega\left(t + \frac{x}{v_1}\right).$$

At the boundary ($x = 0$), unless the rope breaks these two displacements must be equal, and therefore

$$\tau = 1 + R \tag{2.17}$$

Again at the right side of the boundary $\frac{\partial y}{\partial x} = \frac{\tau}{v_2}\sin\omega\left(t - \frac{x}{v_2}\right)$

and at the left side of the boundary $\frac{\partial y}{\partial x} = \frac{1}{v_1}\sin\omega\left(t - \frac{x}{v_1}\right) - \frac{R}{v_1}\sin\omega\left(t + \frac{x}{v_1}\right)$.

At the boundary ($x = 0$), unless there is a kink in the rope, these two must be equal, and therefore

$$\frac{\tau}{v_2} = \frac{1}{v_1} - \frac{R}{v_1} \tag{2.18}$$

Combining this equation with Eq. (2.17) we get

$$R = \frac{v_2 - v_1}{v_2 + v_1} \tag{2.19}$$

and $$\tau = \frac{2v_2}{v_2 + v_1} \tag{2.20}$$

So the reflected wave is in phase if $v_2 > v_1$ but it is out of phase if $v_2 < v_1$, *i.e.*, there is a phase change due to reflection.

If $v_2 = v_1$ (*i.e.*, if there is only one rope) there is no reflected wave as expected.

For a complex wave at the boundary, Eqs. (2.17) and (2.18) will be, $\tau = 1 + R$ and $\frac{\tau}{Z_2} = \frac{1}{Z_1} - \frac{R}{Z_1}$ where $Z_1 = \frac{T}{v_1}$ and $Z_2 = \frac{T}{v_2}$ are the complex impedances.

In that case, the reflection and transmission coefficients may be expressed as

$$R = \frac{Z_1 - Z_2}{Z_1 + Z_2} \tag{2.21}$$

and $$\tau = \frac{2Z_1}{Z_1 + Z_2} \tag{2.22}$$

Note that if $Z_1 = Z_2$ there is no reflection but a complete transmission. For no reflection, we need to match impedances, *e.g.*, for impedance-matching across the two strings having different mass densities m_1 and m_2 we can write the tensions as, $T_2 = \frac{m_1}{m_2} T_1$ so that,

$$Z_2 = \frac{T_2}{v_2} = \sqrt{T_2 m_2} = \sqrt{\frac{m_1}{m_2} T_1 m_2} = \sqrt{T_1 m_1} = Z_1 \tag{2.23}$$

Hence the impedances can match along strings of different mass density. If $Z_1 < Z_2$, the transmission coefficient becomes greater than 1. This means that the amplitude increases when a wave travels from a medium of lower impedance to a medium of higher impedance.

2.3 LONGITUDINAL VIBRATION OF A ROD

Pochhammer in 1876 and Chree in 1889 successfully examined the longitudinal vibrations of a rod. They published their works for an infinite rod with circular cross-section.

To examine such vibrations, consider a uniform solid rod with its length large compared to its lateral dimensions. Let, the axis of the rod is along the direction of propagation of sound. Let, P and Q are two normal plane sections at distances x and $x + \delta x$ from some arbitrary reference as shown in Fig. 2.4.

2.3.1 Wave Equation of Longitudinal Vibration

Let due to the longitudinal waves at some instant, the layers P and Q are displaced by y and $y + \frac{dy}{dx}\delta x$ to P' and Q' respectively. Now, initially the thickness between the layers P and Q is δx and finally the thickness between the layers P' and Q' is $\delta x + \frac{dy}{dx}\delta x$.

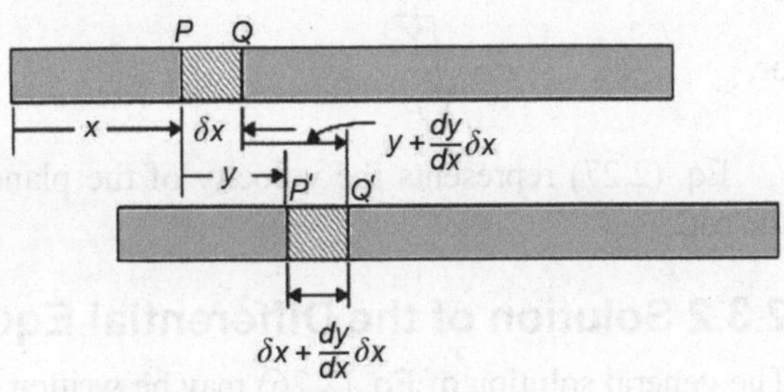

Fig. 2.4: Longitudinal vibration of a rod

So, the change in length of the slice is $\frac{dy}{dx}\delta x$.

$$\therefore \text{ longitudinal strain} = \frac{\frac{dy}{dx}\delta x}{\delta x} = \frac{dy}{dx} \tag{2.24}$$

If F is the stretching force per unit area of P and Y be the Young's modulus of the material of the rod then

$$Y = \frac{F}{\frac{dy}{dx}}$$

or,

$$F = Y\frac{dy}{dx}$$

which acts along the negative X direction on the plane P.

Similarly the force per unit area that acts along the positive X direction on the plane Q is

$$F' = Y\frac{dy}{dx} + \frac{d}{dx}\left(Y\frac{dy}{dx}\right)\delta x$$

Hence the resultant force per unit area δF that acts along the positive X direction on PQ is $Y\frac{d^2y}{dx^2}\delta x$.

This will cause the slice to accelerate and hence from Newton's second law, the equation of motion of the slice of the rod of unit cross section and length δx may be written as,

$$\rho\delta x\frac{d^2y}{dt^2} = Y\frac{d^2y}{dx^2}\delta x \tag{2.25}$$

where ρ is the density of the material of the rod.

So,
$$\frac{d^2y}{dt^2} = \frac{Y}{\rho}\frac{d^2y}{dx^2} \tag{2.26}$$

Comparing this with the general differential wave equation

$$\frac{d^2y}{dt^2} = c^2\frac{d^2y}{dx^2}$$

we get
$$c^2 = \frac{Y}{\rho}$$

or,
$$c = \sqrt{\frac{Y}{\rho}} \tag{2.27}$$

Eq. (2.27) represents the velocity of the plane progressive wave propagating along the length of the rod.

2.3.2 Solution of the Differential Equation

The general solution of Eq. (2.26) may be written as,

$$y = f_1(vt + x) + f_2(vt - x) \tag{2.28}$$

When a solid is excited to longitudinal vibration by any means, the direct waves and the waves reflected from the ends interfere to produce stationary waves, whose characteristics can be determined from the boundary conditions. As in the case of a string, the displacement should be doubly periodic with respect to x and t and hence

$$y = \sum_1^{\infty}\left(A_s\cos snt + B_s\sin snt\right)\sum_1^{\infty}\left(C_r\cos rmx + D_r\sin rmx\right) \tag{2.29}$$

For Eq. (2.29) to be a solution of Eq. (2.26), we must have $s = r$ and $n = mc$. So, Eq. (2.29) becomes,

$$y = \sum_{s=1}^{\infty}\left(A_s\cos smct + B_s\sin smct\right)\left(C_s\cos smx + D_s\sin smx\right) \tag{2.30}$$

Case I: Both ends of the rod are fixed

In this case the boundary conditions are:

i) $y = 0$ at $x = 0$ and

ii) $y = 0$ at $x = l$.

From condition i) using Eq. (2.30) we have

$$\sum_{s=1}^{\infty}(A_s \cos smct + B_s \sin smct)C_s = 0$$

So, $$C_s = 0 \tag{2.31}$$

and from condition ii) using Eq. (2.30) we have

$$\sum_{s=1}^{\infty}(A_s \cos smct + B_s \sin smct)D_s \sin sml = 0$$

Now, $D_s \neq 0$ as in that case $y = 0$ for all x and t.

Thus, $\sin sml = 0 = \sin s\pi$

or, $$m = \frac{\pi}{l} \tag{2.32}$$

So, $$y = \sum_{s=1}^{\infty}\left(A_s \cos\frac{s\pi ct}{l} + B_s \sin\frac{s\pi ct}{l}\right)D_s \sin\frac{s\pi x}{l}$$

or, $$y = \sum_{s=1}^{\infty}\left(A_s' \cos\frac{s\pi ct}{l} + B_s' \sin\frac{s\pi ct}{l}\right)\sin\frac{s\pi x}{l} \tag{2.33}$$

In the above equation, A_s' and B_s' are two new constants.

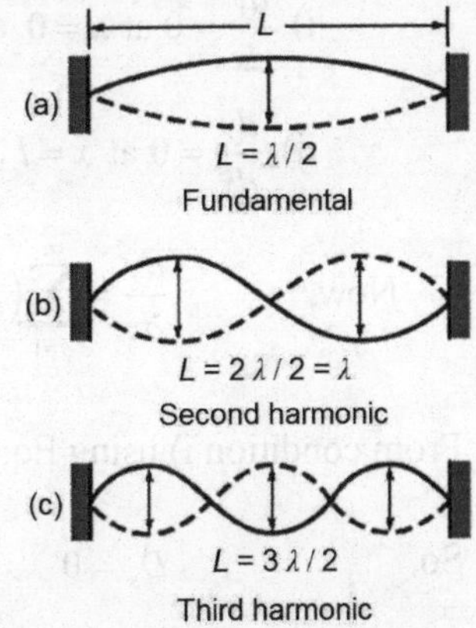

Fig. 2.5: Fundamental and harmonics when both ends of a rod are fixed

From, Eq. (2.33) we have

$$\omega_s = \frac{s\pi c}{l}$$

or, $$\upsilon_s = \frac{sc}{2l} \tag{2.34}$$

which represents different modes of frequencies.

Hence, the fundamental frequency is $\upsilon_1 = \frac{c}{2l} = \frac{1}{2l}\sqrt{\frac{Y}{\rho}}$, frequency of the second harmonic is $\upsilon_2 = \frac{2c}{2l} = \frac{2}{2l}\sqrt{\frac{Y}{\rho}} = 2\upsilon_1$, that of the third harmonic is $\upsilon_3 = \frac{3c}{2l} = \frac{3}{2l}\sqrt{\frac{Y}{\rho}} = 3\upsilon_1$ and so on.

So, a complete series of harmonics will be obtained.

Case II: Both ends of the rod are free

In this case the boundary conditions are:

i) $\frac{dy}{dx} = 0$ at $x = 0$ and

ii) $\frac{dy}{dx} = 0$ at $x = l$.

Now,
$$\frac{dy}{dx} = \sum_{s=1}^{\infty}\left(A_s \cos smct + B_s \sin smct\right) \times sm\left(-C_s \sin smx + D_s \cos smx\right) \tag{2.35}$$

From condition i) using Eq. (2.35) we have $\sum_{s=1}^{\infty}\left(A_s \cos smct + B_s \sin smct\right) smD_s = 0$

So,
$$D_s = 0 \tag{2.36}$$

and from condition ii) using Eq. (2.35) we have $\sum_{s=1}^{\infty}\left(A_s \cos smct + B_s \sin smct\right) smC_s \sin sml = 0$

Now, $C_s \neq 0$ as in that case $y = 0$ for all x and t. Thus, $\sin sml = 0 = \sin s\pi$

or,
$$m = \frac{\pi}{l}$$

So,
$$y = \sum_{s=1}^{\infty}\left(A_s \cos\frac{s\pi ct}{l} + B_s \sin\frac{s\pi ct}{l}\right) C_s \cos\frac{s\pi x}{l}$$

or,
$$y = \sum_{s=1}^{\infty}\left(A_s'' \cos\frac{s\pi ct}{l} + B_s'' \sin\frac{s\pi ct}{l}\right)\cos\frac{s\pi x}{l} \tag{2.37}$$

In the above equation, A_s'' and B_s'' are two new constants.

Case III: Rod is clamped at one end and free at the other

In this case the boundary conditions are:

i) $y = 0$ at $x = 0$ and

ii) $\frac{dy}{dx} = 0$ at $x = l$.

From condition i) using Eq. (2.30) we have $\sum_{s=1}^{\infty}\left(A_s \cos smct + B_s \sin smct\right) C_s = 0$

So,
$$C_s = 0$$

and from condition ii) using Eq. (2.35) we have $\sum_{s=1}^{\infty}\left(A_s \cos smct + B_s \sin smct\right) smD_s \cos sml = 0$

Now, $D_s \neq 0$ as in that case $y = 0$ for all x and t.

Thus,
$$\cos sml = 0 = \cos\frac{s\pi}{2} \text{ with } s = 1,3,5,\ldots$$

or, $$m = \frac{\pi}{2l} \tag{2.38}$$

So, $$y = \sum_{s=1}^{\infty}\left(A_s \cos\frac{s\pi ct}{2l} + B_s \sin\frac{s\pi ct}{2l}\right) D_s \sin\frac{s\pi x}{2l}$$

or, $$y = \sum_{s=1}^{\infty}\left(A_s' \cos\frac{s\pi ct}{2l} + B_s' \sin\frac{s\pi ct}{2l}\right) \sin\frac{s\pi x}{2l} \tag{2.39}$$

From, Eq. (2.39) we have $\omega_s = \dfrac{s\pi c}{2l}$

or, $$\upsilon_s = \frac{sc}{4l} \tag{2.40}$$

which represents different modes of frequencies.

Hence, the fundamental frequency is $\upsilon_1 = \dfrac{c}{4l} = \dfrac{1}{4l}\sqrt{\dfrac{Y}{\rho}}$, frequency of the second harmonic is $\upsilon_2 = \dfrac{2c}{4l} = \dfrac{2}{4l}\sqrt{\dfrac{Y}{\rho}} = 2\upsilon_1$, that of the third harmonic is $\upsilon_3 = \dfrac{3c}{4l} = \dfrac{3}{4l}\sqrt{\dfrac{Y}{\rho}} = 3\upsilon_1$ and so on.

So, a complete series of harmonics will be obtained.

2.4 STANDING WAVE

When two waves travel in the opposite directions with the same amplitude and interfere with each other, then due to their superposition a standing wave is generated. This interference so occurs at the particular points along the medium that seems to be standing still. As the wave pattern is characterized by the points, the pattern is named as a *standing wave pattern* which is created within the medium at certain frequencies of vibration known as *harmonics*. At a frequency other than harmonics, the interference of reflected and incident waves produces a resulting disturbance of the medium which is irregular and non-repeating. The points where the displacement does not change are known as the *nodes* while the points where the displacement changes with maximum amplitude are known as the *antinodes*.

2.4.1 Standing Wave in a Pipe

If a wave propagates through a closed end pipe it will reflect the wave. At the closed end of a pipe there will be a pressure antinode where the pressure fluctuates from minimum to maximum value. So, closed ends of the pipe are responsible for velocity nodes and pressure antinodes.

Though an open end pipe will partly transmit and partly reflect the sound wave as there is a discontinuity in the medium. At the open end of a pipe, there will be a pressure node. In a pipe, at open end, air is free to move in and out. However, the pressure cannot differ from atmospheric pressure. Therefore, at the open ends of the pipe there are velocity antinodes and pressure nodes.

The minimum frequency of the standing wave which occurs in a cavity (*e.g.*, an air pipe) is called the fundamental resonant frequency while the higher ones are known as overtones of the cavity. First, we consider the case of a pipe whose both ends are open. At the ends there have to be pressure nodes.

The first few resonant patterns possible in an open and a closed pipe are shown in Fig. 2.7. For a pipe that is open at both ends (*e.g.*, flute) or closed at both ends, the frequencies and wavelengths are given by the same equations as that of a string fixed at both ends. These are given below,

$$\lambda = \frac{2L}{n} \tag{2.41}$$

$$\upsilon_n = \frac{n}{2L}v = n\upsilon_1 \tag{2.42}$$

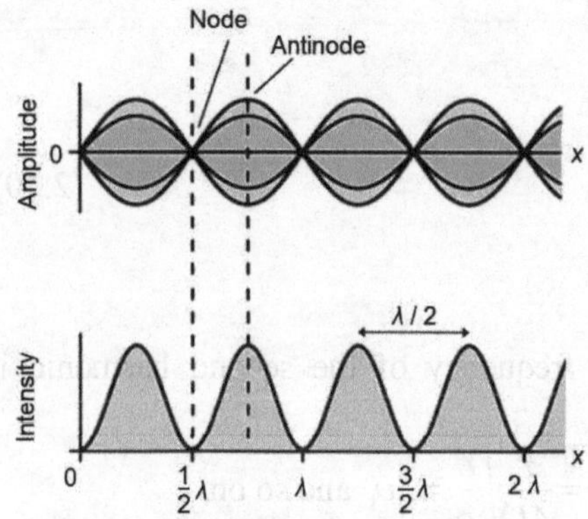

Fig 2.6: Formation of standing waves

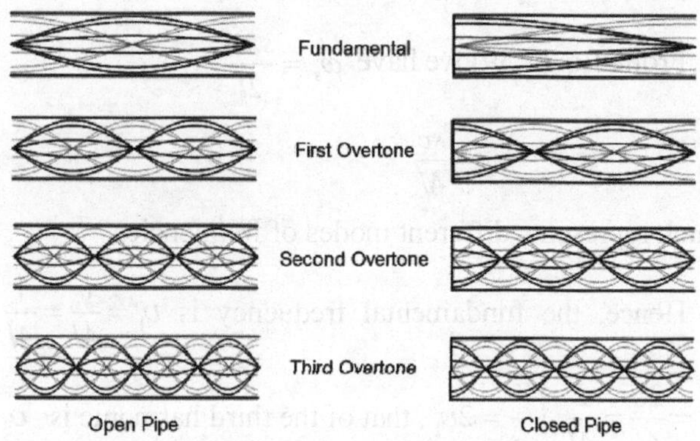

Fig 2.7: Fundamental and overtones in open and closed pipes

For a pipe which is closed at one end and open at the other (*e.g.*, xylophone), there should be a velocity node (and therefore, a pressure antinode) at the closed end, and a pressure node (*i.e.*, velocity antinode) at the open end. For a pipe, that is open at one end and closed at the other, only the odd harmonics (n = 1, 3, 5, ,...) exist and the fundamental has a wavelength that is 4 times the length of the pipe and so for such a pipe

$$\lambda = \frac{4L}{n} \tag{2.43}$$

and

$$\upsilon_n = \frac{n}{4L}v = n\upsilon_1 \tag{2.44}$$

2.4.2 Standing Waves in Finite String

We can choose a string whose both ends are fixed or one end is fixed and one end is free or both ends are free. Let the ends be located at $x = 0$ and L. Generally, for all t, the boundary conditions are $y = 0$ at fixed end and $\partial y/\partial x = 0$ at a free end (as the slope is zero and there is no transverse force on the massless endpoint).

Two fixed ends: At the fixed ends of the string, the boundary conditions are $y\,(0, t) = 0$ and $y\,(L, t) = 0$ for all t and the most general equation for $y(x,t)$ will be,

$$y = \sum_{n=0}^{\infty} B_s \sin(\omega_n t + \varphi_n)\sin k_n x$$

where $k_n = \dfrac{n\pi}{L}$

and $\omega_n = vk_n$.

Thus, the frequency is given by $\upsilon_n = \dfrac{\omega_n}{2\pi} = \dfrac{vk_n}{2\pi} = \dfrac{v\dfrac{n\pi}{L}}{2\pi} = \dfrac{nv}{2L}$ (2.45)

and the wavelength is given by $\lambda_n = \dfrac{2L}{n}$ (2.46)

The possible frequencies are therefore integer multiples of *fundamental frequency* $\upsilon_1 = \dfrac{v}{2L}$.

One fixed end, one free end: When one end of the string is free and the other end is fixed, the boundary conditions are $y(0, t) = 0$ and $\partial y/\partial x = 0$ at $x = L$ for all t. Here, the most general equation for $y(x,t)$ will be,

$$y = \sum_{n=0}^{\infty} B_s \sin(\omega_n t + \varphi_n) \sin k_n x$$

where $k_n = \dfrac{\left(n+\dfrac{1}{2}\right)\pi}{L}$ and $\omega_n = vk_n$.

Dissimilar to the two fixed ends case, $n = 0$ value will give a non-trivial wave.

SCAN ME

to see standing waves on a *vibrating* string

The frequency is given by $\upsilon_n = \dfrac{\omega_n}{2\pi} = \dfrac{vk_n}{2\pi} = \dfrac{v\left(n+\dfrac{1}{2}\right)\pi}{2\pi L} = \dfrac{\left(n+\dfrac{1}{2}\right)v}{2L}$ (2.47)

and the wavelength is given by $\lambda_n = \dfrac{2L}{n+\dfrac{1}{2}}$ (2.48)

So, for the lowest mode ($n = 0$), a quarter of a wavelength fits on the string. The higher modes are then obtained by successively adding on half oscillation (which ensures that $x = L$ is always located at an antinode, with zero slope).

Two free ends: At the free ends of the string, the two boundary conditions are $\partial y/\partial x = 0$ at $x = 0$ and $\partial y/\partial x = 0$ at $x = L$ for all t. In this case, the most general equation for $y(x,t)$ will be,

$$y = \sum_{n=0}^{\infty} B_s \cos(\omega_n t + \varphi_n) \cos k_n x$$

where $k_n = \dfrac{n\pi}{L}$ and $\omega_n = vk_n$.

The frequency is given by $\upsilon_n = \dfrac{\omega_n}{2\pi} = \dfrac{vk_n}{2\pi} = \dfrac{v\dfrac{n\pi}{L}}{2\pi} = \dfrac{nv}{2L}$ (2.49)

and the wavelength is given by $\lambda_n = \dfrac{2L}{n}$ (2.50)

The possible frequencies are therefore integer multiples of the *fundamental frequency* $\upsilon_1 = \dfrac{v}{2L}$.

Example 2.3 The waves on a string move at 50 m/s. The length of the string is 7.5 m. If it is fixed at both the ends, what will be the fundamental standing wave frequency? If one end is kept free, what will be the fundamental standing wave frequency? Is there any possibility to say that the string is either fixed or free at one end by considering the sequence of resonance frequencies?

Solution

A string when fixed at both ends the standing wave resonance frequencies is given by, $\upsilon_n = \frac{nv}{2L}$

So, the fundamental standing wave frequency is

$$\upsilon_1 = \frac{v}{2L} = \frac{50}{2\times 7.5} = 3.33 \text{ Hz.}$$

Again, the standing wave resonance frequencies of a string fixed at only one end is given by, $\upsilon_n = \frac{nv}{4L}$

So, the fundamental standing wave frequency, when one end is kept free is

$$\upsilon_1 = \frac{v}{4L} = \frac{50}{4\times 7.5} = 1.67 \text{ Hz.}$$

For a sequence of resonance frequencies, when the string is fixed at either ends or open at one end, we can infer by considering at the ratio of the frequencies. To get the ratio we can divide it by the highest common divisor. For a string when fixed at both ends, the ratio will have both even and odd numbers. A string, if fixed at one end, will have only odd numbers. It may also be noted that the greatest common factor is the fundamental frequency.

EXAMPLE 2.3

2.5 SUPERPOSITION OF WAVES

2.5.1 Superposition Principle

If two waves or more than two waves are simultaneously present at any single point in space, the displacement at that point becomes the sum of the displacement of each individual wave.

2.5.2 Fourier Theorem

Any function $f(x)$, provided it satisfies a set of conditions, can be expanded in a trigonometric series, *i.e.*, a series of *sines* and *cosines* as

$$f(x) = a_0 + \sum a_n \cos nx + \sum b_n \sin nx \tag{2.51}$$

The requirements of conditional representation of $f(x)$ are that

i) $f(x)$ is single valued in the interval $(-\pi, \pi)$,

ii) $f(x)$ is bounded,

iii) $f(x)$ has at most a finite number of both maxima and minima,

iv) $f(x)$ has a finite number of discontinuities - piecewise continuous and

v) $f(x+2\pi)=f(x)$ for values of x outside of $(-\pi,\pi)$.

The above set of conditions laid down in (i) to (v) is referred to as Dirichlet's conditions. If it is the function of *t*, then $f(t)=a_0+\sum a_n\cos n\omega t+\sum b_n\sin\omega nt$ in the interval $(-T/2,T/2)$.

The equation can be shown into another form as

$$f(x)=a_0+\sum_1^\infty(a_n\cos nx+b_n\sin nx)=a_0+\sum_1^\infty A_n\sin(nx+\delta_n) \tag{2.52}$$

where $a_n=A_n\sin\delta_n$ and $b_n=A_n\cos\delta_n$.

The coefficients a_n, b_n in the Eq. (2.52) are called the Fourier's coefficients and a_0 is a constant defining the average value of the function.

Evaluation of the coefficients: We have,

$$f(x)=a_0+\sum a_n\cos nx+\sum b_n\sin nx.$$

Multiplying both sides by dx and integrating over the entire interval we get,

$$\int_{-\pi}^{\pi}f(x)dx=\int_{-\pi}^{\pi}a_0dx+\int_{-\pi}^{\pi}(\sum a_n\cos nx)dx+\int_{-\pi}^{\pi}(\sum b_n\sin nx)dx$$

$$=a_0\times 2\pi+\sum a_n\int_{-\pi}^{\pi}\cos nxdx+\sum b_n\int_{-\pi}^{\pi}\sin nxdx$$

$$=2\pi a_0$$

$$\therefore \quad a_0=\frac{1}{2\pi}\int_{-\pi}^{\pi}f(x)dx \tag{2.53}$$

Multiplying $f(x)$ by $\cos mx$ and integrating we get

$$\int_{-\pi}^{\pi}f(x)\cos mxdx=a_0\int_{-\pi}^{\pi}\cos mxdx+\sum a_n\int_{-\pi}^{\pi}\cos nx\cos mxdx+\sum b_n\int_{-\pi}^{\pi}\sin nx\cos mxdx$$

Now, $\int_{-\pi}^{\pi}\cos mxdx=0$, $\int_{-\pi}^{\pi}\sin nx\cos mxdx=0$ and $\int_{-\pi}^{\pi}\cos nx\cos mxdx=\pi\delta_{mn}$

where δ_{mn} is known as the Kronecker delta function and is given by,

$$\delta_{mn}=1 \text{ for } m=n$$
$$=0 \text{ otherwise.}$$

$$\therefore \quad \int_{-\pi}^{\pi}f(x)\cos mxdx=\sum a_n\pi\delta_{mn}$$

or, $$\int_{-\pi}^{\pi}f(x)\cos nxdx=\pi a_n$$

$$\therefore \qquad a_n = \frac{1}{\pi}\int_{-\pi}^{\pi} f(x)\cos nx dx \tag{2.54}$$

Similarly we can have,

$$b_n = \frac{1}{\pi}\int_{-\pi}^{\pi} f(x)\sin nx dx \tag{2.55}$$

Applications of Fourier Theorem

i) Square wave: For a square wave, shown in Fig. 2.8 let, $y = k$ for $0 \le t \le \frac{T}{2}$ and $y = 0$ for $\frac{T}{2} \le t \le T$

Now consider, $y = f(t) = a_0 + \sum a_n \cos n\omega t + \sum b_n \sin n\omega t$

So, $$a_0 = \frac{1}{T}\int_0^T f(t)dt = \frac{1}{T}\left[\int_0^{T/2} f(t)dt + \int_{T/2}^{T} f(t)dt\right] = \frac{1}{T}\int_0^{T/2} k dt = \frac{k}{2}$$

$$a_n = \frac{2}{T}\int_0^T f(t)\cos n\omega t dt = \frac{2}{T}\left[\int_0^{T/2} f(t)\cos n\omega t dt + \int_{T/2}^{T} f(t)\cos n\omega t dt\right]$$

$$= \frac{2}{T}\int_0^{T/2} k\cos n\omega t dt = 0$$

and $$b_n = \frac{2}{T}\int_0^T f(t)\sin n\omega t dt = \frac{2}{T}\left[\int_0^{T/2} f(t)\sin n\omega t dt + \int_{T/2}^{T} f(t)\sin n\omega t dt\right]$$

$$= \frac{2}{T}\int_0^{T/2} k\sin n\omega t dt = \frac{2k}{n\omega T}[1 - \cos n\pi]$$

$$= \frac{2k}{n\pi} \text{ for odd } n$$

$$= 0 \quad \text{for even } n$$

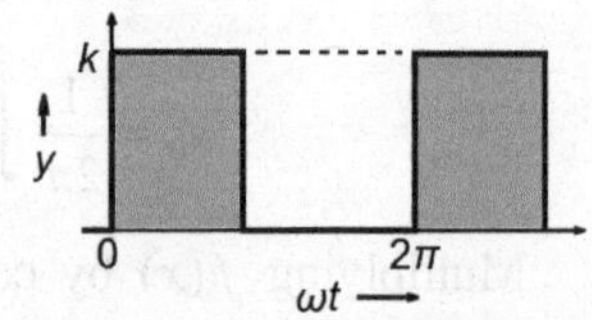

Fig. 2.8: Square waveform

Hence $$y = f(t) = \frac{k}{2} + \frac{2k}{\pi}\left(\sin\omega t + \frac{1}{3}\sin 3\omega t + \frac{1}{5}\sin 5\omega t + ...\right) \tag{2.56}$$

The series shown in Eq. (2.56) when plotted, gives rise to a square waveform (Fig. 2.8).

ii) Saw-tooth wave: For a saw-tooth, shown in Fig. 2.9, wave the displacement reduces to zero from a fixed value as the time is increased from 0 to T.

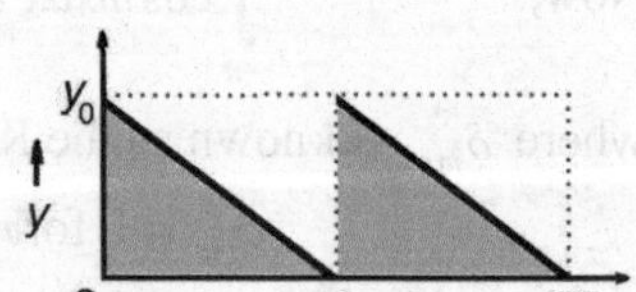

Fig. 2.9: Saw tooth waveform

Let, $$y = f(t) = y_0\left(1 - \frac{t}{T}\right)$$

So, $$a_0 = \frac{1}{T}\int_0^T f(t)dt = \frac{y_0}{T}\int_0^T \left(1 - \frac{t}{T}\right)dt = \frac{y_0}{2}$$

$$b_n = \frac{2}{T}\int_0^T f(t)\sin n\omega t dt = \frac{2y_0}{T}\int_0^T f(t)\sin n\omega t dt = \frac{2y_0}{T}\int_0^T \left(1-\frac{t}{T}\right)\sin n\omega t dt$$

or, $$b_n = \frac{y_0}{n\pi}$$

and, $$a_n = 0$$

Hence $$y = f(t) = \frac{y_0}{2} + \frac{y_0}{\pi}\left(\sin\omega t + \frac{1}{2}\sin 2\omega t + \frac{1}{3}\sin 3\omega t + \ldots\right) \quad (2.57)$$

The series shown in Eq. (2.57) when plotted, gives rise to a saw-tooth waveform (Fig. 2.9).

Fourier theorem can be effectively used to find out the harmonic components from the fundamental to the overtones and or harmonics in a given vibration. It is also useful for determining the quality of a musical note or the designing of suitable filter circuits.

2.6 WAVE VELOCITY AND GROUP VELOCITY

Waves can be taken in group and these groups are termed as wave packets. The velocity with which a wave packet travels is known as *group velocity*. On the other hand, the velocity with which the phase of a wave travels is known as *phase velocity*. It is seen that the relation between group velocity and phase velocity are proportional to each other.

During the wave motion the oscillating particles in the medium do not travel in the medium with the wave. The phase relationship of the particles is being considered as a wave. For the motion of a wave we have to consider the particle velocity, the wave or phase velocity and the group velocity together. The particle velocity is the velocity of simple harmonic motion of the oscillating particles whereas the phase velocity is the velocity with which the plane of same phase travels through the medium. The group velocity on the other hand is the velocity of the group of waves formed due to the superposition of a group of waves.

Now, let us consider a group of waves (Fig. 2.10) formed by two individual progressive waves having equal amplitudes but slightly different frequencies represented by

$$y_1 = a\sin\frac{2\pi}{\lambda_1}(vt - x_1) = a\sin(\omega_1 t - k_1 x) \quad (2.58)$$

and $$y_2 = a\sin\frac{2\pi}{\lambda_2}(vt - x_2) = a\sin(\omega_2 t - k_2 x) \quad (2.59)$$

where $\frac{\omega_1}{k_1}$ and $\frac{\omega_2}{k_2}$ represents respectively the velocity of the individual waves.

The resultant amplitude may be obtained by the principle of superposition as,

$$y = y_1 + y_2 = a\sin(\omega_1 t - k_1 x) + a\sin(\omega_2 t - k_2 x)$$

$$= 2a\sin\left\{\left(\frac{\omega_1+\omega_2}{2}\right)t - \left(\frac{k_1+k_2}{2}\right)x\right\}\cos\left\{\left(\frac{\omega_1-\omega_2}{2}\right)t - \left(\frac{k_1-k_2}{2}\right)x\right\}$$

$$= 2a\sin(\omega t - kx)\cos\left\{\left(\frac{\Delta\omega}{2}\right)t - \left(\frac{\Delta k}{2}\right)x\right\} \quad (2.60)$$

where $\omega = \frac{\omega_1 + \omega_2}{2}$, $k = \frac{k_1 + k_2}{2}$, $\Delta\omega = \omega_1 - \omega_2$ and $\Delta k = k_1 - k_2$.

The resultant wave with amplitude $2a$ has two parts:

(i) A wave of angular frequency ω, propagation constant k and velocity $v_p = \frac{\omega}{k}$ which is modulated in space and time by

(ii) a second wave of angular frequency $\frac{\Delta\omega}{2}$, propagation constant $\frac{\Delta k}{2}$ and velocity $v_g = \frac{\Delta\omega}{\Delta k}$ consisting of a group of waves of type (i).

The velocity with which the modulated pattern moves is known as the group velocity and is given by, $v_g = \frac{\Delta\omega}{\Delta k} = \frac{d\omega}{dk}$ (2.61)

On the other hand the phase velocity is given by $v_p = \frac{\omega}{k}$

Therefore $\frac{dv_p}{dk} = -\frac{\omega}{k^2} + \frac{1}{k}\frac{d\omega}{dk}$

or, $k\frac{dv_p}{dk} = -\frac{\omega}{k} + \frac{d\omega}{dk}$

or, $\frac{d\omega}{dk} = \frac{\omega}{k} + k\frac{dv_p}{dk}$

or, $v_g = v_p + k\frac{dv_p}{dk}$ (2.62)

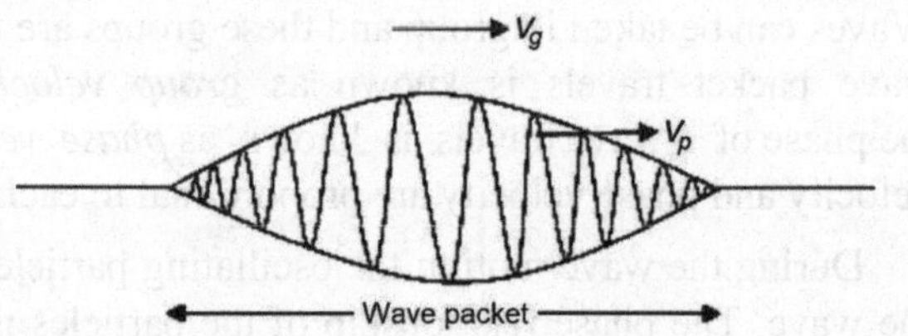

Fig. 2.10: Illustration of group and phase velocities

This gives the relationship between the phase velocity and the group velocity.

2.7 WATER WAVES

Water waves are waves which propagate on the water surface, with gravity and surface tension as the restoring forces. That is why water with a free surface is usually considered as a dispersive medium.

The general idea of dispersive waves originated from the problems of water waves. The problems are of great interest in the Maritime and Offshore settings.

We consider an inviscid incompressible fluid of constant density (ρ). Let, the spatial domain is given in (x_1, x_2, y) and the components of the velocity vector $\vec{u}$ are (u_1, u_2, v).

Let the external force on the fluid is $\vec{F} = -\rho g \hat{j}$

Now, for irrotational flow of the fluid, $\omega = \nabla \times u = 0$.

Also, let the velocity potential φ is associated with velocity as $\vec{u} = \vec{\nabla}\varphi$.

Thus, the inviscid incompressible equations are $\vec{\nabla}.\vec{u} = 0$

and $$\frac{d\vec{u}}{dt} = \frac{\partial\vec{u}}{\partial t} + (\vec{u}.\vec{\nabla})\vec{u} = -\frac{1}{\rho}\vec{\nabla}p - g\hat{j}$$

so that $$\frac{\partial \vec{u}}{\partial t}+\vec{\nabla}\left(\frac{u^2}{2}\right)+\vec{\omega}\times\vec{u}=-\frac{1}{\rho}\vec{\nabla}p-g\hat{j} \tag{2.63}$$

On integrating, we get

$$\frac{p-p_0}{\rho}=-\varphi t-\frac{1}{2}\left|\vec{\nabla}\varphi\right|^2-gy \tag{2.64}$$

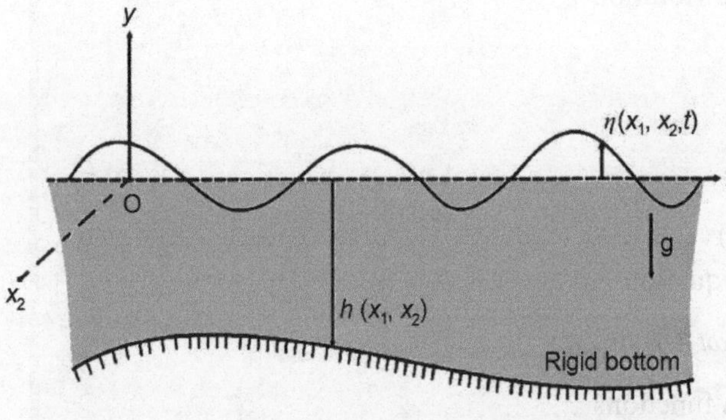

Fig. 2.11: Flow domain and boundaries: Water waves

where $\vec{u}=\vec{\nabla}\varphi$.

Choosing an interface,

$$f(x_1,x_2,y,t)=0$$

with $y=\eta(x_1,x_2,t)$ such that

$$f(x_1,x_2,y,t)\equiv\eta(x_1,x_2,t)-y$$

Kinematic free surface condition will be

$$\frac{D_\eta}{D_t}=\eta_t+u_1\eta_{x_1}+u_2\eta_{x_2}=v$$

and we obtain dynamic free surface condition as $p=p_0$.

Boundary conditions at the free surface (on $y=\eta$) is

$$\eta_t+\varphi_{x_1}\eta_{x_1}+\varphi_{x_1}\eta_{x_2}-\varphi_y=0$$

and $$\varphi_t+\frac{1}{2}\left|\vec{\nabla}\varphi\right|^2+g\eta=0.$$

At the bottom $\left[y=-h_0(x_1,x_2)\right]$ we obtain kinematic bottom boundary condition

$$\varphi_{x_1}h_{0x_1}+\varphi_{x_1}h_{0x_2}-\varphi_y=0 \text{ on } y=-h_0$$

and for a horizontal flat bottom, $\varphi_y=0$ on $y=-h_0$.

Water waves are classified into three main categories:

Shallow water or long waves, if $\frac{h_0}{\lambda}<\frac{1}{20}$

Intermediate water waves, if $\frac{1}{20}<\frac{h_0}{\lambda}<\frac{1}{2}$ and

Deep water waves, if $\frac{h_0}{\lambda}>\frac{1}{2}$.

UNIT SUMMARY

- ***Transverse wave on a string***

  Wave equation of transverse vibration

  $$\frac{\partial^2 y}{\partial t^2} = \frac{T}{m}\frac{\partial^2 y}{\partial x^2}$$

  $$v = \sqrt{\frac{T}{m}}$$

  $$y = f_1(vt + x) + f_2(vt - x)$$

  Solution of the differential equation

  $$y(x,t) = D\sin\frac{\omega x}{v}\left(A\cos\omega t + B\sin\omega t\right)$$

  Eigen frequencies and eigen functions

  $$\frac{\omega l}{v} = n\pi$$

  $$y(x,t) = \sum_{n=1}^{\infty} y_n(x,t) = \sum_{n=1}^{\infty} \sin\frac{n\pi x}{l}\left(A_n\cos\frac{n\pi vt}{l} + B_n\sin\frac{n\pi vt}{l}\right)$$

  Reflection and transmission of waves: impedance matching

  $$R = \frac{Z_1 - Z_2}{Z_1 + Z_2}$$

  $$\tau = \frac{2Z_1}{Z_1 + Z_2}$$

  $$Z_2 = \sqrt{T_2 m_2} = \sqrt{T_1 m_1} = Z_1$$

- ***Longitudinal vibration of a rod***

  The wave equation of longitudinal vibration of a rod

  $$\frac{d^2 y}{dt^2} = \frac{Y}{\rho}\frac{d^2 y}{dx^2}$$

  Velocity of propagation

  $$c = \sqrt{\frac{Y}{\rho}}$$

- ***Standing wave***

  Standing waves in pipes

  Open at both ends

  $$\lambda = \frac{2L}{n}$$

$$\upsilon_n = \frac{n}{2L} v = n\upsilon_1$$

Open at one end and closed at the other

$$\lambda = \frac{4L}{n}$$

$$\upsilon_n = \frac{n}{4L} v = n\upsilon_1$$

Standing sound waves in finite string

Two fixed ends

$$\upsilon_n = \frac{\omega_n}{2\pi} = \frac{vk_n}{2\pi} = \frac{v\frac{n\pi}{L}}{2\pi} = \frac{nv}{2L}$$

$$\lambda_n = \frac{2L}{n}$$

One end fixed, one end free

$$\upsilon_n = \frac{\omega_n}{2\pi} = \frac{vk_n}{2\pi} = \frac{v\left(n+\frac{1}{2}\right)\pi}{2\pi L} = \frac{\left(n+\frac{1}{2}\right)v}{2L}$$

$$\lambda_n = \frac{2L}{n+\frac{1}{2}}$$

Two ends free

$$\upsilon_n = \frac{\omega_n}{2\pi} = \frac{vk_n}{2\pi} = \frac{v\frac{n\pi}{L}}{2\pi} = \frac{nv}{2L}$$

$$\lambda_n = \frac{2L}{n}$$

- ***Superposition of waves***

  Superposition principle: If two waves or more than two waves are simultaneously present at any single point in space, the displacement at that point becomes the sum of the displacement of each individual wave.

  Fourier Theorem

  $$f(x) = a_0 + \sum a_n \cos nx + \sum b_n \sin nx$$

- ***Wave velocity and group velocity***

  $$v_g = \frac{\Delta\omega}{\Delta k} = \frac{d\omega}{dk}$$

  $$v_p = \frac{\omega}{k}$$

$$v_g = v_p + k\frac{dv}{dk}$$

- ***Water waves***

$$\frac{\partial \vec{u}}{\partial t} + \vec{\nabla}\left(\frac{u^2}{2}\right) + \vec{\omega}\times\vec{u} = -\frac{1}{\rho}\vec{\nabla}p - g\hat{j}$$

EXERCISES

Multiple Choice Questions

2.1 What type of waves are sound waves?

(a) latitudinal waves (b) longitudinal waves
(c) latitudinal mechanical waves (d) longitudinal mechanical waves

2.2 A disturbance in the medium which causes the particles of the medium to undergo vibratory motion about their mean position in equal intervals of time is known as

(a) wave (b) energy (c) vibration (d) amplitude

2.3 Water waves and sound waves are the examples of

(a) transverse waves (b) longitudinal waves
(c) mechanical waves (d) electromagnetic waves

2.4 Which one of the following cannot represent a traveling wave?

(a) $y = f(x - vt)$ (b) $y = y_0 \sin(kx - \omega t)$ (c) $y = y_0 \log(kx - \omega t)$ (d) $y = f(x^2 - vt^2)$

2.5 When a compression is incident on rigid wall it is reflected as

(a) Compression with a phase change of π (b) Compression with no phase change
(c) Rarefaction with a phase change of π (d) Rarefaction with no phase change

2.6 In the longitudinal waves, the direction of vibration of particle in a medium is

(a) perpendicular to propagation of wave (b) parallel to propagation of wave
(c) different from each other (d) variable from time to time

2.7 With the propagation of a longitudinal wave through a material medium, the quantities transmitted in the propagation direction is

(a) energy, momentum and mass (b) energy
(c) energy and mass (d) energy and linear momentum

2.8 Phase difference between two successive troughs in the transverse wave is

(a) 2π (b) π (c) π/2 (d) π/4

2.9 Which of the following phenomenon cannot take place with sound wave?

(a) reflection (b) interference (c) diffraction (d) polarization

2.10 If v is the velocity of the wave and ω is the angular velocity, then the propagation constant (k) of the wave is given by

(a) $k = \frac{v}{\omega}$ (b) $k = \frac{\omega}{v}$ (c) $k = 2\pi n$ (d) $k = \frac{\lambda}{2\pi}$

2.11 Transverse waves can propagate

(a) in a gas but not in a metal (b) in a metal but not in gas

(c) neither in a gas nor in a metal (d) either in a gas or in a metal

2.12 A progressive wave is one which

(a) has vibrations that are perpendicular to the direction of wave propagation

(b) has vibrations that are parallel to the direction of wave propagation

(c) transfers energy in the direction of wave propagation

(d) transfers energy and particles in the direction of wave propagation

2.13 What kind of wave is formed in organ pipes?

(a) transverse stationary waves (b) electromagnetic waves

(c) mechanical waves (d) longitudinal stationary waves

2.14 Which of the following is also known as pressure waves?

(a) transverse waves (b) longitudinal waves (c) mechanical waves (d) stationary waves

2.15 While moving from deep water to shallow water

(a) frequency of water waves increases (b) frequency of water waves decreases

(c) frequency of water waves remain the same (d) none of these

Answers of Multiple Choice Questions

2.1 (d), 2.2 (a), 2.3 (c), 2. 4 (c), 2.5 (a), 2.6 (b), 2.7 (d), 2.8 (a), 2.9 (d), 2.10 (b), 2.11 (b), 2.12 (c), 2.13 (d), 2.14 (b), 2.15 (c)

Short and Long Answer Type Questions

Category I

2.1 Note down the differences between transverse and longitudinal waves.

2.2 Show that the velocity of transverse vibration of a string is $v = \sqrt{\frac{T}{m}}$.

2.3 Establish the differential equation for the transverse vibration of a string and hence solve it to find a possible solution. Using the boundary conditions calculate the normal frequencies of the vibrating string.

2.4 Mention the importance of impedance matching for the reflection and transmission of waves at a boundary.

2.5 Show that the amplitude of waves increases when a wave travels from lower impedance to a higher impedance medium.

2.6 Show that the velocity of longitudinal vibration of a rod is $c=\sqrt{\frac{Y}{\rho}}$.

2.7 Explains a water wave. Classify its types. Explain the kinematic and dynamic free surface conditions for a water wave.

2.8 Using Fourier theorem show that a saw-tooth wave can be represented by

$$y=f(t)=\frac{y_0}{2}+\frac{y_0}{\pi}\left(\sin\omega t+\frac{1}{2}\sin 2\omega t+\frac{1}{3}\sin 3\omega t+.....\right)$$

2.9 Establish the differential equation for the transverse vibration of a rod and hence solve it to find a possible solution. Using the boundary conditions, calculate the modes of frequencies of the rod when the rod is i) fixed at both ends ii) clamped at one end while free at the other and ii) free at both ends.

2.10 What is group velocity and phase velocity? Establish their relationships.

Category II

2.11 Using Fourier theorem show that a square wave can be represented by

$$y=f(t)=\frac{k}{2}+\frac{2k}{\pi}\left(\sin\omega t+\frac{1}{3}\sin 3\omega t+\frac{1}{5}\sin 5\omega t+.....\right)$$

2.12 Mention the conditions for long, intermediate and deep water waves?

2.13 Give the physical meaning of superposition of waves.

2.14 Explain briefly how the different type of waves can be generated.

2.15 Find an expression for the fundamental, first overtone and second overtone for the vibration of air column in an open air pipe.

2.16 Find an expression for the fundamental, first overtone and second overtone for the vibration of air column in a closed air pipe.

2.17 Explain Fourier theorem with its statement. What basic assumptions make the theorem valid?

Numerical Problems

2.1 Find the solution for wave equation $\frac{\partial^2 u}{\partial t^2}=a^2\frac{\partial^2 u}{\partial x^2}$ for a fixed string with the boundary condition $u(x,0)=f(x)$ and $\frac{\partial u(x,0)}{\partial t}=g(x)$ where $f(x)$ and $g(x)$ are some functions such that $f(0)=f(L)=g(0)=g(L)=0$.

2.2 Transverse waves travel through a string where the tension equals 5 N with a speed of 20 m/s. What tension would be required for a wave speed of 30 m/s?

2.3 Three successive resonance frequencies for a certain string are 175 Hz, 245 Hz, and 315 Hz. Find whether this string has an antinode or a node at one end? What is the fundamental frequency? Which harmonics are these resonance frequencies?

If the speed of transverse waves on this string is 125 m/s, find the length of the string?

[*Ans*: antinode; 35 Hz; 5th, 7th, 9th harmonics; 0.893 m]

2.4 Show that a square wave may be represented by the function

$$y = f(t) = \frac{k}{2} + \frac{2k}{\pi}\left(\sin\omega t + \frac{1}{3}\sin 3\omega t + \frac{1}{5}\sin 5\omega t + ...\right).$$

KNOW MORE

Dispersion relations consider the effect of dispersion on the properties of waves in any medium. A dispersion relation is associated with the wavelength or wave number of a wave to its frequency. From the dispersion relation, we can calculate the phase velocity and group velocity of waves in the medium, as a function of frequency. In addition to the geometry-dependent and material-dependent dispersion relations, the Kramers–Kronig relations consider the frequency dependence of wave propagation and the corresponding attenuation.

Activity

Waves are formed in many shapes. All waves share basic properties but some of them can be distinguished from others based on some characteristics. It is common to categorize different type of waves based on these distinguishing characteristics.

Interesting facts

Water waves are good examples of waves that involve a combination of both longitudinal and transverse motions. As wave travels through the waver, the particles travel in clockwise circles. The radius of the circles diminishes as the depth into the water increases.

Analogy

Categorizing waves leads to three distinct categories: transverse waves, longitudinal waves, and surface waves. Dispersion may be the result either by geometric boundary conditions (waveguides, shallow water) or by interaction of the waves with the transmitting medium. Elementary particles, considered as matter waves, have a nontrivial dispersion relation even in the absence of geometric constraints and other media.

History

1564–1642: Modern study of waves is said to have originated with Galileo Galilei.

1636: French mathematician Marin Mersenne investigated the vibration of stretched strings.

1650: First bell-in-vacuum experiment was performed by Athanasius Kircher.

1740: G. L. Bianconi demonstrated that the speed of sound in air increases with temperature.

Timelines

1776: Dispersion of waves on water was investigated by Pierre-Simon Laplace

1926–27: The universality of the Kramers–Kronig relations became apparent with further observations on the dispersion relation. This can further be interconnected with the scattering theory of different types of waves and particles.

Applications (Real Life / Industrial)

Waves propagating through a solid medium can be either transverse or longitudinal waves. But waves moving through the bulk of a fluid (*e.g.*, a liquid or a gas) are always longitudinal. Transverse waves need a relatively rigid medium to transmit their energy. It is for this reason that only longitudinal waves are found to move through the bulk of liquids such as our oceans. Earthquakes are capable of producing both transverse and longitudinal waves which travel through the solid structures of the Earth. When seismologists started to investigate earthquake waves they found that only longitudinal waves were capable of propagating through the core of the Earth. That is why geologists believe that the Earth's core consists of a liquid of molten iron.

Case Study (Environmental / Sustainability / Social / Ethical Issues)

A sound wave moving through air is a classic example of a longitudinal wave. As a sound wave propagates from the lips of a speaker to the ear of a listener, particles of air vibrate back and forth in the same and the opposite direction of energy transport. Each individual particle pushes on its neighboring particle so as to push it forward. This back and forth motion of particles in the direction of energy transport produces regions within the medium where the particles are pressed together and in the other regions the particles are spread apart. Longitudinal waves can be quickly identified by the presence of such regions. This process continues along the chain of particles until the sound wave goes the ear of the listener.

Inquisitiveness and Curiosity Topics

Earthquakes and other geologic disturbances sometimes result in the production of seismic waves. Seismic waves are waves of energy which are transported through the earth and over its surface by both transverse and longitudinal waves.

REFERENCES AND SUGGESTED READINGS

1. F. A. Jenkins and H. E. White, *Fundamentals of Optics*, New York: McGraw-Hill, 1957.
2. R. G. Dean and R. A. Dalrymple, *Water Wave Mechanics for Engineers and Scientists, Advanced Series on Ocean Engineering*, 2, World Scientific, Singapore, 1991.
3. A. D. D. Craik, *The Origins of Water Wave Theory*, Annual Review of Fluid Mechanics, 36: 1–28, 2004.
4. J. S. Toll, *Causality and the Dispersion Relation*: *Logical Foundations*, Phys. Rev, 104 (6): 1760–1770, 1956.
5. R. A. Serway, C. J. Moses and C. A. Moyer, *Modern Physics*, Philadelphia: Saunders, 1989.
6. Crawford Jr., S. Frank, *Waves* (Berkeley Physics Course, Vol. 3), McGraw-Hill, 1968.
7. E. W. Weisstein, *Wave Velocity*, Science World, Retrieved 2009-05-30.
8. A. E. H. Love, *A Treatise on the Mathematical Theory of Elasticity*, New York: Dover, 1892.
9. https://scholar.harvard.edu/files/david-morin/files/waves_interference.pdf
10. https://nptel.ac.in/courses/115/105/115105083/
11. https://nptel.ac.in/courses/115/106/115106119/

3 Propagation of Light and Geometric Optics

UNIT SPECIFICS

We have given emphasis on the following topics in this unit:

- Fermat's principle of stationary time;
- Applications of Fermat's principle in explaining laws of reflection and refraction;
- Light as an electromagnetic wave and Fresnel equations;
- Reflectance and transmittance;
- Total internal reflection;
- Brewster's angle and evanescent wave;
- Mirrors, lenses and optical instruments based on mirrors and lenses;
- Transfer formula and the matrix method.

The practical applications of the topics are discussed for generating further curiosity and creativity as well as improving problem solving capacity.

Besides giving a large number of multiple choice questions as well as questions of short and long answer types marked in two categories following lower and higher order of Bloom's taxonomy, assignments through a number of numerical problems, a list of references and suggested readings are given in the unit so that one can go through them for practice. It is important to note that for getting more information on various topics of interest some QR codes have been provided in different sections which can be scanned for relevant supportive knowledge.

After the related practicals, based on the content, there is a "Know More" section. This section has been carefully designed so that the supplementary information provided in this part becomes beneficial for the users of the book. This section largely focuses the preliminary activity giving examples of some interesting facts, analogy, history of the development of the subject highlighting the salient observations and finding. It further covers timelines starting from the development of the concerned topics up to the recent time. Applications of the subject matter for our day-to-day real life as well as industrial applications on variety of aspects are discussed including various case studies related to environmental, sustainability, social and ethical issues and finally inquisitiveness and curiosity topics of the unit are critically considered.

RATIONALE

This unit on the Propagation of Light and Geometric Optics will help students to get a primary idea about the principle of propagation of light from the geometrical point of view. The Fermat's principle and its use or justifying the laws of reflection and refraction of light are discussed at length. It further considers light as an electromagnetic wave and different related laws applicable to mirrors and lenses, including various optical instruments, based on these principles.

Light wave can propagate either in one mode or in multiple simultaneous modes. Optical fibers with a small core permit only a single mode while larger fibers permit multiple modes. However, light can behave in a variety of ways like a wave or a packet of photons and travels in a straight line, showing the behavior of rectilinear propagation of light. Geometrical optics essentially deals with the propagation of light in a straight line and the methods of geometrical optics allow understanding the conditions for the formation of an optical image of an object. It provides an explanation for many phenomena related to the transmission of optical radiation in various media.

PRE-REQUISITES

Mathematics: Differential equations (Class XII)
Physics: Optics (Class XII)

UNIT OUTCOMES

List of outcomes of this unit is as follows:

U3-O1: Illustrate various laws and principles in optics
U3-O2: Describe various types of mirrors and lenses
U3-O3: Outline total internal reflection from the concepts of critical angle as well as Brewster's angle
U3-O4: Apply the concept of geometric optics in different optical instruments
U3-O5: Apply Fresnel's equations for reflection and transmission coefficient calculation
U3-O6: Develop transfer formula using matrix method

Unit-3 Outcome	EXPECTED MAPPING WITH COURSE OUTCOMES (1- Weak Correlation; 2- Medium Correlation; 3- Strong Correlation)					
	CO-1	CO-2	CO-3	CO-4	CO-5	CO-6
U3-O1	3	1	-	-	-	1
U3-O2	1	1	-	-	-	1
U3-O3	2	-	-	-	-	2
U3-O4	2	-	-	-	-	2
U3-O5	2	-	-	-	-	2
U3-O6	2	-	-	-	-	2

3.1 INTRODUCTION

When light wave is allowed to fall on a surface separating two media, then a part of the incident wave is reflected back to the first medium *i.e.*, returns in the first medium from the interface, while another part is absorbed in the second medium along a different direction. So when a ray of light travels in one homogeneous medium and is incident obliquely on the surface of second medium, it undergoes a change in the direction and this phenomenon is called the refraction of light. This is illustrated in Fig. 3.1 (a) and (b) respectively for refraction from denser to rarer as well as that from rarer to denser medium.

From the figure it is obvious that when light rays travel from denser to rarer medium the refracted ray of light goes away from the normal (MN) and when light rays travel from rarer to denser medium the refracted ray of light moves toward the normal. Here *i*, the angle made by the incident ray (PQ) with the normal (MN), is the angle of incidence and *r*, the angle made by the refracted ray (QR) with the normal (MN), is the angle of refraction.

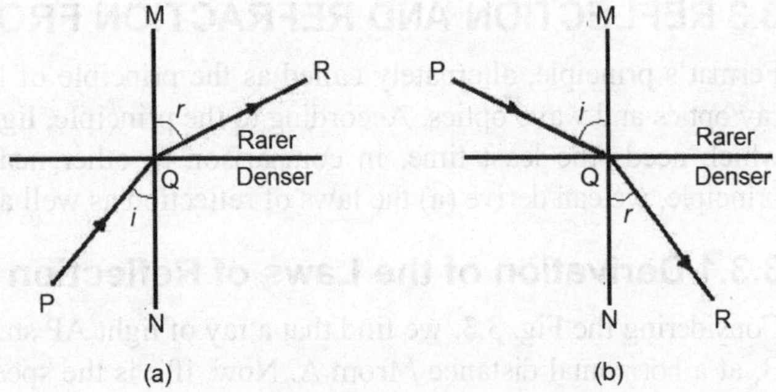

Fig. 3.1: Refraction from (a) denser to rarer and (b) rarer to denser medium

3.2 LAWS OF REFLECTION AND REFRACTION

When light wave meets a boundary of two media, the forward propagation of the wave is disrupted due to scattering and under that situation the incident light is scattered backwards. This is what is known as the reflection of light. It is seen that when the incident light is directed at an angle to the normal, the ray changes its direction. This is what is known as the refraction of light. This change in direction is caused by the difference in speed as the ray travels from one medium to another.

3.2.1 Laws of Reflection

We assume that a ray PO falls on a plane mirror M_1M_2 (Fig. 3.2). Laws of reflection say that the incident ray (PO), the reflected ray (OQ) and the normal (ON) to the mirror will lie in the same plane. Furthermore, the angle of reflection *r* is same as the angle of incidence *i*,

i.e., $\angle i = \angle r$

Here both angles are to be measured with respect to the normal (ON) drawn on the mirror.

3.2.2 Laws of Refraction

When light wave is incident on a interface separating two media (Fig. 3.1), then the direction of the refracted ray is related to that of the corresponding incident ray by the laws of refraction. They are:

1. First law: Incident ray, refracted ray and the normal to the refracting surface at the point of incidence will lie on the same plane.

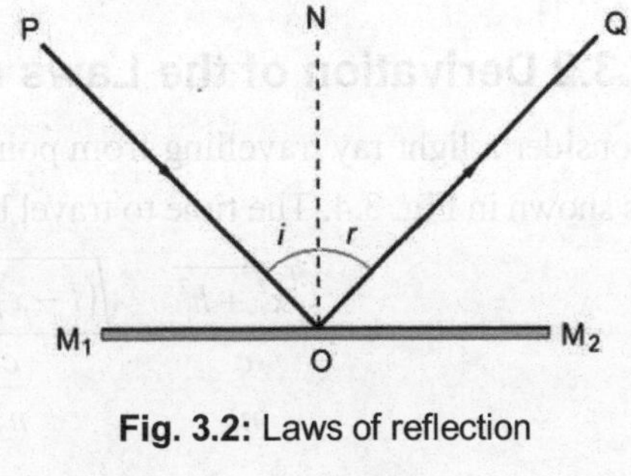

Fig. 3.2: Laws of reflection

2. Second law: For a particular set of media and for a given wavelength of light, the ratio of the *sine* of angle of incidence (sin *i*) and the *sine* of angle of refraction (sin *r*) always exhibits a constant value. It is called Snell's law of refraction and the constant of proportionality, denoted by *n* is known as the refractive index of the medium. *n* is constant for a particular wavelength and set of media.

Mathematically Snell's law may be written as,

$$n = \frac{\sin i}{\sin r} \tag{3.1}$$

3.3 REFLECTION AND REFRACTION FROM FERMAT'S PRINCIPLE

Fermat's principle, alternately called as the principle of least time, provides the connection between ray optics and wave optics. According to the principle, light travels between two points along the path which needs the least time, in comparison to other nearby paths. From the well known Fermat's principle, we can derive (a) the laws of reflection as well as (b) the laws of refraction.

3.3.1 Derivation of the Laws of Reflection

Considering the Fig. 3.3, we find that a ray of light AP starting from A, reflects at P before arriving at B, at a horizontal distance *l* from A. Now, if *c* is the speed of light, the time required by the light to travel between the two points A and B is

$$t = \frac{\sqrt{x^2 + h_1^2}}{c} + \frac{\sqrt{(l-x)^2 + h_2^2}}{c}$$

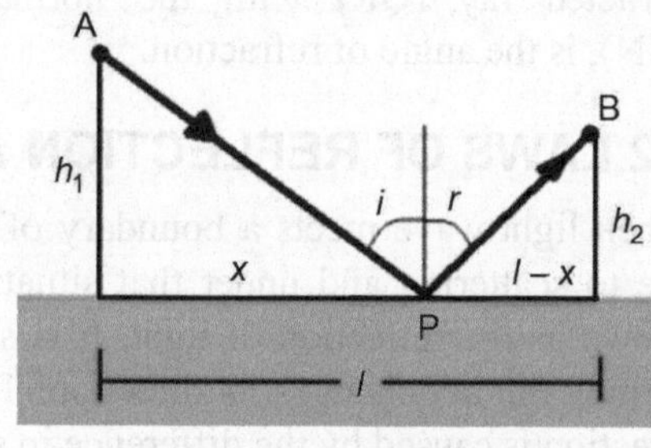

Fig. 3.3: Laws of reflection from Fermat's principle

To minimize the time of travel, we set the derivative of the travel time *t* with respect to *x* as equal to zero. Thus,

$$0 = \frac{dt}{dx} = \frac{x}{c\sqrt{x^2 + h_1^2}} - \frac{(l-x)}{c\sqrt{(l-x)^2 + h_2^2}}$$

or, $$\frac{x}{c\sqrt{x^2 + h_1^2}} = \frac{(l-x)}{c\sqrt{(l-x)^2 + h_2^2}}$$

or, $$\sin i = \sin r$$

or, $$i = r$$

which is the second law of reflection. Hence the laws of reflection are verified from the Fermat's principle.

3.3.2 Derivation of the Laws of Refraction

Consider a light ray travelling from point A to point B in two media with refracting indices n_1 and n_2, as shown in Fig. 3.4. The time to travel between the two points A and B is

$$t = \frac{\sqrt{x^2 + h_1^2}}{\frac{c}{n_1}} + \frac{\sqrt{(l-x)^2 + h_1^2}}{\frac{c}{n_2}} \quad \left[\frac{c}{n} \text{ represents the velocity of light propagation}\right]$$

To minimize the time of travel, we set the derivative of the travel time t with respect to x as equal to zero. Thus,

$$0 = \frac{dt}{dx} = \frac{n_1 x}{c\sqrt{x^2 + h_1^2}} - \frac{n_2 (l - x)}{c\sqrt{(l - x)^2 + h_2^2}}$$

$$\therefore \quad \frac{n_1 x}{c\sqrt{x^2 + h_1^2}} = \frac{n_2 (l - x)}{c\sqrt{(l - x)^2 + h_2^2}}$$

or, $\quad n_1 \sin i = n_2 \sin r$

which is Snell's law of refraction. So, the laws of refraction are verified from Fermat's principle.

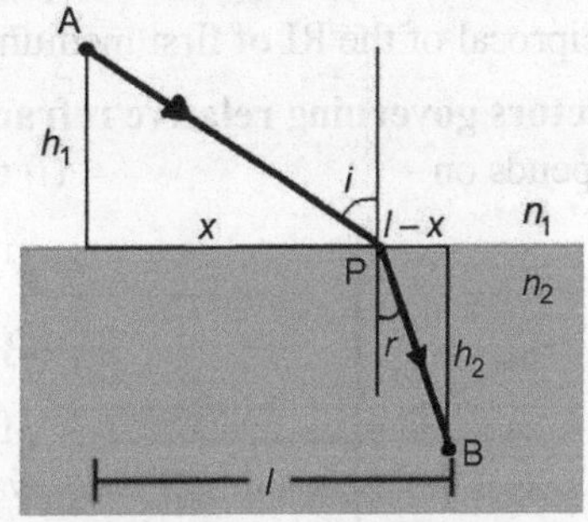

Fig. 3.4: Laws of refraction from Fermat's principle

3.4 REFRACTIVE INDEX

Refractive index (RI) of a medium may be classified into two categories:

(i) relative refractive index and (ii) absolute refractive index.

3.4.1 Relative Refractive Index

When light wave of a particular wavelength is incident from one homogeneous medium to another homogeneous medium then the ratio of *sine* of the angle of incidence and that of the angle of refraction is the relative refractive index of the second medium with respect to the first medium. Thus from Fig. 3.1(b) we have,

$$\frac{\sin i}{\sin r} = \frac{n_2}{n_1} = {}_1n^2 = \text{constant} \tag{3.2}$$

where ${}_1n^2$ is the RI of the denser medium with respect to the rarer medium. From Fig. 3.1 (a) the law may be written as,

$$\frac{\sin i}{\sin r} = \frac{n_1}{n_2} = {}_2n^1 = \text{constant} \tag{3.3}$$

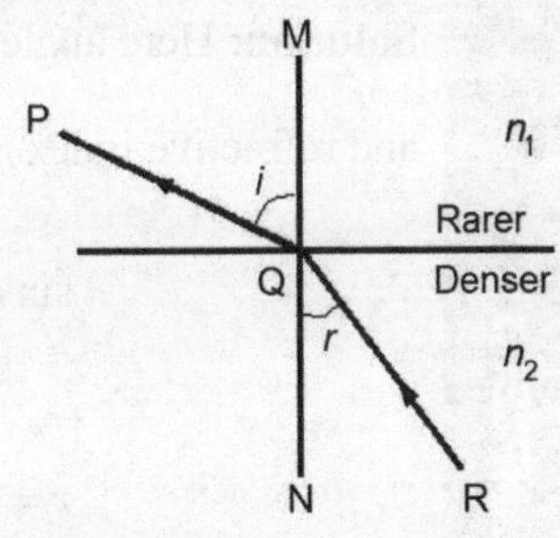

Fig. 3.5: Reverse path of Fig 3.1 (b)

where ${}_2n^1$ is the RI of the rarer medium with respect to the denser medium.

If we consider that the light ray PQ is incident in medium 2 (denser medium) at an angle of incidence r and it is refracted in medium 1 (rarer medium) at an angle of refraction i, then by considering the reversibility of the path of light as in Fig. 3.1 (b), and using the Snell's law we have,

$$\frac{\sin i}{\sin r} = \frac{n_1}{n_2} = {}_2n^1 = \text{constant} \tag{3.4}$$

Now, if we multiply Eqs. (3.2) and (3.4) for the two reversible paths, then we have,

$$1 = {}_1n^2 \times {}_2n^1$$

or, $$ {}_1n^2 = \frac{1}{{}_2n^1} \tag{3.5}$$

i.e., the RI of the second medium (medium 2) with respect to first medium (medium 1) is the reciprocal of the RI of first medium with respect to second medium and vice versa.

Factors governing relative refractive index of a medium: The relative refractive index of a medium depends on –

(i) the wavelength of light used and

(ii) the nature of the media.

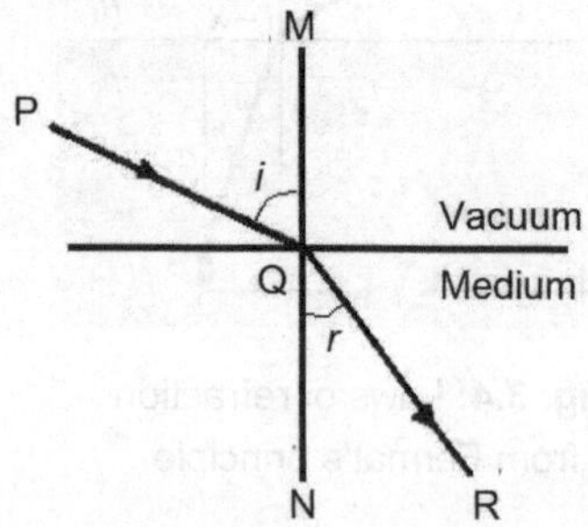

Fig. 3.6: Absolute refractive index of a medium

3.4.2 Absolute Refractive Index

If a ray of light, having a particular wavelength is incident from vacuum to a particular medium then the ratio of *sine* of the angle of incidence to that of the angle of refraction is called the absolute refractive index of the medium corresponding to that wavelength.

So, from Fig. 3.6 using Snell's law we get,

$$\frac{\sin i}{\sin r} = n = \text{constant} \tag{3.6}$$

where n is called the absolute refractive index of the medium *i.e.*, the refractive index of the medium with respect to vacuum ($n_{vac} = 1$, as $i = r$). The absolute refractive index of air is 1.

EXAMPLE 3.1

Example 3.1: A light ray passing through air is making incidence on a glass surface at an angle of incidence of 30^0. Find the refracting angle. Given RI of glass is 1.5.

Solution: Here angle of incidence $i = 30^0$

and refractive index $n = \dfrac{\sin i}{\sin r}$

$$\therefore \qquad \sin r = \frac{\sin i}{n} = \frac{\sin 30^0}{1.5} = \frac{1}{2\times\frac{3}{2}} = \frac{1}{3}$$

$$\therefore \qquad r = \sin^{-1}\left(\frac{1}{3}\right) = 19.47^0$$

Thus the angle of refraction is $= 19.47^0$.

Factors governing absolute refractive index of a medium: The absolute refractive index of a medium depends on –

(i) the wavelength of light used and

(ii) the nature of the medium.

The relationship between absolute refractive index (n) of a medium and wavelength of light used is given by Cauchy's relation,

$$n = A + \frac{B}{\lambda^2} \tag{3.7}$$

where A and B are constants.

3.5 RELATION OF ABSOLUTE AND RELATIVE RI

If n_1 and n_2 be the absolute refractive indices for two media 1 and 2 and ${}_1n^2$ be the relative refractive index of medium 2 with respect to medium 1 then,

$$ {}_1n^2 = \frac{n_2}{n_1} \tag{3.8} $$

Thus, the relative refractive index of one medium with respect to another medium is the ratio of the absolute refractive indices of the second and first medium respectively.

3.6 RI AND SPEED OF LIGHT

The speed of light is different in different media. For some fixed medium, the speed of different colors of light is different *i.e.*, speed of red light is different from that of the green light in a particular medium (say glass). Again for fixed wavelength of light, for different medium, the speed of light is different *i.e.*, for example the velocity of red light in water is different from that in glass.

3.6.1 Absolute RI and Speed of Light

The ratio of velocity of light in vacuum to the velocity of light in a particular medium is called the absolute refractive index of the medium and is given by,

$$ n = \frac{\text{velocity of light in vacuum}}{\text{velocity of light in a medium}} = \frac{c}{v} \tag{3.9} $$

From Eq. (3.9), it is clear that for vacuum ($v = c$) the absolute refractive index is

$$ n_{vac} = \frac{c}{c} = 1 \tag{3.10} $$

Now since the velocity of light in a medium is always less than that of light in vacuum (*i.e.*, $v < c$), so from Eq. (3.9) we have

$$ n_{med} > 1 \tag{3.11} $$

i.e., the absolute refractive index of a medium is always greater than 1.

Example 3.2: What is the speed of light in glass, if the refractive index of glass is 1.5? Given speed of light in air is 3×10^8 m/s.

Solution: We know, refractive index of a medium is,

$$ n = \frac{\text{velocity of light in vacuum}}{\text{velocity of light in a medium}} $$

or, $$ n = \frac{c}{v} $$

or, $$ v = \frac{c}{n} = \frac{3\times10^8}{1.5} \text{ m/s.} $$

So, the speed of light in glass is 2×10^8 m/s.

EXAMPLE 3.2

3.6.2 Relative RI and Speed of Light

The relative RI of one medium (medium 2) with respect to another (medium 1) is given by,

$$ {}_1n^2 = \frac{n_2}{n_1} = \frac{c/v_2}{c/v_1} = \frac{v_1}{v_2} \quad \text{[Using Eq. (3.9)]} \tag{3.12}$$

or, $nv = $ constant

So, for a given wavelength of light, the product of n and v is constant and it is not depending on the nature of the media *i.e.*, for two media water and glass,

$$ n_{\text{water}} \times v_{\text{water}} = n_{\text{glass}} \times v_{\text{glass}} \tag{3.13}$$

or, $$\frac{v_{\text{glass}}}{v_{\text{water}}} = \frac{n_{\text{water}}}{n_{\text{glass}}}$$

or, $$v \propto \frac{1}{n} \tag{3.14}$$

So, the speed of light for different media, for a particular wavelength varies inversely as the absolute refractive index of the media.

Again, using $v = \gamma\lambda$, where γ is the frequency and λ the wavelength, we get from Eq. (3.13)

$$ n_1\gamma_1\lambda_1 = n_2\gamma_2\lambda_2 \tag{3.15}$$

Now, $\gamma_1 = \gamma_2$, as the frequency of the refracted ray is the same as that of the incident ray. Thus from Eq. (3.15) we get,

$$ n\lambda = \text{constant} \tag{3.16}$$

So, for a particular wavelength of light passing from one medium (water) to another medium (glass) the product of n and λ is constant *i.e.*,

$$ n_{\text{water}} \times \lambda_{\text{water}} = n_{\text{glass}} \times \lambda_{\text{glass}}$$

or, $$\frac{n_{\text{glass}}}{n_{\text{water}}} = \frac{\lambda_{\text{water}}}{\lambda_{\text{glass}}}$$

or, $$n \propto \frac{1}{\lambda} \tag{3.17}$$

So, the refractive index of the material of a medium increases with decrease in wavelength and *vice-versa*.

Using Eq. (3.16) in Eq. (3.13) we get,

$$\frac{v_1}{v_2} = \frac{n_2}{n_1} = \frac{\lambda_1}{\lambda_2} \tag{3.18}$$

or, $$\frac{v_1}{v_2} = \frac{\lambda_1}{\lambda_2} \tag{3.19}$$

or, $$v \propto \lambda \tag{3.20}$$

Hence, speed of light in a particular medium increases as the wavelength increases and *vice-versa*.

3.7 GENERALIZED SNELL'S LAW

The interface of two media of varying refractive indices can be artificially structured for introducing an abrupt shift of phase in the light path. This represents a function of the position along the interface.

For a particular wavelength of light when it passes from one medium to another medium then the relationship between the *sine* of the angle of incidence and that of the angle of refraction is given by,

$$\frac{\sin i}{\sin r} = \frac{n_2}{n_1}$$ [n_1 and n_2 respectively represents the absolute RI of media 1 and 2]

or, $$n_1 \sin i = n_2 \sin r \quad (3.21)$$

Now, if the light travels from denser to rarer medium then $n_1 > n_2$ and from Eq. (3.21) we get,

$$\sin i < \sin r$$

i.e., $$i < r$$

So, when light travels from denser to rarer medium, the refracted ray moves away from the normal. This is shown in Fig. 3.7 (a).

Again if light waves travel from rarer to denser medium, *i.e.*, $n_1 < n_2$ then from Eq. (3.21) we get,

$$\sin i > \sin r$$

or, $$i > r$$

So, when light travels from rarer to denser medium, the refracted ray will bend towards the normal. This is shown in Fig. 3.7 (b).

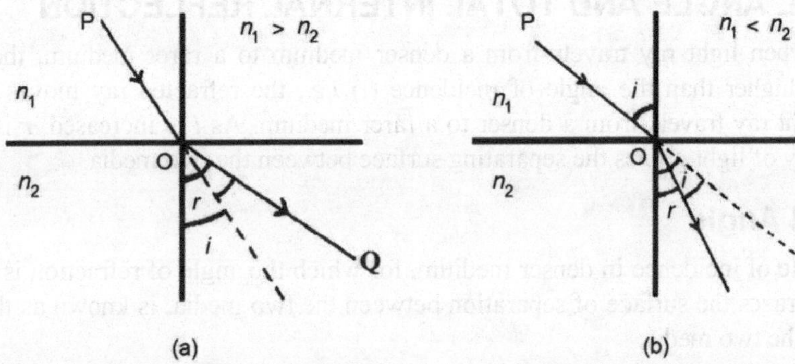

Fig. 3.7: Refraction from (a) denser to rarer and (b) rarer to denser medium

Example 3.3: If the absolute RI of water and glass are $\frac{4}{3}$ and $\frac{3}{2}$ respectively then find RI of glass with respect to air.

Solution: Here $n_g = \frac{4}{3}$ and $n_w = \frac{4}{2}$.

Thus, the relative refractive index of glass with respect to air is,

$${}_wn^g = \frac{n_g}{n_w} = \frac{9}{8}.$$

EXAMPLE 3.3

3.7.1 Snell's Law in Terms of Speed of Light

From Eq. (3.21) we get,

$$\frac{\sin i}{\sin r} = \frac{n_2}{n_1} = \frac{\frac{c}{v_2}}{\frac{c}{v_1}}$$

where v_1 and v_2 are the velocities of a particular wavelength of light in medium 1 and 2 respectively.

or, $$v_2 \sin i = v_1 \sin r \tag{3.22}$$

Eq. (3.22) is the Snell's law in terms of speed of light.

3.7.2 Snell's Law in Terms of Wavelength

From Eq. (3.22) we get,

$$\frac{\sin i}{\sin r} = \frac{v_1}{v_2} = \frac{\gamma\lambda_1}{\gamma\lambda_2}$$

where γ is the frequency of the given color of light.

or, $$\lambda_2 \sin i = \lambda_1 \sin r \tag{3.23}$$

Here, λ_1 and λ_2 are the wavelengths of the given color of light in medium 1 and 2 respectively.

3.8 CRITICAL ANGLE AND TOTAL INTERNAL REFLECTION

We know that when light ray travels from a denser medium to a rarer medium, then the angle of refraction (r) is higher than the angle of incidence (i) *i.e.*, the refracted ray moves away from the normal if the light ray travels from a denser to a rarer medium. As i is increased, r is also increased and ultimately ray of light grazes the separating surface between the two media.

3.8.1 Critical Angle

The limiting angle of incidence in denser medium, for which the angle of refraction is 90º *i.e.*, the ray after refraction grazes the surface of separation between the two media, is known as the critical angle of incidence for the two media.

If we apply generalized Snell's law for the ray of light, then we have,

$$n_1 \sin\theta_c = n_2 \sin 90^\circ$$

or, $$\sin\theta_c = \frac{n_2}{n_1}$$

or, $$\theta_c = \sin^{-1}\left(\frac{n_2}{n_1}\right) \tag{3.24}$$

If the rarer medium is air, then choosing $n_1 = n$ and $n_2 = 1$, from Eq. (3.24) we get

$$\theta_c = \sin^{-1}\left(\frac{1}{n}\right) \tag{3.25}$$

SCAN ME

for more on total internal reflection

Example 3.4: If the refractive index of a medium with respect to air is 2 then what is the value of critical angle at the interface of two media?

Solution: We know, $n = \dfrac{1}{\sin\theta_c}$

So, $$\theta_c = \sin^{-1}\left(\frac{1}{n}\right) = \sin^{-1}\left(\frac{1}{2}\right) = 30^0$$

Thus, the critical angle at the interface of the two media is 30^0.

EXAMPLE 3.4

Example 3.5: If the absolute refractive index of glass and water are $\dfrac{4}{3}$ and $\dfrac{3}{2}$ respectively then find the critical angle for incidence of light from glass to water.

Solution: For critical angle of incidence of a light ray from glass to water, we have from Snell's law

$$_g n^w = \frac{n_w}{n_g} = \frac{\sin\theta_c}{\sin 90^0}$$

or, $$\frac{4}{3} / \frac{3}{2} = \sin\theta_c$$

or, $$\sin\theta_c = \frac{8}{9} = 0.889$$

$\therefore$ $$\theta_c = \sin^{-1}(0.889) = 62.75^0.$$

EXAMPLE 3.5

Example 3.6: If the speed of light in a medium is 1.5×10^8 m/s then find the critical angle for the medium with respect to air.

Solution: The absolute refractive index of the medium is,

$$n = \frac{c}{v} = \frac{3 \times 10^8}{1.5 \times 10^8} = 2.$$

Hence the critical angle for the light for the medium with respect to air is,

$$\theta_c = \sin^{-1}\left(\frac{1}{n}\right) = \sin^{-1}\left(\frac{1}{2}\right) = 30^0.$$

EXAMPLE 3.6

Factors governing critical angle: The critical angle depends on –

i) the pair of media and ii) the wavelength of the light used.

From Eq. (3.25) it is clear that, when n increases, θ_c decreases and as from Eq. (3.17) n is inversely proportional to λ, thus the critical angle θ_c increases with the increase in wavelength λ of the light used *i.e.*,

$$\theta_{c_{violet}} < \theta_{c_{red}}$$

On the other hand, the critical angle θ_c decreases with the increase in refractive indices and since $n_{violet} > n_{red}$

$$\therefore \quad \theta_{c_{violet}} < \theta_{c_{red}}$$

So, for a given pair of media, critical angle for red color of light is greater than that for violet color of light.

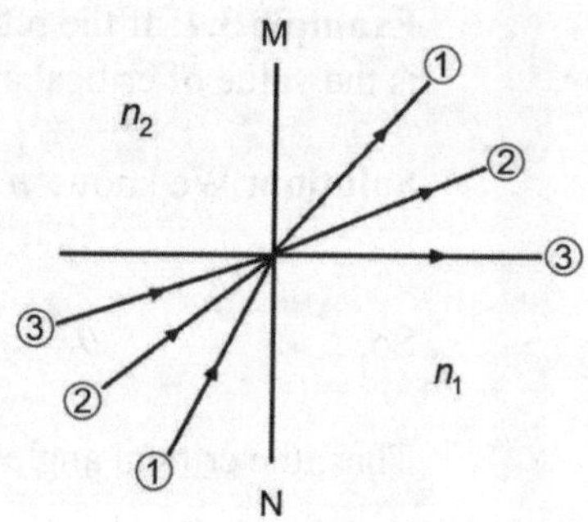

Fig. 3.8: Refraction at critical angle

3.8.2 Total Internal Reflection

When the incident angle is increased more than the critical angle, then there is no refracted ray in rarer medium, for incidence of light from a denser medium.

In this case, the entire incident ray is reflecting back to the denser medium from the interface. This phenomenon is termed as the total internal reflection.

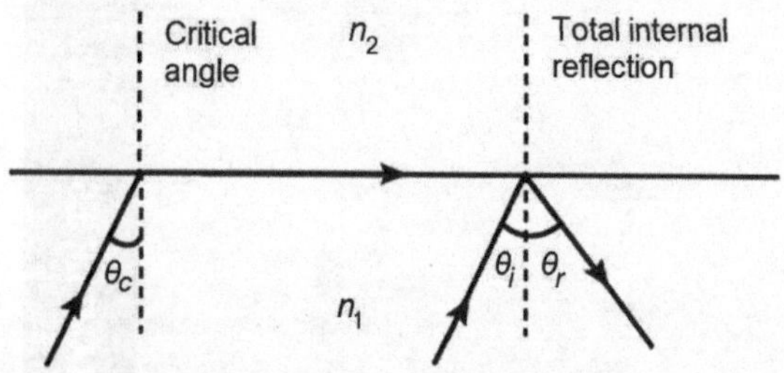

Fig. 3.9: Total internal reflection

So, total internal reflection is a phenomenon where a ray of light travelling from denser to rarer medium is sent back to the denser medium if it is incident on the interface at an angle more than the critical angle. This is illustrated in Fig. 3.9.

Factors governing total internal refection: For total internal reflection to take place, the following conditions must be satisfied –

i) The incident ray of light must have to make incidence from denser to rarer medium.

ii) For a fixed interface, the angle of incidence must be higher than the critical angle.

EXAMPLE 3.7

Example 3.7: A light ray passing through air is incident on a medium of absolute refractive index n. If the angle in between the reflected and refracted light ray is 90^0 then obtain the angle of incidence.

Solution: If i is the angle of incidence and r is the angle of refraction then,

$$i + r = 90^0$$

Now, from Snell's law we have,

$$\frac{\sin i}{\sin r} = n$$

or,
$$\frac{\sin i}{\sin\left(90^0 - i\right)} = n$$

or,
$$\tan i = n$$

$\therefore$ $i = \tan^{-1} n$, which is the required angle of incidence.

Example of total internal reflection: Some examples of total internal reflection are -

a) Formation of mirage, both inferior (for hot countries) and superior (for cold countries).

b) Sparkling of diamond.

c) Shinning of air bubble inside water.

Mirage: During summer, the air close to the ground surface becomes more heated than the air which is far away from the ground. As the RI of air increases with density, hotter air becomes less dense, and has lower RI than the cooler air. As a result, a ray of light from a tall tree successively bends away from the normal and undergoes total internal reflection, when the incident angle for the air near the ground becomes greater than the critical angle. This is shown in Fig. 3.10. To an observer at far, the light seems to be coming from below the ground.

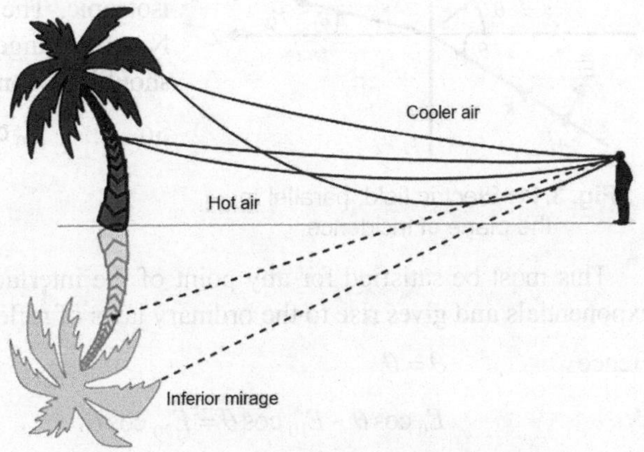

Fig. 3.10: Formation of mirage due to total internal reflection

Naturally the observer imagines that light is being reflected from the ground surface, by a pool of water close to the tall object. These inverted images of distant objects are responsible for producing optical illusion to the observer. This phenomenon is known as the mirage. This type of mirage is particularly common in deserts with very high temperature. In general, during summers at the time of moving in a bus or a car, a distant patch of road is found to be wet, in a highway. But, no evidence can be seen when you reach that spot. This is also due to mirage.

3.9 FRESNEL EQUATION

To know about the fraction of the amplitude of the incident wave, which is reflected or transmitted, we need to have an idea of Fresnel equation. This equation represents the reflection and transmission of light, when incident on an interface between two different optical media. To derive the equation, we will consider light as an electromagnetic wave, consisting of the electric and magnetic vectors.

3.9.1 Electric Field Parallel to the Plane of Incidence

Let the electric vector $\vec{E}$ is perpendicular to the propagation vector $\vec{k}$ that lies in the X-Z plane (plane of incidence) as indicated in Fig. 3.11. Let the incident, reflected and transmitted waves are represented as

$$\vec{E} = \vec{E}_0 e^{i(\vec{k}.\vec{r}-\omega t)}, \ \vec{H} = \frac{\vec{k}\times\vec{E}}{\omega\mu_1} \tag{3.26}$$

$$\vec{E}_1 = \vec{E}_{10} e^{i(\vec{k}_1.\vec{r}-\omega t)}, \ \vec{H}_1 = \frac{\vec{k}_1\times\vec{E}_1}{\omega\mu_1} \tag{3.27}$$

$$\vec{E}_2 = \vec{E}_{20} e^{i(\vec{k}_2.\vec{r}-\omega t)}, \ \vec{H}_2 = \frac{\vec{k}_2\times\vec{E}_2}{\omega\mu_2} \tag{3.28}$$

where μ_1 and μ_2 are respectively the permeability of medium 1 and medium 2.

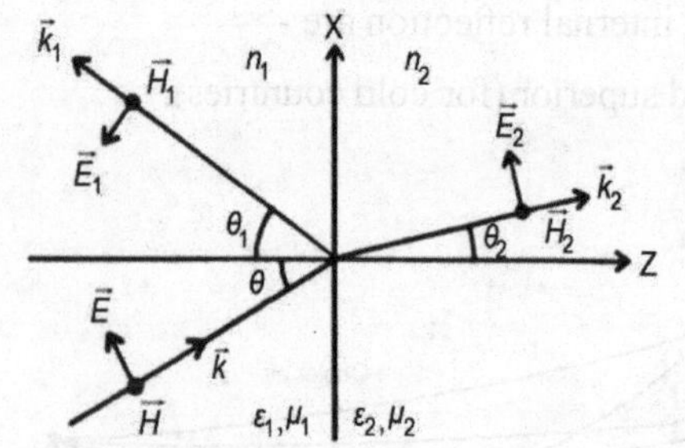

Fig. 3.11: Electric field, parallel to the plane of incidence

Here the $\vec{E}$ vectors for the reflected and refracted waves are so chosen that they are in the plane of incidence, considering the media on two sides of the interface are isotropic. The $\overline{H}$ fields are chosen parallel to the interface. Now the tangential component of the concerned electric field should be continuous in nature across the interface ($z = 0$).

So, $$E_0 \cos\theta e^{i(\vec{k}.\vec{r}-\omega t)} - E_{10} \cos\theta_1 e^{i(\vec{k}_1.\vec{r}-\omega t)} = E_{20} \cos\theta_2 e^{i(\vec{k}_2.\vec{r}-\omega t)} \quad (3.29)$$

This must be satisfied for any point of the interface, at any instant, leading to the cancelation of exponentials and gives rise to the ordinary laws of reflection and refraction.

Hence $$\theta = \theta_1$$

$\therefore$ $$E_0 \cos\theta - E_{10} \cos\theta = E_{20} \cos\theta_2 \quad (3.30)$$

In a similar way the continuity of tangential component of $\overline{H}$ fields at the interface gives

$$H_0 + H_{10} = H_{20} \quad (3.31)$$

Now, $$\overline{H} = \frac{\vec{k} \times \vec{E}}{\omega\mu_1} = \sqrt{\frac{\varepsilon_1}{\mu_1}} \frac{\vec{k} \times \vec{E}}{k} \qquad [\because k = \omega\sqrt{\mu_1\varepsilon_1}]$$

So, $H_0 = \sqrt{\frac{\varepsilon_1}{\mu_1}} E_0$,where ε_1 is the permittivity of medium 1.

Similarly, $H_{10} = \sqrt{\frac{\varepsilon_1}{\mu_1}} E_{10}$ and $H_{20} = \sqrt{\frac{\varepsilon_2}{\mu_2}} E_{20}$ [ε_2 is the permittivity of medium 2]

Hence, $$\sqrt{\frac{\varepsilon_1}{\mu_1}} E_0 + \sqrt{\frac{\varepsilon_1}{\mu_1}} E_{10} = \sqrt{\frac{\varepsilon_2}{\mu_2}} E_{20}$$

or, $$\sqrt{\frac{\varepsilon_1}{\mu_1}} (E_0 + E_{10}) = \sqrt{\frac{\varepsilon_2}{\mu_2}} E_{20} \quad (3.32)$$

For dielectric medium,

$$\mu_1 = \mu_2 = \mu_0.$$

So, $$n_1 (E_0 + E_{10}) = n_2 E_{20} \quad (3.33)$$

where $$n_1 = \sqrt{\frac{\varepsilon_1}{\varepsilon_0}} \text{ and } n_2 = \sqrt{\frac{\varepsilon_2}{\varepsilon_0}}$$

Solving Eq. (3.30) and Eq. (3.33) we get, reflection coefficient

$$r = \frac{E_{10}}{E_0} = \frac{n_2 \cos\theta - n_1 \cos\theta_2}{n_2 \cos\theta + n_1 \cos\theta_2} \quad (3.34)$$

and transmission coefficient

$$t = \frac{E_{20}}{E_0} = \frac{2n_1 \cos\theta}{n_2 \cos\theta + n_1 \cos\theta_2} \tag{3.35}$$

Now, from Snell's law we have $\dfrac{\sin\theta}{\sin\theta_2} = \dfrac{n_2}{n_1}$

$$\therefore \quad r = \frac{\dfrac{n_2}{n_1}\cos\theta - \cos\theta_2}{\dfrac{n_2}{n_1}\cos\theta + \cos\theta_2} = \frac{\dfrac{\sin\theta}{\sin\theta_2}\cos\theta - \cos\theta_2}{\dfrac{\sin\theta}{\sin\theta_2}\cos\theta + \cos\theta_2} = \frac{\sin\theta\cos\theta - \sin\theta_2\cos\theta_2}{\sin\theta\cos\theta - \sin\theta_2\cos\theta_2}$$

$$\text{or,} \quad r = \frac{\sin 2\theta - \sin 2\theta_2}{\sin 2\theta - \sin 2\theta_2} = \frac{2\cos(\theta+\theta_2)\sin(\theta-\theta_2)}{2\sin(\theta+\theta_2)\cos(\theta-\theta_2)} = \frac{\tan(\theta-\theta_2)}{\tan(\theta+\theta_2)} \tag{3.36}$$

$$\text{and} \quad t = \frac{2\cos\theta}{\dfrac{n_2}{n_1}\cos\theta + \cos\theta_2} = \frac{2\cos\theta}{\dfrac{\sin\theta}{\sin\theta_2}\cos\theta + \cos\theta_2} = \frac{2\sin\theta_2\cos\theta}{\sin\theta\cos\theta + \sin\theta_2\cos\theta_2}$$

$$\text{or,} \quad t = \frac{4\sin\theta_2\cos\theta}{\sin 2\theta + \sin 2\theta_2} = \frac{2\sin\theta_2\cos\theta}{2\sin(\theta+\theta_2)\cos(\theta-\theta_2)} \tag{3.37}$$

Eqs. (3.36) and (3.37) are known as the Fresnel's equation when the electric vector is parallel to the plane of incidence.

3.9.2 Electric Field Perpendicular to the Plane of Incidence

Let the electric vector $\vec{E}$ is perpendicular to the plane of incidence as indicated in Fig. 3.12. Here the $\vec{E}$ vectors for the reflected and refracted waves are so chosen that they are normal to the plane of incidence, considering the media on two sides of the interface are isotropic. The $\vec{H}$ fields are chosen in the plane of incidence. Thus in this case from the continuity of tangential component of $\vec{E}$ fields at the interface, we have

$$E_0 + E_{10} = E_{20} \tag{3.38}$$

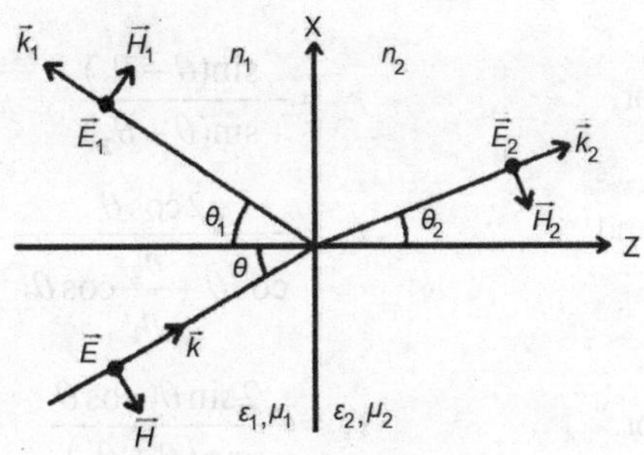

Fig. 3.12: Electric field, perpendicular to the plane of incidence

whereas the continuity of tangential component of $\vec{H}$ fields at the interface gives

$$-H_0\cos\theta + H_{10}\cos\theta_1 = -H_{20}\cos\theta_2 \tag{3.39}$$

$$\text{Now,} \quad H_0 = \sqrt{\frac{\varepsilon_1}{\mu_1}}E_0 \; ; \; H_{10} = \sqrt{\frac{\varepsilon_1}{\mu_1}}E_{10} \text{ and } H_{20} = \sqrt{\frac{\varepsilon_2}{\mu_2}}E_{20}$$

$$\text{Hence,} \quad -\sqrt{\frac{\varepsilon_1}{\mu_1}}E_0\cos\theta + \sqrt{\frac{\varepsilon_1}{\mu_1}}E_{10}\cos\theta = -\sqrt{\frac{\varepsilon_2}{\mu_2}}E_{20}\cos\theta_2 \qquad [\text{As, } \theta = \theta_1]$$

or, $$\sqrt{\frac{\varepsilon_1}{\mu_1}}\left(-E_0+E_{10}\right)\cos\theta=-\sqrt{\frac{\varepsilon_2}{\mu_2}}E_{20}\cos\theta_2 \qquad (3.40)$$

For dielectric medium, $\mu_1=\mu_2=\mu_0$

So, $$n_1\left(-E_0+E_{10}\right)\cos\theta=-n_2E_{20}\cos\theta_2 \qquad (3.41)$$

where $n_1=\sqrt{\frac{\varepsilon_1}{\varepsilon_0}}$ and $n_2=\sqrt{\frac{\varepsilon_2}{\varepsilon_0}}$.

Solving we get, reflection coefficient

$$r_\perp=\frac{E_{01}}{E_0}=\frac{n_1\cos\theta-n_2\cos\theta_2}{n_1\cos\theta+n_2\cos\theta_2} \qquad (3.42)$$

and transmission coefficient

$$t_\perp=\frac{E_{02}}{E_0}=-\frac{2n_1\cos\theta}{n_1\cos\theta+n_2\cos\theta_2} \qquad (3.43)$$

Now, from Snell's law we have

$$\frac{\sin\theta}{\sin\theta_2}=\frac{n_2}{n_1}$$

So, $$r_\perp=\frac{\cos\theta-n_2\cos\theta_2}{\cos\theta+n_2\cos\theta_2}=\frac{\cos\theta-\dfrac{\sin\theta}{\sin\theta_2}\cos\theta_2}{\cos\theta+\dfrac{\sin\theta}{\sin\theta_2}\cos\theta_2}=\frac{\sin\theta_2\cos\theta-\sin\theta\cos\theta_2}{\sin\theta_2\cos\theta-\sin\theta\cos\theta_2}$$

or, $$r_\perp=-\frac{\sin(\theta-\theta_2)}{\sin(\theta+\theta_2)} \qquad (3.44)$$

and $$t_\perp=\frac{2\cos\theta}{\cos\theta+\dfrac{n_2}{n_1}\cos\theta_2}=\frac{2\cos\theta}{\cos\theta+\dfrac{\sin\theta}{\sin\theta_2}\cos\theta_2}=\frac{2\sin\theta_2\cos\theta}{\sin\theta_2\cos\theta+\sin\theta\cos\theta_2}$$

or, $$t_\perp=-\frac{2\sin\theta_2\cos\theta}{\sin\left(\theta+\theta_2\right)} \qquad (3.45)$$

Eqs. (3.44) and (3.45) are known as the Fresnel's equation when the electric vector is perpendicular to the plane of incidence. From Fresnel's equations we can describe a number of different properties of light.

3.10 BREWSTER'S LAW

For any dielectric interface from a medium with RI n_1 to a medium with RI n_2, there is a certain incident angle θ_p at which r_p approaches to 0 and the p-polarized (electric field parallel to the plane of incidence) incident wave is refracted purely and consequently all the reflected light becomes s-polarized (electric field perpendicular to the plane of incidence). This angle is called as *Brewster's angle*. From Eq. (3.36) we have $r=0$ if $\theta+\theta_2=\frac{\pi}{2}$. Hence under this condition no parallel light is reflected. However under this condition $r_\perp \neq 0$, *i.e.*, the perpendicular light is partly reflected. So, an unpolarized light incident at an angle $\theta=\theta_p$, satisfying the condition $\theta+\theta_p=\frac{\pi}{2}$ will be plane polarized by reflection.

In this case Snell's law reduces to-

$$\frac{\sin\theta_p}{\sin\left(\frac{\pi}{2}-\theta_p\right)}=\frac{n_2}{n_1}$$

or, $$\tan\theta_p=\frac{n_2}{n_1}=n \tag{3.46}$$

or, $$\theta_p=\tan^{-1} n \tag{3.47}$$

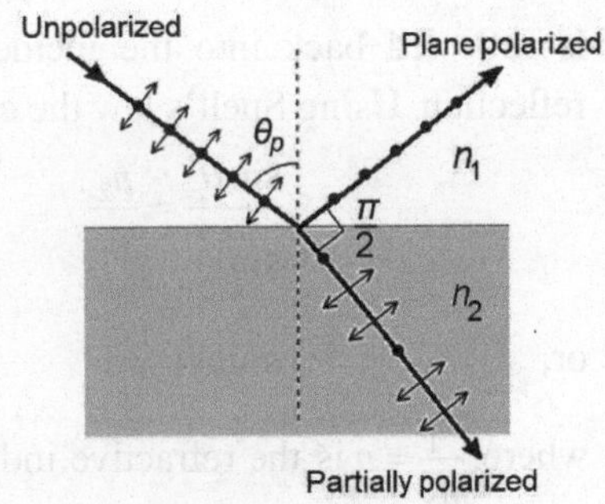

Fig. 3.13: Illustration of Brewster's law

The angle θ_p is known as the Brewster's angle or the *angle of polarization*.

From Eq. (3.47) it is obvious that the tangent related to the angle of polarization becomes same as the RI of the reflecting medium with respect to the incident medium. This is known as the *Brewster's law*.

3.10.1 Amplitude Coefficients and Phase Change

For normal incidence, $\theta=\theta_2=0$

Thus, $$r_\perp=-r=\frac{n_1-n_2}{n_1+n_2} \tag{3.48}$$

and $$t_\perp=t=\frac{2n_1}{n_1+n_2} \tag{3.49}$$

For grazing incidence, $\theta\to 90^0$

Thus, $$r_\perp=-r=-1 \tag{3.50}$$

and $$t_\perp=t=0 \tag{3.51}$$

3.11 TOTAL INTERNAL REFLECTION FROM FRESNEL EQUATION

If a ray of light is incident on the surface between materials having different refractive indices, the reflection of a part of light occurs. This Fresnel reflection depends on both the difference between refractive indices of the materials and on the angle of incidence. When $n_1>n_2$, total internal reflection comes into play. In this case $\theta_2>\theta$ and $r_\perp$ as given by Eq. (3.42) is always positive starting from the value $\frac{n_1-n_2}{n_1+n_2}$ at $\theta=0$ and then increases with θ and reaches the value 1 when $\theta=\theta_2$ (also regarded as the critical angle θ_c).

The critical angle is the value of the angle of incidence θ for which the angle of reflection θ_2 becomes 90^0. In a similar way, if we look into Eq. (3.34) for the parallel incidence of the $\vec{E}$ waves, we have reflection coefficient $r = \frac{n_1 - n_2}{n_1 + n_2}$ at $\theta = 0$ and then it becomes zero at $\theta = \theta_p$ (known as the polarizing angle) and reaches the value 1 at $\theta = \theta_c$.

As r and $r_\perp$ increases with θ, t and $t_\perp$ should decrease with θ. For $\theta \geq \theta_c$, the entire incident energy is reflected back into the incident medium leading to a phenomena known as the total internal reflection. Using Snell's law the critical angle θ_c may be given as,

$$\frac{\sin\theta_c}{\sin\frac{\pi}{2}} = \frac{n_2}{n_1}$$

or,
$$n\sin\theta_c = 1 \tag{3.52}$$

where, $\frac{n_1}{n_2} = n$ is the refractive index of the denser medium with respect to the rarer medium. For any incident angle,

$$n\sin\theta = \sin\theta_2$$

and
$$\cos\theta_2 = \sqrt{1-\sin^2\theta_2} = \sqrt{1-n^2\sin^2\theta}$$

If $\theta > \theta_c$, $n\sin\theta > 1$ and $\cos\theta_2 = i\sqrt{n^2\sin^2\theta - 1}$.

So,
$$r_\perp = \frac{n\cos\theta - i\sqrt{n^2\sin^2\theta - 1}}{n\cos\theta + \sqrt{n^2\sin^2\theta - 1}} \tag{3.53}$$

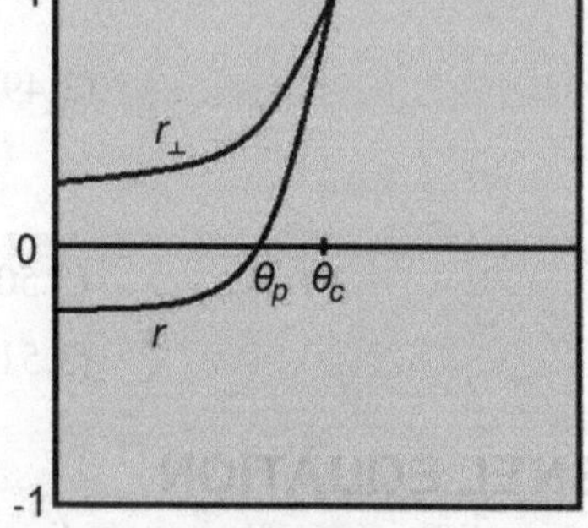

Fig. 3.14: Reflection coefficient

Obviously the magnitude of the reflection coefficient will be 1 and as it is complex, it indicates a phase change due to reflection. Similarly, we can show that the coefficient $r_\perp$ is also having magnitude 1 and is also complex. Thus the entire incident energy is found to reflect back into the denser medium. This is indeed due to the total internal reflection. The change of phase between the incident and the reflected wave in the denser medium is not equal to π. As a result, some resultant $\vec{E}$ field exists at the denser side of the boundary. So, to satisfy the boundary condition, there must be a disturbance at the rarer side whose amplitude should decay with the depth of penetration into the rarer medium. This is known as the *evanescent wave* which on the average cannot carry energy across the boundary.

3.12 MIRRORS AND LENSES

A lens refracts light coming from an object and creates either a virtual or a real image of it. Virtual images are produced at a location beyond the lens, where the paths of the primary rays intersect, if projected backward from their direction.

A real image is produced at a location where the light converges originally.

Mirrors also reflect light coming from an object and create images in a way similar to a lens, depending on where an object is located in relation to a mirror.

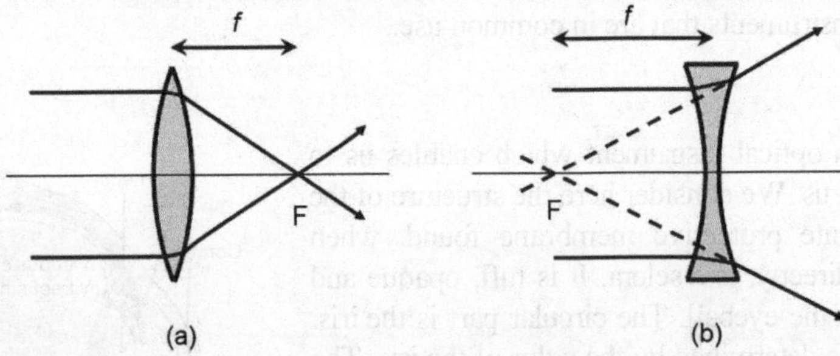

Fig. 3.15: Refraction through (a) convex and (b) concave lenses

Convex lens: A convex lens is thicker at middle than on its outer edge. As a result of the middle portion being the thickest part, light travelling through the lens converges into a single point. Parallel rays of light after refraction meet at a single point beyond the convex lens. How an image appears in a convex lens depends on the distance and position of the object being viewed.

When an object is situated within the focal distance, it appears similar to a *virtual image i.e.*, its right side is up and greater than the actual object. An image beyond focal range will appear upside down and such an image can be smaller, larger or as the same size as the original image.

Concave lens: Concave lens are thicker on the ends and thinner at the middle. They divert light rays away from a focal point and only create virtual or smaller images.

Plane mirror: The shape of the mirror is flat and it sends out light in multiple directions either by reflection or emission. The point where the reflected light beams intersect is where the image is formed. The image formed by a plane mirror will always be of the same size as the original object.

Convex mirror: A convex mirror, similar to the outside portion of a bowl, works like a concave lens. From its middle, it curves light away. This mirror can produce only smaller and virtual images.

Concave mirror: A concave mirror, similar to the inside of a bowl, works much like a convex lens. It bends light more away in the middle.

How, the images will appear is dependent on the proximity of the objects to the mirror.

At certain distances of the object, in front of a mirror or lens, the image will appear as virtual whereas other locations of the object will cause an image to appear larger, inverted, real or erect.

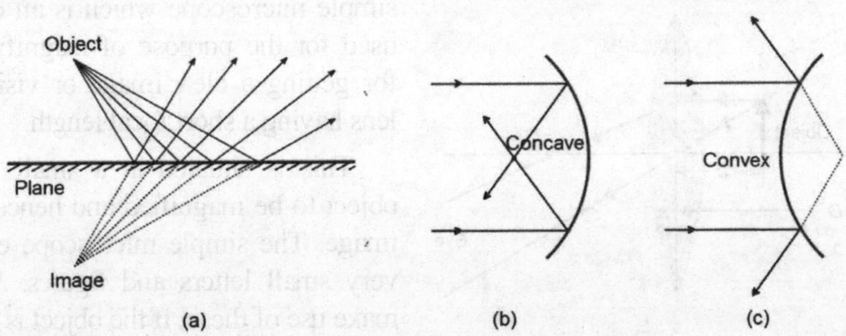

Fig. 3.16: Reflection through (a) plane mirror, (b) concave mirror and (c) convex mirror

3.13 OPTICAL INSTRUMENTS

A number of optical instruments have been designed by utilizing reflecting and refracting properties of mirrors and lenses. Our eye is an important optical instrument. Microscopes and telescopes are some other examples of optical instruments that are in common use.

3.13.1 The Eye

The human eye is an optical instrument which enables us to see all objects around us. We consider here the structure of the human eye. The white protective membrane found, when looked into the eye directly, is a sclera. It is tuff, opaque and fibrous outer layer of the eyeball. The circular part is the iris. The color of the eye is determined by the color of the iris. The center transparent area of the iris is the pupil. The iris works like the shutter of the camera. It absorbs majority of light falling on it and permits it to go through the pupil. The amount of light that goes inside the inner part of the eye depends on the size of the pupil. In bright light, the iris contracts the pupil to restrict the light, whereas in low light it widens the pupil to emit more light into the eye. The eyeball is spherical in shape.

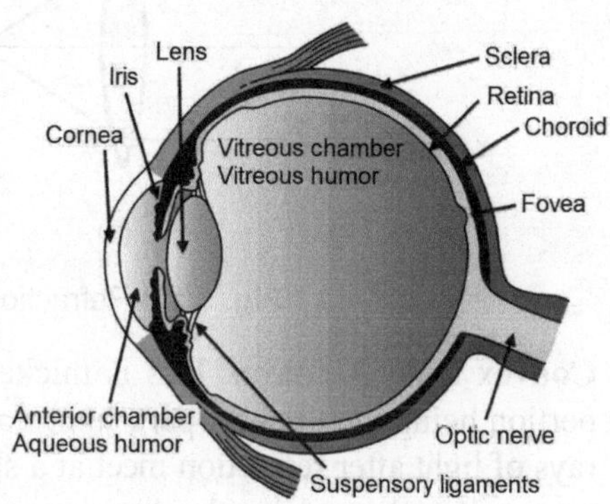

Fig. 3.17: The human eye

The retina of the eye is able to detect the light and its color because of the presence of senses known as rods and cones. Light passing through the human eye is first refracted by the cornea which is then incident on the iris. The lens is located just behind the iris and light after being refracted through the pupil falls on it and forms a sharp image. Formation of an image, exactly on the retina, enables us to see the object clearly.

Defects in Human Eye: The ability to focus diminishes with the age of a person and this defect is *Presbyopia*. This defect is corrected by using converging lenses. The other defect is *Hypermetropia*. It is observed in the people of any age. The person who is suffering from this defect will have a normal vision while observing at the far object. But the vision is blurred for nearby objects. This is corrected by using a convex lens.

3.13.2 The Microscope

Microscope is an optical instrument to view small object. Two preliminary such instruments are:

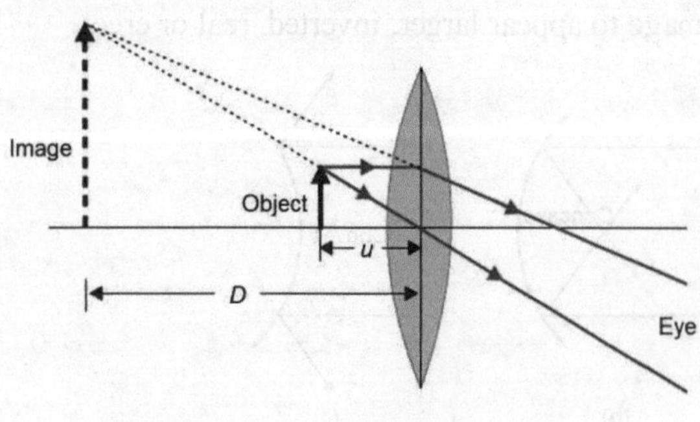

Fig. 3.18: The simple microscope

i) Simple Microscope: We first consider the simple microscope which is an optical instrument, used for the purpose of magnifying small objects for getting a clear image or vision. It is a convex lens having a short focal length.

This is situated at a small distance from the object to be magnified and hence it forms a virtual image. The simple microscope enables us to view very small letters and figures. Watchmakers also make use of these. If the object is at the focal length, the image is formed at infinity (Fig. 3.18).

However, if the object is at a distance slightly less than the focal length of the lens, the image is virtual and situated at a distance closer than infinity. The magnification produced by a simple microscope for the image formed at the near point D is,

$$m = 1 + \frac{D}{f} \tag{3.54}$$

and the angular magnification for the image at infinity is,

$$m = \frac{D}{f} \tag{3.55}$$

Simple microscope has a limitation of maximum magnification (≤ 9) when we consider realistic focal lengths.

ii) Compound Microscope: Now we consider the case of a compound microscope. For much larger magnifications, one generally uses two lenses, so that one lens can compound the effect of the other lens to generate a compound microscope (Fig. 3.19).

The components of a compound microscope are eye-piece, objective lens, fine adjustment screw and rough adjustment screw. The lens nearer to the object is the objective which forms a real, inverted, magnified image of the object. This behaves as the object for the second lens close to the eye and is called as the eye-piece. This essentially magnifies to produce the final virtual image. With a compound microscope, we get very large values of magnification. This microscope is used to observe microscopic objects. It is comprised of two convex lenses, both of which are used for magnification.

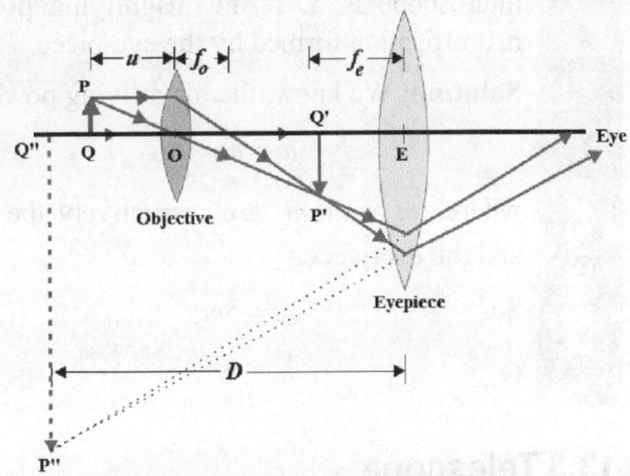

Fig. 3.19: The compound microscope

The resultant magnification of a compound microscope is the combination of the magnifications produced by the objective (m_o) and the eye-piece (m_e) and is given by

$$m = m_o \times m_e = \frac{L}{f_o} \times \frac{D}{f_e} = \frac{LD}{f_o f_e} \tag{3.56}$$

So, to produce large magnification, the focal lengths of the objective and eye-piece should be as small as possible.

Example 3.8: A convex lens having focal length 2.5 cm is utilized as a simple microscope. Find the magnifying power of the image at the distance of distinct vision. If it is used as an eye-piece of a compound microscope with objective of magnifying power 25, calculate the magnifying power of the compound microscope.

Solution: Distance of distinct vision = D = 25 cm.

Now, the magnification produced by simple microscope, for the formation of image at the near point D is $m = 1 + \frac{D}{f}$.

EXAMPLE 3.8

EXAMPLE 3.8

So, $m = 1 + \frac{25}{2.5} = 11.$

If it is used as an eye-piece of a compound microscope then

$$m_e = 11.$$

For the compound microscope the magnifying power of the objective is

$$m_o = 25.$$

So, the magnifying power of the said microscope will be

$$m = m_o \times m_e = 11 \times 25 = 275.$$

EXAMPLE 3.9

Example 3.9: The magnification produced by the objective of a compound microscope is 5. If the magnifying power of the microscope is 30, find the magnification formed by the eye-piece.

Solution: We know, the magnifying power of a compound microscope is

$$m = m_o \times m_e$$

where, m_o and m_e are respectively the magnifications formed by the objective and the eye-piece.

So, $30 = 5m_e$

or, $m_e = 6.$

3.13.3 Telescope

It is utilized to give angular magnification of long ranged objects. Its objective has a large focal length (f_o) and a larger aperture compared to the eye-piece focal length (f_e) and aperture. The overall magnification produced by a telescope is,

$$m = \frac{f_o}{f_e} \tag{3.57}$$

where $f_o + f_e = L$ is the total length of the telescope tube.

The telescope is of two types. One is the reflecting type and the other one is the refracting type. Reflecting telescopes are such telescopes which do not use lenses at all. They use mirrors for focusing the light together. The mirror used for this purpose is a concave mirror which bend the light together, by reflecting the light instead of bending it. There is no chromatic aberration in a mirror. One obvious problem with a reflecting telescope is that the objective mirror focuses light inside the telescope tube. Refracting telescopes (Fig. 3.20) use two lenses to focus the light and make the object to look like closer than it really is. Both of these lenses are convex in nature and helps in bending light inwards. Refracting telescopes can be used both for terrestrial and astronomical observations. Terrestrial telescopes have, in addition, a pair of inverting lenses to make the final image erect. The main consideration with an astronomical telescope is the resolving power.

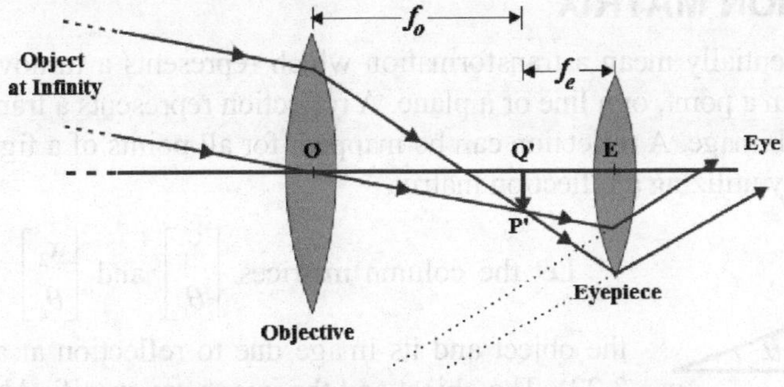

Fig. 3.20: Refracting telescope

3.14 TRANSLATION MATRIX

Translation matrix permits to change a default coordinate system and map the original coordinate system (x, y) with a new coordinate system (x', y'). It depends on how we change the coordinate system which effectively rotate, scale, translate or shear the object in a particular fashion.

Let a ray of light passes from a point A, specified by a column matrix $\begin{bmatrix} x_1 \\ \theta_1 \end{bmatrix}$ to another point B specified by another column matrix $\begin{bmatrix} x_2 \\ \theta_2 \end{bmatrix}$ through a translation d in air (Fig. 3.21). Since the translation of the ray is in the direction of the ray, d is negative and θ_1 is also negative.

Here, $\theta_1 = \theta_2$

Also from the figure, $\dfrac{x_2 - x_1}{-d} = \tan(-\theta_1)$

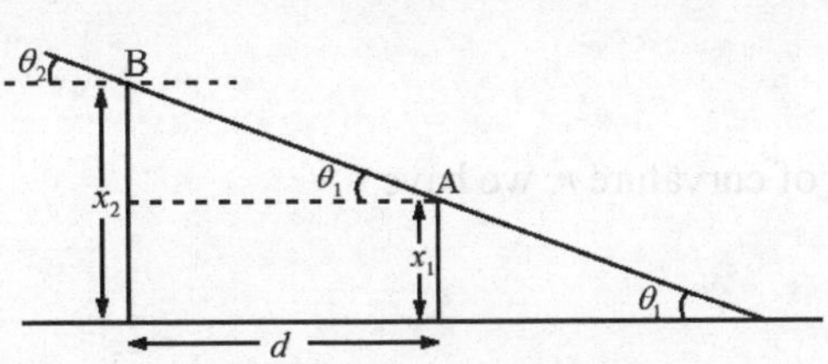

Fig. 3.21: Translation matrix

or, $x_2 - x_1 = d \tan \theta_1$

or, $x_2 = x_1 + d \tan \theta_1$

For paraxial ray *i.e.* for small value of θ_1, $\tan \theta_1 \simeq \theta_1$

$$\therefore \quad x_2 = x_1 + d\theta_1 = 1 \times x_1 + d \times \theta_1 \tag{3.58}$$

$$\text{or,} \quad \theta_2 = \theta_1 = 0 \times x_1 + 1 \times \theta_1 \tag{3.59}$$

$\therefore$ Eqs. (3.58) and (3.59) may be written in the matrix form as

$$\begin{bmatrix} x_2 \\ \theta_2 \end{bmatrix} = \begin{bmatrix} 1 & d \\ 0 & 1 \end{bmatrix} \begin{bmatrix} x_1 \\ \theta_1 \end{bmatrix} \tag{3.60}$$

$\therefore$ Translation matrix may be defined as

$$[T] = \begin{bmatrix} 1 & d \\ 0 & 1 \end{bmatrix} \tag{3.61}$$

3.15 REFLECTION MATRIX

By reflection we essentially mean a transformation which represents a turnover of a coordinate that may be reflected from a point, or a line or a plane. A reflection represents a transformation that maps a figure to its reflected image. A reflection can be mapped, for all points of a figure, to an image across a line of symmetry by utilizing a reflection matrix.

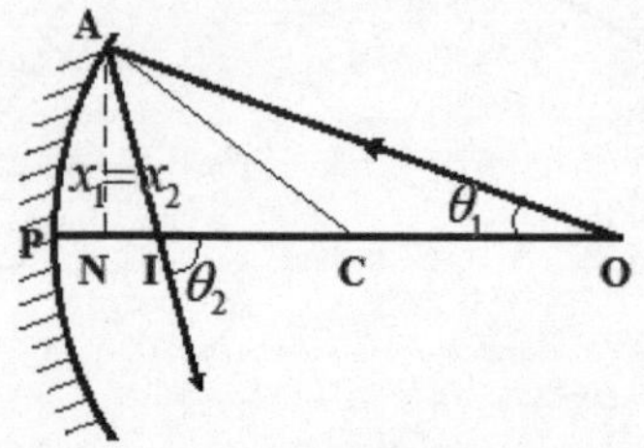

Fig. 3.22: Reflection matrix

Let the column matrices $\begin{bmatrix} x_1 \\ \theta_1 \end{bmatrix}$ and $\begin{bmatrix} x_2 \\ \theta_2 \end{bmatrix}$ respectively represents the object and its image due to reflection at a curved reflector (Fig. 3.22). The object and the image are specified by the points O and I in figure. For paraxial ray *i.e.* for small value of θ_1 the points P and N will be very close to each other and hence from the figure we get

$$\frac{x_1}{NO} \simeq \frac{x_1}{PO} = \tan(-\theta_1)$$

or $$\frac{x_1}{u} = -\theta_1$$

or, $$u = -\frac{x_1}{\theta_1}$$

In a similar way we can have

$$\frac{x_2}{NI} \simeq \frac{x_2}{PI} = \tan(-\theta_2)$$

or, $$\frac{x_2}{v} = -\theta_2$$

or, $$v = -\frac{x_2}{\theta_2}$$

Now, for the reflection at a curved surface of radius of curvature r, we have

$$\frac{1}{v} + \frac{1}{u} = \frac{2}{r}$$

or, $$-\frac{\theta_2}{x_2} - \frac{\theta_1}{x_1} = \frac{2}{r}$$

or, $$\theta_2 = -\theta_1 - \frac{2}{r}x_1 \tag{3.62}$$

$\because$ $$x_1 = x_2$$

As the object and image height are the same,

$$x_2 = x_1 \tag{3.63}$$

$\therefore$ Eq. (3.62) and Eq. (3.63) may be written in the matrix form as

$$\begin{bmatrix} x_2 \\ \theta_2 \end{bmatrix} = \begin{bmatrix} 1 & 0 \\ -\frac{2}{r} & -1 \end{bmatrix} \begin{bmatrix} x_1 \\ \theta_1 \end{bmatrix} \tag{3.64}$$

$\therefore$ Reflection matrix may be defined as

$$[R] = \begin{bmatrix} 1 & 0 \\ -\frac{2}{r} & -1 \end{bmatrix} \tag{3.65}$$

For a plane surface, $r = \propto$. Hence the reflection matrix for a plane reflector will be,

$$[R] = \begin{bmatrix} 1 & 0 \\ 0 & -1 \end{bmatrix} \tag{3.66}$$

3.16 REFRACTION MATRIX

This matrix is made by multiplying both the refraction and translation matrices. The principal planes, the front and back surface powers and the equivalent focal length are obtained from the system matrix.

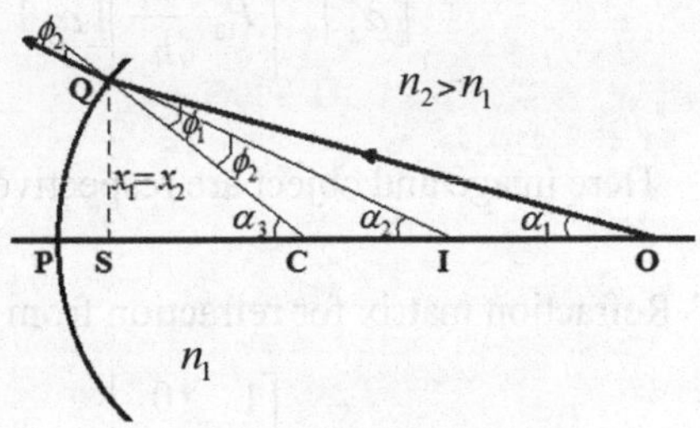

Fig. 3.23: Refraction matrix

Let us consider a concave refracting surface of radius of curvature r, separating the two media of refractive indices n_1 and n_2 ($n_2 > n_1$), forms the virtual image at I for the object situated at O. Let the angle of incidence and the angle of refraction at the curved surface are respectively ϕ_1 and ϕ_2. From Fig. 3.23 we have,

$$\phi_1 + \alpha_1 = \alpha_3$$

and $$\phi_1 - \phi_2 + \alpha_1 = \alpha_2$$

Again from the Snell's law we have,

$$n_1 \sin \phi_1 = n_2 \sin \phi_2$$

or, $$n_1 \phi_1 = n_2 \phi_2 \qquad [\because \phi_1, \phi_2 \to 0]$$

or, $$\phi_2 = \frac{n_1}{n_2} \phi_1$$

$\therefore$ $$\alpha_2 = \phi_1 - \frac{n_1}{n_2}\phi_1 + \alpha_1 = \left(1 - \frac{n_1}{n_2}\right)\phi_1 + \alpha_1 = \left(1 - \frac{n_1}{n_2}\right)(\alpha_3 - \alpha_1) + \alpha_1$$

or, $$\alpha_2 = \alpha_3 \frac{n_2 - n_1}{n_2} + \alpha_1 \frac{n_1}{n_2}$$

Here, α_1, α_2 and α_3 are all negative. For paraxial rays, as the value of α_3 is small so,

$$\alpha_3 = \frac{x_1}{r} \qquad [\because \text{CP} \simeq \text{CS}]$$

$$\therefore \qquad \alpha_2 = \frac{n_2 - n_1}{n_2}\frac{x_1}{r} + \frac{n_1}{n_2}\alpha_1 = P_{12}x_1 + \frac{n_1}{n_2}\alpha_1 \tag{3.67}$$

where,
$$P_{12} = \frac{n_2 - n_1}{n_2 r} \tag{3.68}$$

As the object and image height are the same,

$$\therefore \qquad x_2 = x_1 \tag{3.69}$$

Hence, Eq. (3.68) and Eq. (3.69) may be written in the matrix form as

$$\begin{bmatrix} x_2 \\ \alpha_2 \end{bmatrix} = \begin{bmatrix} 1 & 0 \\ P_{12} & \dfrac{n_1}{n_2} \end{bmatrix} \begin{bmatrix} x_1 \\ \alpha_1 \end{bmatrix} = [R_{12}] \begin{bmatrix} x_1 \\ \alpha_1 \end{bmatrix} \tag{3.70}$$

Here image and object are respectively characterized by the column matrix $\begin{bmatrix} x_1 \\ \alpha_1 \end{bmatrix}$ and $\begin{bmatrix} x_2 \\ \alpha_2 \end{bmatrix}$.

$\therefore$ Refraction matrix for refraction from medium 1 to medium 2 may be defined as

$$[R_{12}] = \begin{bmatrix} 1 & 0 \\ P_{12} & \dfrac{n_1}{n_2} \end{bmatrix} \tag{3.71}$$

UNIT SUMMARY

- ***Laws of reflection and refraction***

  Law of reflection

  $i = r$

  Law of refraction

  $$n = \frac{\sin i}{\sin r} \qquad n_1 \sin i = n_2 \sin r$$

- ***Refractive indices***

  Relative refractive index

  $$\frac{\sin i}{\sin r} = \frac{n_2}{n_1} = {}_1n^2 = \text{constant}$$

  $$\frac{\sin i}{\sin r} = \frac{n_1}{n_2} = {}_2n^1 = \text{constant}$$

  Absolute refractive index

  $$\frac{\sin i}{\sin r} = n = \text{constant}$$

- ***Relationship between absolute and relative refractive indices of a medium***

  $${}_1n^2 = \frac{n_2}{n_1}$$

- ***Relationship between refractive index and speed of light***

  The absolute refractive index of a medium and speed of light

  $$n = \frac{c}{v}$$

  Relative refractive index of a medium and speed of light

  $nv =$ constant

  $n\lambda =$ constant

  $$\frac{v_1}{v_2} = \frac{\lambda_1}{\lambda_2}$$

- ***Generalized Snell's law***

  Snell's law in terms of speed of light

  $v_2 \sin i = v_1 \sin r$

  Snell's law in terms of wavelength

  $\lambda_2 \sin i = \lambda_1 \sin r$

- ***Total internal reflection and critical angle***

  Critical angle

  $$\theta_c = \sin^{-1}\left(\frac{1}{n}\right)$$

Total internal reflection

i) The incident ray of light must go from denser to rarer medium.

ii) The angle of incidence must be greater than the critical angle.

- ***Fresnel equation***

Electric field is parallel to the plane of incidence

$$r = \frac{\tan(\theta - \theta_2)}{\tan(\theta + \theta_2)}$$

$$t = \frac{2\sin\theta_2 \cos\theta}{2\sin(\theta + \theta_2)\cos(\theta - \theta_2)}$$

Electric field is perpendicular to the plane of incidence

$$r_\perp = -\frac{\sin(\theta - \theta_2)}{\sin(\theta + \theta_2)}$$

$$t_\perp = -\frac{2\sin\theta_2 \cos\theta}{\sin(\theta + \theta_2)}$$

- ***Brewster's law from Fresnel's equation***

$$\theta_p = \tan^{-1} n$$

Amplitude coefficient and phase change

For normal incidence, $\theta = \theta_2 = 0$

$$r_\perp = -r = \frac{n_1 - n_2}{n_1 + n_2} \text{ and } t_\perp = t = \frac{2n_1}{n_1 + n_2}$$

For grazing incidence, $\theta \to 90^0$

$$r_\perp = -r = -1 \text{ and } t_\perp = t = 0$$

- ***Total internal reflection from Fresnel's equation***

$$n\sin\theta_c = 1$$

$$\cos\theta_2 = \sqrt{1 - \sin^2\theta_2} = \sqrt{1 - n^2\sin^2\theta}$$

If $\theta > \theta_c$, $\cos\theta_2 = i\sqrt{n^2\sin^2\theta - 1}$.

$$r_\perp = \frac{n\cos\theta - i\sqrt{n^2\sin^2\theta - 1}}{n\cos\theta + \sqrt{n^2\sin^2\theta - 1}}$$

- ***Optical instruments***

The Microscope

Simple Microscope $m = 1 + \frac{D}{f}$

Compound Microscope $m = m_o \times m_e = \frac{LD}{f_o f_e}$

Telescope

$$m = \frac{f_o}{f_e}$$

- ***Translation matrix***

$$[T] = \begin{bmatrix} 1 & d \\ 0 & 1 \end{bmatrix}$$

- ***Reflection matrix***

$$[R] = \begin{bmatrix} 1 & 0 \\ -\frac{2}{r} & -1 \end{bmatrix}$$

- ***Refraction matrix***

$$[R_{12}] = \begin{bmatrix} 1 & 0 \\ P_{12} & \frac{n_1}{n_2} \end{bmatrix}$$

EXERCISES

Multiple Choice Questions

3.1 When light travels from one medium to another, separated by a sharp boundary, the characteristic which does not change is

(a) velocity (b) amplitude (c) wavelength (d) frequency

3.2 A ray of light passes from vacuum into a medium of refractive index n. If the angle of incidence being twice the angle of refraction then the angle of incidence is

(a) $\cos^{-1}\frac{n}{2}$ (b) $2\sin^{-1}\frac{n}{2}$ (c) $\sin^{-1}\frac{n}{2}$ (d) $2\cos^{-1}\frac{n}{2}$

3.3 The refractive index of diamond is 2 and the velocity of light in vacuum is 3×10^8 m/s. The velocity of light in diamond is,

(a) 10^8 m/s (b) 1.5×10^8 m/s (c) 2×10^8 m/s (d) none of these

3.4 A ray of light passes from vacuum into a medium of refractive index *n*. If the reflected ray and the refracted ray are at right angle to each other then the angle of incident is,

(a) $\tan^{-1} n$ (b) $\sin^{-1} n$ (c) $\cos^{-1} n$ (d) $\sin n$

3.5 The absolute refractive index of water is $\frac{4}{3}$ and that of glass is $\frac{3}{2}$. The refractive index of glass with respect to water will be,

(a) 2 (b) $\frac{9}{8}$ (c) $\frac{8}{9}$ (d) none of these

3.6 Refractive index of a material

(a) does not depend on the material (b) does not depend on the color of light

(c) does not depend on the angle of incidence (d) none of these

3.7 Critical angle will be largest when light goes from

(a) water to air (b) glass to air (c) glass to water (d) diamond to air

3.8 It is not possible of visualize total internal reflection when light travels from

(a) water to air (b) glass to air (c) glass to water (d) diamond to air

3.9 When the angle of incidence in the denser medium is equal to the critical angle then angle of refraction is

(a) 0º (b) 90º (c) 180º (d) none of these

3.10 Velocity of light is minimum in

(a) water (b) glass (c) diamond (d) free space

3.11 The refractive index of a material for which the critical angle to air is 30º is,

(a) 0 (b) $1/2$ (c) $\sqrt{3}/2$ (d) 2

3.12 Mirages are due to

(a) reflection (b) total internal reflection

(c) refraction (d) polarization

3.13 When a glass rod is kept inside a beaker containing water, it seems to disappear due to the reason that

(a) they have the same color (b) they have the same density

(c) they have the same refractive index (d) total internal reflection

3.14 The refractive index of a medium is maximum for ______ color

(a) yellow (b) green (c) red (d) violet

3.15 A ray of light is refracted from a medium to another with speeds c_1 and c_2 respectively. If i be the angle of incidence in the first medium and r be the angle of refraction in the second medium, the refractive index of medium 1 with respect to medium 2 is,

(a) $\frac{\sin i}{\sin r}$ (b) $\frac{c_1}{c_2}$ (c) $\frac{\sin r}{\sin i}$ (d) none of these

3.16 The critical angle for a pair of liquid of refractive indices 3/2 and 6/5 is

(a) $\sin^{-1}(5/4)$ (b) $\sin^{-1}(4/5)$ (c) $\sin^{-1}(6/5)$ (d) $\tan^{-1}(4/5)$

3.17 If the refractive index of a medium is 2 then the critical angle for the medium-air combination is

(a) 15º (b) 30º (c) 45º (d) 60º

3.18 When a ray of light passes from one medium to another, which of the following does not change?

(a) wavelength (b) intensity (c) frequency (d) speed

3.19 A light ray is refracted from one medium of refractive index 2 to another medium of refractive index 1.5. If the wavelength of light in the first medium is 600 nm, the wavelength of light in the second medium will be,

(a) 400 nm (b) 600 nm (c) 800 nm (d) none of these

3.20 In optical fibre the principle used is

(a) scattering (b) refraction

(c) interference (d) total internal reflection

3.21 For a plane reflector, the reflection matrix is

(a) $\begin{bmatrix} 1 & 0 \\ 0 & -1 \end{bmatrix}$ (b) $\begin{bmatrix} -1 & 0 \\ 0 & 1 \end{bmatrix}$ (c) $\begin{bmatrix} 0 & 0 \\ 1 & -1 \end{bmatrix}$ (d) $\begin{bmatrix} 1 & -1 \\ 0 & 0 \end{bmatrix}$

3.22 Maximum magnification of a simple microscope is limited to

(a) 19 (b) 5 (c) 9 (d) 10

3.23 The magnification produced by the objective of a compound microscope is 10. If the magnifying power of the microscope is 50, the magnification produced by the eye-piece will be

(a) 500 (b) 5 (c) 6 (d) 0.2

Answers of Multiple Choice Questions

3.1 (d), 3.2 (d), 3.3 (b), 3.4 (a), 3.5 (b), 3.6 (c), 3.7 (c), 3.8 (a), 3.9 (b), 3.10 (c), 3.11 (d), 3.12 (b), 3.13 (c), 3.14 (d), 3.15 (c), 3.16 (b), 3.17 (b), 3.18(c), 3.19 (c), 3.20 (d), 3.21 (a), 3.22 (c), 3.23 (b)

Short and Long Answer Type Questions

Category I

3.1 State the laws of refraction of light.

3.2 Explain the terms total internal reflection and critical angle with diagram.

3.3 Mention the conditions for total internal reflection.

3.4 Find the difference between relative refractive index and absolute refractive index.

3.5 Explain the Fermat's principle and give its statement.

3.6 Derive the laws of reflection from Fermat's principle.

3.7 In how many classes lenses can be classified? Explain with diagram.

3.8 Derive the laws of refraction from Fermat's principle.

3.9 Name the factors which can influence the refractive index of a medium.

3.10 Show that the absolute refractive index of a medium is always greater than 1.

3.11 Show that the speed of light for different media and for a particular wavelength varies inversely as the absolute refractive index of the media.

3.12 Show that the speed of light in a particular medium increases as the wavelength increases.

3.13 Establish Snell's law in terms of velocity and wavelength.

3.14 Explain the role of total internal reflection during the formation of image.

3.15 Give the drawback of ordinary laws of reflection and refraction. What is Fresnel's equation?

3.16 Obtain Fresnel's equation when the electric vector is parallel to the plane of incidence.

3.17 Obtain Fresnel's equation when the electric vector is perpendicular to the plane of incidence.

3.18 Obtain Brewster's law from Fresnel's equation.

3.19 Explain total internal reflection from Fresnel's equations. When total internal reflection will occur?

3.20 In how many classes mirrors can be classified? Explain with diagram.

Category II

3.21 A ray of light passes from a point A specified by a column matrix $[x_1, \theta_1]$ to another point B specified by another column matrix $[x_2, \theta_2]$ through a translation d in air. Obtain the translation matrix during the process.

3.23 A ray of light passes from a point A specified by a column matrix $[x_1, \theta_1]$ to another point B specified by another column matrix $[x_2, \theta_2]$ by reflection at a curved surface. Obtain the reflection matrix during the process considering u and v as the object and image distance with respect to the reflector. From this equation obtain the reflection matrix for a plane reflector also.

3.24 A ray of light passes from a point to another point after refraction at a curved surface. Obtain the refraction matrix during the process considering u and v as the object and image distance with respect to the refractor.

3.25 Justify how the human eye serves as an effective optical instrument.

3.26 With a neat diagram explain the working of a simple microscope.

3.27 With a neat diagram explain the working of a compound microscope and a refracting telescope.

Numerical Problems

3.1 What will be the speed of light in glass (RI = 1.5) if the speed of light in air is 3×10^8 m/s. [*Ans*: 2×10^8 m/s]

3.2 A ray of light passes from glass to water in such a way that the refracted ray just grazes the surface of separation. If the RI of glass and water are 1.5 and 1.3 respectively with respect to air then find the angle of incidence. [*Ans*: 62.9°]

3.3 If the core and cladding of an optical fibre are made of flint glass and crown glass with RI of 1.55 and 1.51 respectively, calculate the NA and acceptance angle of the optical fibre. [*Ans*: 0.352, 20.6°]

3.4 A ray of light in water is incident on a plate at an angle of 45°. What is the angle of refraction in the plate? Given absolute RI for the plate is 1.88 and that for water is 1.33. [*Ans*: 30°]

3.5 Calculate the value of the critical angle for a pair of media with refractive indices 3 and 2. [*Ans*: 41.8°]

3.6 The focal lengths of the objective and the eye-piece of compound microscope are respectively 10 mm and 15 mm. An object, 1.1 cm away from the objective, produces the final image at infinity. Calculate the distance between the objective and the eye-piece. [*Ans:* 12.5 cm]

3.7 The focal lengths of the objective and the eye-piece of compound microscope are respectively 11 cm and 2 cm. An object, at what position from the objective, produces the final image at a distance of 25 cm from the eye-piece, if the distance between objective and eye-piece is 15 cm. [*Ans:* −1.08 cm]

PRACTICALS

1. To verify the laws of reflection

Aim

Raw data, primarily in the form of pencil tracings of both incident and reflected rays, are to be found out to verify the laws of reflection.

Apparatus

Laser pointer, flat back-silvered mirror, 0.25 inch thick target, graph paper, protractor, binder clips, pen and pencil

Procedure

First Part

1. Test the laser pointer and be sure that the binder clips can hold it horizontally while keeping the pointer on.
2. By using a pen, draw a straight line on the graph paper and to label it as "mirror". On the line mark a central point where the grid of the graph intersects the line and label it as the "reflection point"
3. Next, using pen and protractor, draw the normal line which should be normal to the mirror line. Label this normal line as "normal".
4. Next, using pencil and protractor, draw the incoming angle at the reflection point. Choose angle between 20 degree and 50 degree from the normal and label this line as the "incident ray".

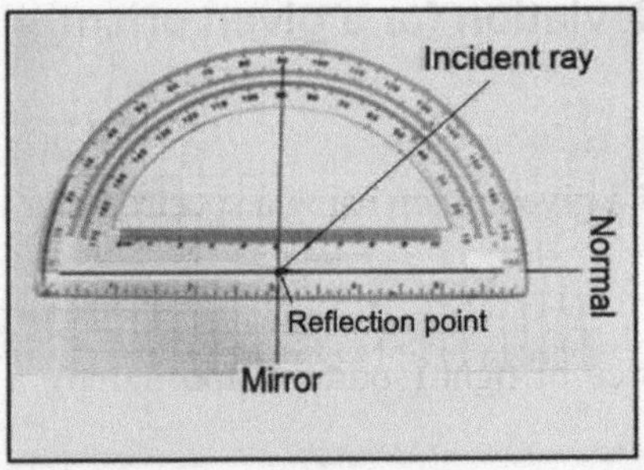

Fig. (i)

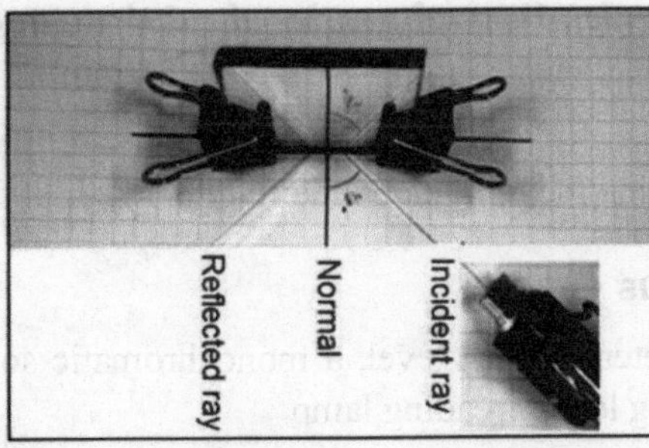

Fig. (ii)

5. Using protractor, find the angle between the normal and the incident ray. Note the measured angle on the graph.

6. Draw a dotted line for indicating the hypothesis ray which represents the "reflected ray" when the incident ray strikes the mirror. Likely, this reflected ray starts at the point on the mirror line and will go straight out at an angle from normal which is equal to the angle of incidence.
7. Now, put the mirror in a manner so that its backside is on the mirror line. On the graph paper then trace a pencil line along the front side of the mirror which will mark the edge of the glass surface.

Second Part

1. It may be noted that the shiny reflective surface is at the back of the mirror, *i.e.*, behind the glass. It will reflect most of the light to produce the brightest reflected ray. The glass surface will reflect some part of light and making them dim reflected rays. The glass edge line is critical to help for a sense of reflected ray observations.
2. Now, turn the laser pointer on. Use the binder clip for maintaining the laser pointer in this position. Be sure that during the experiment, the laser beam follows the incident line and it exhibits a reflected beam on the other end of the normal line.
3. Angle the laser pointer in such a manner that both the incident and reflected rays glides the surface of the graph paper. After adjusting the laser pointer it should not be disturbed until the recording of all the data on the graph paper.
4. Identify properly and carefully to mark the reflected ray for at least three places along the reflected beam of the laser. Never disturb the position of the pointer. After completing the observation, turn the pointer off.
5. Using protractor, line up the reflected ray's to draw a straight line which will intersect the mirror line. Find the angle between the normal and the reflected ray.

It is to mention that the laser reflected ray does not line up with the hypothesis ray that starts at the mirror line point. However, if the reflected ray is extended toward the mirror, it intersects the normal line. Now, place the protractor eye at the point of intersection of the normal and reflected ray and align the protractor to the normal. The reflection ray angle will be equal (or very nearly equal) to the incident ray, even though it is shifted.

2. Determination of angle of minimum deviation for a given prism

Aim

To determine the refractive index of the material of a given prism using a spectrometer

Apparatus

Spectrometer, a spirit level, a monochromatic source of light (sodium vapor lamp), a glass prism, a magnifying lens, a reading lamp

Theory

If a light beam strikes the surface of a transparent material (like glass, quartz crystal), a portion of it is transmitted while another portion is reflected. The transmitted light ray has low value of deviation (δ)

of the path, from the incident ray, due to refraction. If a ray of light enters a prism, it suffers refraction in a manner as exhibited in Fig. (i). In this figure, ABC is the principal section of the prism, OP is the incident ray and OR is the emergent ray. The angles, $i_1, r_1, i_2, r_2, \delta_1, \delta_2$ have their usual significance and δ_m represents the angle of minimum deviation. The RI of the prism can then be obtained from the equation,

$$\mu = \frac{\sin\left(\frac{A+\delta_m}{2}\right)}{\sin\frac{A}{2}} \qquad (i)$$

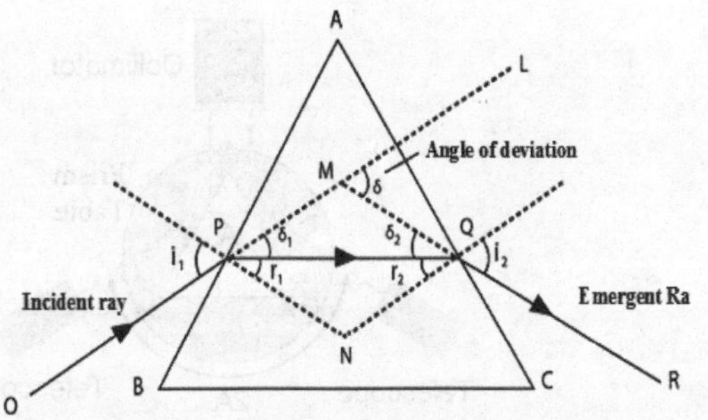

Fig. (i)

Procedure

Telescope adjustment

1. At first, the spectrometer as well as the prism table is to be adjusted in horizontal position. This is done by the leveling screws [Fig. (ii)].
2. Then the telescope is to be turned properly towards a distant object for getting a prominent and distinct image.
3. The slit is then illuminated by using a sodium vapor lamp. The slit and the collimator are to be properly adjusted for getting a narrow and vertical image of the slit.
4. The telescope is then turned properly for receiving the direct ray. This can be confirmed by ensuring that the vertical slit is in coincidence with the vertical crosswire.

Fig. (ii)

For measuring the angle of the prism

1. Find the least count (l. c.) of the given spectrometer.
2. Keep the prism on the prism table of the spectrometer in a manner such that its refracting angle A is faced towards the collimator and also the refracting edge (A) of the prism is to be properly placed at the centre of the prism table.

   Under the situation, some portion of the light after falling on each face of the prism will be reflected. This reflected light is to be observed by the telescope [Fig. (iii)].
3. The telescope is shifted at one side for receiving the light which is reflected from the face AB.

The crosswires are properly focused on the image of the slit and then readings of the two verniers are to be noted. Telescope is then moved to the other side for receiving the reflected light from the face AC of the prism. The crosswires are again focused on the image of the slit at this position for taking readings of the two verniers.

4. The angle through which the telescope is shifted, *i.e.*, the difference of the two positions of the telescope provides twice of the refracting angle of the prism. So half of this angle will give the required angle of the prism.

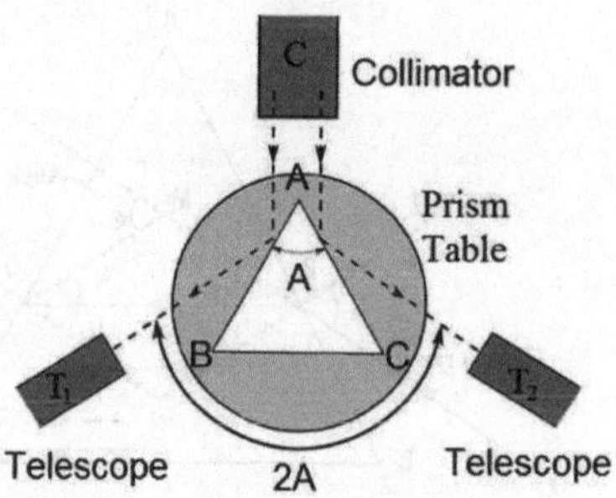

Fig. (iii)

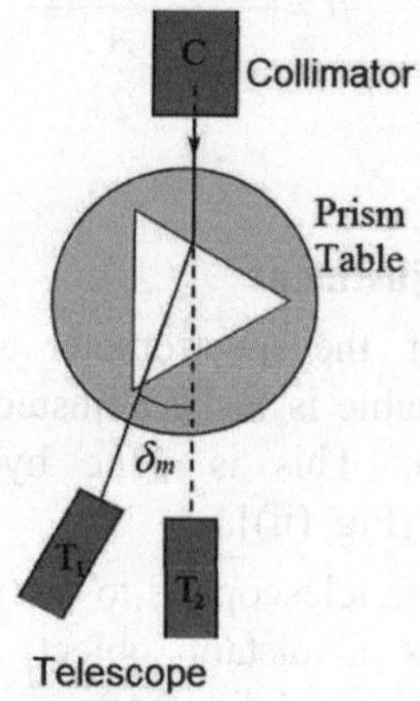

Fig. (iv)

For measuring the angle of minimum deviation

1. Place the prism on the prism table of the spectrometer in a manner such that the centre of the prism coincides with the centre of the table. Light is allowed to fall on one of the polished faces and it will emerge out of the other polished face after refraction. The telescope is turned properly to view the refracted image of the slit on the other face [Fig. (iv)].
2. Vernier table is now slowly rotated so that the slit image is moved in the direction of reducing deviation angle.
3. It is observed that at some particular position, the image becomes stationary for some time. Vernier table is to be fixed at that particular position. Using the fine slider for adjusting the telescope, adjust the slit so that it coincides with the crosswire. Note the main scale and vernier scale readings for the two verniers marked as vernier I and vernier II.
4. Next the prism is to be taken away from the prism table and thereby the telescope is to be turned in such a way that it becomes parallel to the collimator. Keep a record of this direct ray reading.
5. Measure the difference that occurs between direct ray and deviated ray, which represents the angle for minimum deviation (δ_m).
6. RI of the prism material can then be found out by using the working formula of Eq. (*i*).

Results

Least count (l. c.) of the spectrometer = …

Wavelength of light used = … cm

Table 1: Determination of the angle of prism

Sl. No.	Vernier	Telescope reading for reflection						Difference $\theta = a - b$	$A = \theta/2$
		First face			Second face				
		MSR	VSR	TR (a)	MSR	VSR	TR (b)		
1.	V_1								
	V_2								
2.	V_1								
	V_2								
3.	V_1								
	V_2								
4.	V_1								
	V_2								

MSR = Main Scale Reading, VSR = Vernier Scale Reading, TR= Total Reading = MSR + VSR × l. c.

Table 2: Determination of the angle of minimum deviation

Sl. No.	Vernier V_1			Vernier V_2			Mean δ_m
	Telescope reading		Difference δ_m	Telescope reading		Difference δ_m	
	Minimum Deviation	Direct reading		Minimum Deviation	Direct reading		
1. (i)							
1. (ii)							
2. (i)							
2. (ii)							

Calculation

Refractive index of the material of the prism = …

Percentage error = …

Viva-Voce Questions

1. What is meant by monochromatic light?
2. What do you mean by angle of a prism?
3. What is angle of minimum deviation?
4. What is refractive index? What are its units? Give a statement of Snell's law of refraction.
5. Mention the relation between angle of incidence and angle of deviation.
6. How does the deviation angle vary with the wavelength of light used?
7. On what factor does the deviation angle of a prism depend?
8. What type of eye-piece is used in the telescope of a spectrometer?
9. Will the minimum angle of deviation change, if the prism is kept under water?

KNOW MORE

Geometrical optics or ray optics is a section of optics which essentially deals with the propagation of light in terms of rays. The ray in geometrical optics is useful for showing the paths along which the propagation of light occurs under certain circumstances.

Activity

Our lives are completely filled with light. Through vision which is the most valued part of our senses, light can evoke spiritual emotions. Through our vision we view a magnificent sunset or glimpse a rainbow that is breaking through the clouds. Light can amuse us in a theatre or warn to stop at an intersection. It has many uses beyond vision. Light can carry telephone signals through glass fibres or can cook a meal using a solar oven. Starting from photosynthesis in plants to the sun warming a cold-blooded animal, its supply of energy is always vital.

Interesting facts

Fermat's principle in geometrical optics offers a fascinating view. It explains how the photons know which the fastest way is; it reminds of other principles in Physics like principle of least action. However, geometrical optics does not account for some important optical effects like interference and diffraction. The various techniques come under geometrical optics are particularly useful to describe geometrical aspects of imaging, including optical aberrations.

Analogy

The visible light is the type of electromagnetic waves to which our eyes respond. But this knowledge still leaves many questions regarding the nature of light and vision. Some of them are: What is color, and how do our eyes detect it? Why do diamonds sparkle? How does light travel? How do lenses and mirrors form images? These are a few of the questions that can be obtained by the study of optics. In fact, optics is the branch of Physics that deals with the behavior of visible light and other electromagnetic waves. More specifically, optics is concerned with the generation and propagation of light and its interaction with matter.

History

Until the end of the 19th century, light was considered to be a stream of particles. These particles were either emitted by the object being viewed or emanated from the eyes of the viewer. It was Isaac Newton (1642-1727) who can be considered as the chief architect of the particle theory of light. He believed the particles left the object and stimulated the sense of sight when entering the eyes. This was a general view also at the end of the 18th century.

Timelines

Christian Huygens (1629-1695): Huygens first argued that light might have some sort of a wave motion.

Thomas Young (1801): Young provided the first clear demonstration of the wave nature of light. He demonstrated that light rays interfere with each other and this type of behavior could not be explained by particles concept.

During the nineteenth century: Many other developments led to the general acceptance of the wave theory of light. Maxwell asserted that light was a form of high frequency electromagnetic wave and Hertz confirmed Maxwell's predictions.

Applications (Real Life / Industrial)

It is appropriate to categorize optics into two major parts based on the size of the objects that light encounters. When light interacts with an object that is several times larger than the light's wavelength, its observable behavior is similar to that of a ray. It does not prominently display its wave characteristics and this part of optics is concerned with the "geometric optics." On the other hand, when light interacts with smaller objects it has very prominent wave characteristics, such as constructive and destructive interference. At this situation, wave optics will be concentrated.

Case Study (Environmental / Sustainability / Social / Ethical Issues)

It was a great concern for the scientists about the nature of light. The question was whether light behaves like particle or wave or both. Some of the initial experiments were unable to explain the wave nature of light. The photoelectric effect was a major phenomenon which was not possible to explain from the consideration of waves.

When light strikes a metal surface, electrons are ejected from the surface under certain condition, where the kinetic energy of the ejected electron is independent of the frequency of the light.

Christian Huygens (1629-1695)

Inquisitiveness and Curiosity Topics

Huygens in his wave theory assumed that light propagates in the form of wave motion rather than a stream of particles. Huygens's principle is a geometric construction for finding the position of a new wave at some point. It was based on the knowledge of the wavefront that preceded it. Another very interesting fact is the index of refraction which decreases with increasing wavelength, in general. Due to this property violet light bends more than red light when passing into a refracting material.

REFERENCES AND SUGGESTED READINGS

1. J. E. Greivenkamp, *Field Guide to Geometrical Optics*, SPIE Field Guides, 2004.
2. H. D. Young, *University Physics*, 8th Edition, Addison-Wesley, 1992.
3. E. W. Marchand, *Gradient Index Optics*, New York, NY, Academic Press, 1978.
4. E. Hecht, *Optics*, 2nd Edition, Addison Wesley, 1987.
5. A. Schuster, *An Introduction to the Theory of Optics*, London: Edward Arnold, 1904 (online).
6. M. Born and E. Wolf, *Principles of Optics: Electromagnetic Theory of Propagation, Interference and Diffraction of Light*, Elsevier, 2013.
7. M. Kline and I. W. Kay, *Electromagnetic Theory and Geometrical Optics*, Interscience Publishers 1965.

8. R. K. Luneburg, *Mathematical Theory of Optics*, Brown University Press 1944 [mimeographed notes], University of California Press, 1964.
9. S. Borowitz, *Fundamentals of Quantum Mechanics, Particles, Waves, and Wave Mechanics*, W.A. Benjamin, 1967.
10. L. D. Landau and E. M. Lifshitz, *The Classical Theory of Fields*, Butterworth-Heinemann, 1975.
11. http://www.neo-classical-physics.info/uploads/3/0/6/5/3065888/sommerfeld_-_geometrical_optics.

4 Wave Optics

UNIT SPECIFICS

The following topics are considered in this unit:

- Huygens' principle;
- Superposition of waves;
- Interference of light by wavefront splitting and amplitude splitting;
- Young's double slit experiment and Newton's rings experiment;
- Michelson interferometer and Mach-Zehnder interferometer;
- Fraunhofer diffraction from a single slit, double slit, grating and circular aperture;
- Rayleigh criterion of resolution and its application to vision;
- Diffraction gratings and their resolving power.

The practical applications of the topics are discussed for generating further curiosity and creativity as well as improving problem solving capacity.

In addition to a large number of multiple choice questions and also of short and long answer types as marked in two categories we have followed lower and higher order of Bloom's taxonomy. Assignments have been provided through a good number of numerical problems. A list of references and suggested readings are carefully designed at the end of the unit so that one can go through them for proper practice and understanding. It is important to note that for getting more information on various topics of interest some QR codes have been provided in different sections which can be scanned for relevant supportive knowledge.

After the related practicals on some selected topics, we have placed a section under the heading "Know More". This section provides the supplementary information which will be helpful for curious students. This portion largely highlights the activity, analogy, history of development of the subject covering some salient observations and findings. The timelines clearly exhibit the development of the concerned topics up to the recent time. Applications of the subject matter including various industrial uses on variety of aspects as well as different case studies in the concerned areas have been highlighted which include environmental, sustainability, social and ethical issues in addition to inquisitiveness and curiosity topics for the unit.

RATIONALE

This unit on Wave Optics explains Huygens' principle on wave theory and its applications. Superposition of waves, interference and diffraction patterns in single and double slit and Newton's rings experiment, different types of interferometers as well as diffraction gratings and their resolving powers are discussed elaborately to provide a deep insight on wave optics.

Classical optics is categorized into two branches: geometrical optics and physical optics. While in geometrical optics light is considered to move in straight lines, in physical optics light is taken as an electromagnetic wave. In wave optics light travels as a wave around objects. It can interfere and diffract. It is utilized in different areas of science like physics, astronomy, engineering, oceanography and fiber optics. Popular applications of interferometry in industry include the measurement of refractive index, small displacements and surface irregularities.

PRE-REQUISITES

Mathematics: Differential Calculus (Class XII)
Physics: Wave Optics (Class XII)

UNIT OUTCOMES

List of outcomes of this unit is as follows:

U4-O1: Employ Young's interference experiment to locate constructive and destructive interference fringes

U4-O2: Express the wave nature of light using superposition principle and Huygens' principle

U4-O3: Apply principles of thin film interference in Newton's rings experiment

U4-O4: Obtain maxima and minima conditions for the diffraction at a single slit, double slit, grating and circular aperture

U4-O5: Apply interference in design and working of interferometers

U4-O6: Apply Rayleigh criterion to have resolving power of different optical instruments

Unit-4 Outcome	EXPECTED MAPPING WITH COURSE OUTCOMES (1- Weak Correlation; 2- Medium Correlation; 3- Strong Correlation)					
	CO-1	CO-2	CO-3	CO-4	CO-5	CO-6
U4-O1	3	3	-	-	-	2
U4-O2	2	1	-	-	-	2
U4-O3	2	3	-	-	-	2
U4-O4	2	3	-	-	-	2
U4-O5	1	-	-	-	-	2
U4-O6	1	-	-	-	-	2

4.1 INTRODUCTION

Whenever a disturbance is created in any medium, wave is generated. The disturbances from a source of light spread in every direction and the points situated at equal distances from the disturbance are with the same phase. Basically, wave may be regarded as the propagation of a disturbance. For example, if a particle is dropped on the surface of water, the water level is disturbed and waves are generated from that point and spread over the surface within a very short period of time.

The state of vibration of a vibrating particle in a medium is called the phase of the wave. In vacuum or in some other homogeneous medium, the points situated at equal distance from the source of light will vibrate in the same phase and at any instant the locus of all such points having the same phase of vibration is called a wavefront.

Depending on the shape of the emitted source, wavefronts are broadly classified into three classes *viz*: *Spherical, Cylindrical* and *Plane wavefronts.*

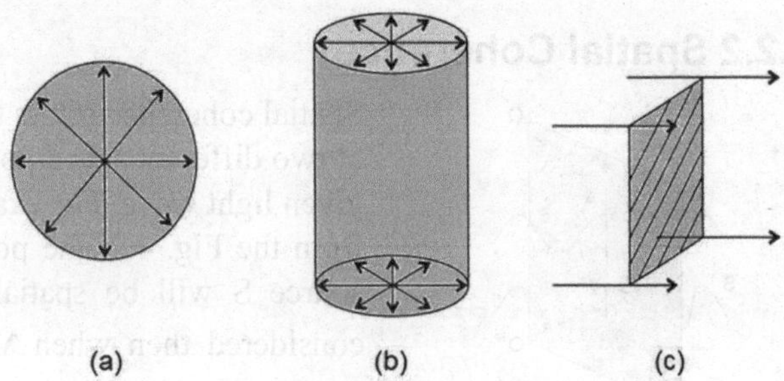

Fig. 4.1: (a) Spherical, (b) cylindrical and (c) plane wavefronts

Spherical Wavefronts: If the source producing a wave is a point source, then the locus of the points, after a finite interval of time will lie on the surface of a sphere and hence the wavefront produced is termed as a spherical wavefront [Fig. 4.1 (a)].

Cylindrical Wavefronts: For a linear source of light, the points which are equidistant from the source will lie on the curved surface of a cylinder and hence the wavefront produced from such sources will be cylindrical [Fig. 4.1 (b)].

Plane Wavefronts: A small part of any spherical or cylindrical wavefront originating from a distant source will appear as a plane and hence at its location of observation it may be regarded as a plane wavefront [Fig. 4.1 (c)].

4.2 COHERENT SOURCES

Two sources of light emitting waves having same frequency with same or nearly same amplitude but a constant phase difference are called as the *coherent sources*. In general it is not possible to have two independent coherent sources as the phase difference of the two independent sources does not remain constant. To generate coherent sources usually, we use a single source and divide it into two parts either by *division of wavefront* or by *division of amplitude.*

Division of Wavefronts: In this case, light from a single source is divided into two beams by some way and the separated beams are subsequently brought together to form the coherent sources.

The method is applicable in Young's double slit experiment, Fresnel bi-prism experiment *etc.*

Division of Amplitudes: In this case, a single beam is divided into two parts by means of partial reflection in a thin film or semi-transparent mirror and the separated beams are superposed subsequently.

This is applicable in Michelson interferometer or in Newton's rings apparatus *etc.*

There are two types of coherence *viz. temporal coherence* and *spatial coherence.*

4.2.1 Temporal Coherence

When the phase difference of any two points or a single point in space lying along the direction of propagation of the wave does not change in between a fixed interval of time, the beam of light is called temporally coherent. In Michelson-Morley experiment or in Newton's rings experiment temporally coherent source is produced. The time interval for which the phase difference remains constant is known as coherence time τ_C and thus the spatial distance for which the source is coherent is the coherence length, given as

$$l_C = c\tau_C \tag{4.1}$$

Thus for perfectly coherent monochromatic plane wave the coherence time τ_C is infinity, while for a completely incoherent source τ_C is zero.

4.2.2 Spatial Coherence

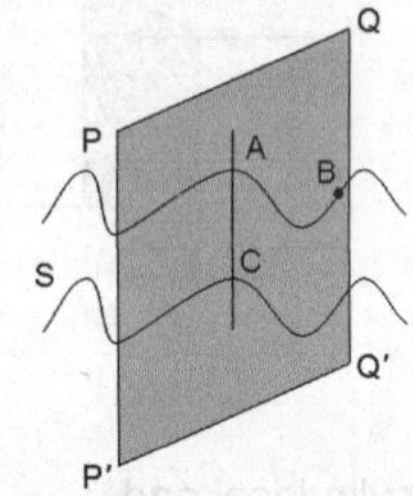

Fig. 4.2: Illustration of spatial coherence

Spatial coherence refers to the correlations between the two light fields at two different space points at any instant of time on a wavefront of a given light wave. For example, if S be the source of wave then we find from the Fig. 4.2, the points A and C which are equidistant from the source S will be spatially coherent but if the points A and B are considered then when $AB \ll l_C$, then there will be a definite phase relationship between the points A and B and they will be coherent to a high degree. But if $AB \gg l_C$, there will not be any coherence between the points A and B.

EXAMPLE 4.1

Example 4.1: If the coherence length of a monochromatic light is 3×10^{-2} m, find its coherence time.

Solution: Coherence time $(\tau_c) = \dfrac{l_c}{c} = \dfrac{3\times10^{-2}}{3\times10^{8}}$

$= 10^{-10}$ s.

4.3 HUYGENS' THEORY OF WAVE PROPAGATION

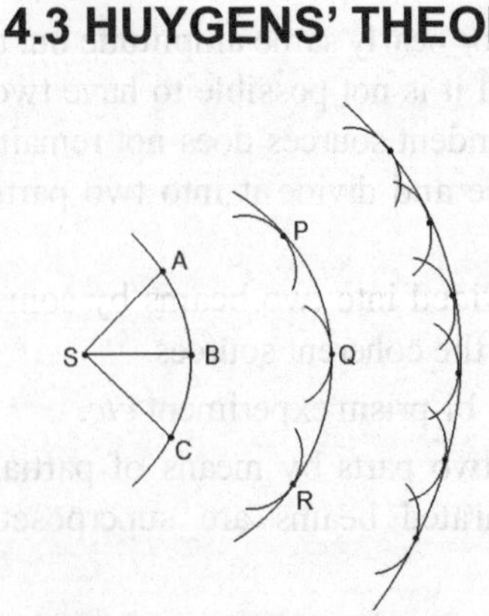

Fig. 4.3: Huygens' theory of wave propagation

Huygens proposed that the sum of the secondary waves produced as a consequence of the disturbances determine the form of the new wave. This theory of light is called the Huygens' theory of wave propagation. According to this theory, each point on a primary wavefront acts as a secondary wavelet of disturbance and spread their disturbance in all directions in a similar manner as the original wavelet. The new position of the wavefront is termed as the secondary wavelet.

Let us consider that S be a source originating waves in an isotropic homogeneous medium at $t = 0$ (Fig. 4.3). If v be the speed of the wave in that medium then in time Δt the wave will travel a distance $v\Delta t$. Let, SA = SB = SC = $v\Delta t$.

Thus ABC will represent the primary wavelet at time Δt. Now, these points A, B and C will act as secondary sources and the wave generated from these secondary sources after a finite interval of time will be represented by the wavelets P, Q and R. The common tangent drawn to these wavelets will represent the *secondary wavefront* at this instant and the process repeats itself to produce the next secondary wavefront. Intensity of the secondary wave at any instant of propagation is proportional to $(1+\cos\theta)$, where θ is the angle between the wave normal and the line joining the point of the secondary wavelet and source.

So, when $\theta=0$ the factor is $(1+1)=2$ and when $\theta=\pi$ the factor is $(1-1)=0$ and thus the intensity of the secondary wave in the backward direction is zero. Hence, according to Huygens' theory the amplitude of the secondary wavelet is highest in the forward direction and is zero in the backward direction.

4.4 INTERFERENCE OF LIGHT

When monochromatic waves of light from two sources proceed almost in the same direction and superpose at a point either in same or in opposite phase, the intensity of light at that point will be maximum or minimum accordingly as the waves meet the point in the same or in the opposite phase. This phenomenon is known as *interference of light.*

4.4.1 Conditions for Sustained Interference

For the attainment of sustained interference pattern, the following conditions must be satisfied:

(i) The two sources of light must be coherent *i.e.*, there must be a constant phase relation between the interfering waves.

(ii) For the formation of constructive interference (at maxima) the path difference between two light waves is always even multiple of $\lambda/2$ and for destructive interference (at minima) it is odd multiple of $\lambda/2$.

(iii) The two sources must lie very close to each other; else overlapping of bright and dark fringe will destroy the interference pattern.

(iv) The two sources must emit light of same wavelength but the amplitude may slightly differ.

(v) The two sources must be very narrow.

(vi) For interference with polarized light, they must be in the same state of polarization.

SCAN ME

to know more on interference and its effect

4.5 YOUNG'S DOUBLE-SLIT EXPERIMENT

Thomas Young proposed through some definite experimental findings that light is a wave, subjected to the superposition principle. His experimental achievement was to demonstrate the constructive and destructive interference of light and these observations of interference effects confirmed the presence of overlapping waves.

In this experiment (Fig. 4.4) S is a source of monochromatic light. A and B are two narrow parallel slits, equidistant from S. Thus they will act as coherent sources. The wavefronts from A and B will spread out in all directions and after superposition they form alternate bright and dark interference fringes on the screen S_1S_2. At the centre of the screen, the intensity of light is maximum (as $AO = BO$) and it is called central maximum. Let, the wavelength of the wave associated with the coherent sources A and B be λ and let, P be any point on the screen.

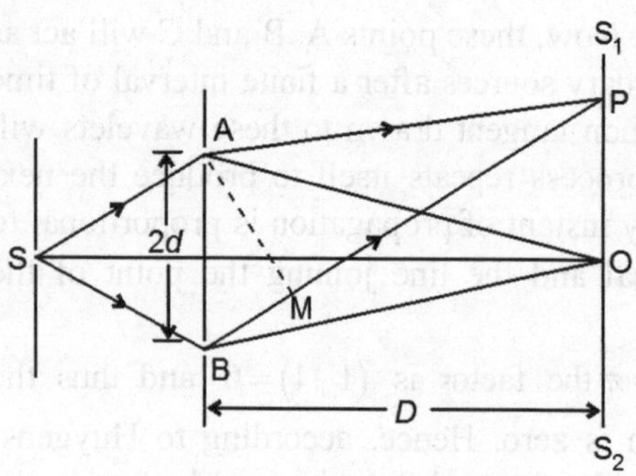

Fig. 4.4: Young's double slit experiment

Now the resultant disturbance at P due to the waves at A and B is the sum of the displacements at P owing to the individual waves.

Now, the path difference between two interfering waves is,

$$BP - AP = BM = x$$

Let the individual disturbance at P due to the waves from A and B be,

$$y_1 = a\sin\omega t$$

and $$y_2 = a\sin(\omega t + \phi)$$

where ϕ is the phase difference corresponding to the path difference BM of the individual waves.

Thus the resultant disturbance at P is given by,

$$y = y_1 + y_2 = a\sin\omega t + a\sin(\omega t + \phi) = a\sin\omega t + a\sin\omega t\cos\phi + a\cos\omega t\sin\phi$$

$$= a\sin\omega t(1 + \cos\phi) + a\cos\omega t\sin\phi$$

Let $$a(1+\cos\phi) = A\cos\theta \quad \text{(i)}$$

$$a\sin\phi = A\sin\theta \quad \text{(ii)}$$

Then $$y = A\sin\omega t\cos\theta + A\cos\omega t\sin\theta$$

or, $$y = A\sin(\omega t + \theta) \quad (4.2)$$

Hence the resultant displacement at the point P is SHM of amplitude *A* and may be obtained from Eqs. (i) and (ii) as,

$$A^2 = a^2(1+\cos\phi)^2 + a^2\sin^2\phi = a^2(1 + 2\cos\phi + \cos^2\phi) + a^2\sin^2\phi = 2a^2(1+\cos\phi)$$

or, $$A^2 = 4a^2\cos^2\frac{\phi}{2} \quad (4.3)$$

Hence the intensity (*I*) at the point P due to the superposed wave is,

$$I = A^2 = 4a^2\cos^2\frac{\phi}{2} \quad (4.4)$$

4.5.1 Constructive and Destructive Interference

a) Constructive interference: The intensity of light at point P will be maximum if, $\cos^2\frac{\phi}{2} = 1$

or, $$\cos\frac{\phi}{2} = \pm 1$$

It is possible only when $\phi = 2n\pi$, where $n = 0, 1, 2, 3, \ldots$

So, $\phi = 0, 2\pi, 4\pi, \ldots.$

In terms of path difference, the condition is,

$$x = 2n\frac{\lambda}{2} \quad (\text{where } \phi = \frac{2\pi}{\lambda}x)$$

Hence, $I_{max} = A^2 = 4a^2$

So, for constructive interference (maximum intensity) the phase difference is even integral multiple of π *i.e.*, the path difference is an even integral multiple of $\frac{\lambda}{2}$.

b) Destructive interference: The intensity of light at point P will be minimum if, $\cos^2\frac{\phi}{2} = 0$

or, $\cos\frac{\phi}{2} = 0$

It is possible only when $\phi = (2n+1)\pi$, where $n = 0, 1, 2, \ldots$

So, $\phi = \pi, 3\pi, 5\pi, \ldots.$

In terms of path difference, the condition is,

$$x = (2n+1)\frac{\lambda}{2} \quad (\text{where } \phi = \frac{2\pi}{\lambda}x)$$

Hence, $I_{min} = A^2 = 0$

So, for destructive interference (minimum intensity) the phase difference is an odd integral multiple of π *i.e.*, the path difference is an odd integral multiple of $\frac{\lambda}{2}$.

4.5.2 Conservation of Energy in Interference

In absence of interference the intensity of the interfering waves is,

$$I = I_1 + I_2 = a^2 + a^2 = 2a^2$$

When the two light waves from coherent sources interfere constructively,

$$I_{max} = (a+a)^2 = 4a^2$$

and for destructive interference,

$$I_{min} = (a-a)^2 = 0$$

and thus the average intensity is,

$$I_{av} = \frac{4a^2 + 0}{2} = 2a^2$$

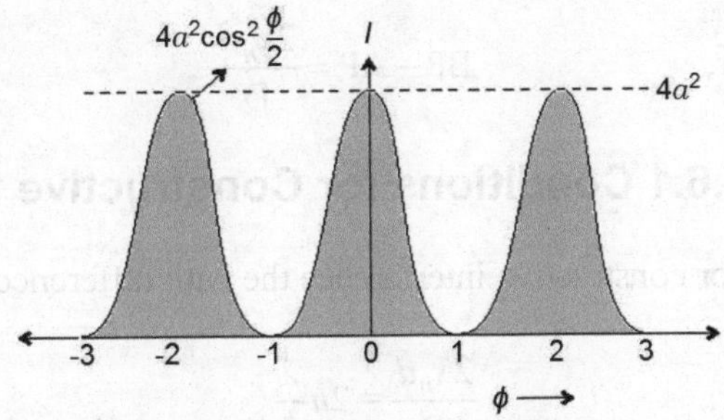

Fig. 4.5: Variation of intensity

Thus the energy is conserved. Fig. 4.5 shows the variation of intensity $I = 4a^2\cos^2\frac{\phi}{2}$ with ϕ.

EXAMPLE 4.2

Example 4.2: If amplitudes of two coherent light waves are in the ratio is $1:4$, find the ratio of maximum and minimum intensity in the pattern of interference produced.

Solution: Let for the two coherent sources, a_1 and a_2 be amplitudes of light waves.

$\therefore \quad a_1 : a_2 = 1 : 4$

or, $\quad a_2 = 4a_1$

$\therefore$ the ratio of the maximum to minimum intensity is

$$\frac{I_{max}}{I_{min}} = \frac{(a_1 + a_2)^2}{(a_1 - a_2)^2} = \frac{25}{9}$$

i.e., $\quad I_{max} : I_{min} = 25 : 9.$

4.6 FRINGE WIDTH

Fringe width means the separation between two consecutive bright and dark fringes. It depends on the wavelength of light, separation between the slits and also on the distance between the screen and slit.

Let us consider two coherent sources A and B sending monochromatic wave of wavelength λ. The point O on the screen S_1S_2 is equidistant from A and B and hence waves from A and B arrive at O at the same time and thus central bright band is produced there (Fig. 4.4). Let the nth bright band be formed at P at a distance x_n from the point O. Now, if D be the distance between the sources and the screen and $2d$ be the separation between the coherent sources then, we have from Fig. 4.4,

$$BP^2 = D^2 + (x_n + d)^2 \text{ and } AP^2 = D^2 + (x_n - d)^2$$

$$\therefore \quad BP^2 - AP^2 = \{D^2 + (x_n + d)^2\} - \{D^2 + (x_n - d)^2\}$$

or, $$BP - AP = \frac{4x_n d}{BP + AP}$$

In practice P is very close to O and hence, $BP \approx AP \approx D$

SCAN ME to see Young's double slit interference simulation

$$\therefore \quad BP - AP = \frac{4x_n d}{2D}$$

or, $$BP - AP = \frac{2x_n d}{D} \tag{4.5}$$

4.6.1 Conditions for Constructive Interference

For constructive interference the path difference should be even multiple of $\frac{\lambda}{2}$ *i.e.* $BP - AP = 2n\frac{\lambda}{2}$

or, $$\frac{2x_n d}{D} = 2n\frac{\lambda}{2}$$

$$\therefore \quad x_n = n\frac{\lambda D}{2d} \tag{4.6}$$

This gives the position of the nth bright fringe.

Similarly, the distance of the $(n + 1)^{th}$ bright fringe from the central fringe is,

$$x_{n+1} = (n+1)\frac{\lambda D}{2d} \tag{4.7}$$

Thus the distance between the nth and $(n + 1)^{th}$ bright fringe, called as the fringe width β, is given by

$$\beta = x_{n+1} - x_n = (n+1)\frac{\lambda D}{2d} - n\frac{\lambda D}{2d}$$

or,
$$\beta = \frac{\lambda D}{2d} \tag{4.8}$$

This gives the separation between two consecutive bright fringes.

4.6.2 Conditions for Destructive Interference

For destructive interference the path difference should be odd multiple of $\frac{\lambda}{2}$ *i.e.*, $BP - AP = (2n+1)\frac{\lambda}{2}$

or,
$$\frac{2x_n' d}{D} = (2n+1)\frac{\lambda}{2}$$
[where x_n' represents the position of the nth dark fringe]

∴
$$x_n' = \frac{(2n+1)\lambda D}{4d} \tag{4.9}$$

This gives the position of the nth dark fringe from the central fringe. In a similar way, the distance of the $(n+1)^{th}$ dark fringe from the central fringe is given by,

$$x_{n+1}' = \frac{(2n+3)\lambda D}{4d} \tag{4.10}$$

Thus the distance between the nth and $(n+1)^{th}$ dark fringe is given by,

$$\beta' = x_{n+1}' - x_n' = \frac{(2n+3)\lambda D}{4d} - \frac{(2n+1)\lambda D}{4d}$$

or,
$$\beta' = \frac{\lambda D}{2d} \tag{4.11}$$

Thus the distances between consecutive bright and dark fringes are the same *i.e.* they are of equal width and hence the fringe width is given by, $\beta = \beta' = \frac{\lambda D}{2d}$.

Example 4.3: In the experiment of Young's double slit, the distance between the two coherent sources is 1.15 mm. Calculate the fringe width that would be observed on a screen placed at a distance of 85 cm from the source. The wavelength of the light used is 5893 Å.

Solution: We know, fringe width $\beta = \frac{\lambda D}{2d}$.

Here $\lambda = 5893$ Å $= 5893 \times 10^{-8}$ cm, $2d = 1.15$ mm $= 0.115$ cm, $D = 85$ cm.

∴
$$\beta = \frac{\lambda D}{2d} = \frac{5893 \times 10^{-8} \times 85}{0.115} = 0.43 \text{ cm.}$$

EXAMPLE 4.3

EXAMPLE 4.4

Example 4.4: In Young's double slit experiment the distance between two slits is 0.5 mm, the wavelength of light used is 5000 Å and the separation between the source and the screen is 50 cm. Calculate the fringe width.

Solution: We have, fringe width $\beta = \dfrac{\lambda D}{2d}$

Here, $2d = 0.5$ mm = 0.05 cm, $\lambda = 5000$Å $= 5000 \times 10^{-8}$ cm and $D = 50$ cm.

$\therefore$ fringe width $\beta = \dfrac{\lambda D}{2d} = 0.05$ cm.

EXAMPLE 4.5

Example 4.5: In Young's double slit experiment the slits are 0.2 mm apart and the screen is 1.5 m away. It is observed that the distance between the central bright fringe and the fourth dark fringe is 1.8 cm. Find the wavelength of light used.

Solution: For $(n+1)^{\text{th}}$ dark fringe $x_n = \dfrac{D(2n+1)\lambda}{4d}$

For fourth dark fringe, $n = 3$.

Also, $x_4 = 1.8$ cm, $2d = 0.2$ mm, $D = 1.5$ m.

$$\therefore \quad 1.8 = \frac{1.5\times10^2\,(2\times3+1)\lambda}{0.4\times10^{-2}}$$

$$\text{or,} \quad \lambda = \frac{1.8\times0.4\times10^{-2}}{7\times1.5\times10^2} = 6.857\times10^{-5}\ \text{cm} = 6857\ \text{Å}.$$

EXAMPLE 4.6

Example 4.6: In a Young's double slit experiment, the fringes are formed at a distance of 1 m from the double-slit of separation 0.12 mm. Calculate the distance of 5th dark band from the center of screen. Given wavelength of light = 5000 Å.

Solution: For $(n+1)^{\text{th}}$ dark rings $x_n = \left(n + \dfrac{1}{2}\right)\dfrac{D\lambda}{2d}$

Here, $2d = 0.012$ cm, $D = 1$ m $= 100$ cm, $\lambda = 5 \times 10^{-5}$ cm.

$$\text{So,} \quad x_5 = \left(4 + \frac{1}{2}\right)\frac{100\times5\times10^{-5}}{0.012} = 3.75\ \text{cm}.$$

4.7 SHAPE OF THE INTERFERENCE FRINGES

From Fig. 4.4 we have, $BP^2 = D^2 + (x_n + d)^2$ and $AP^2 = D^2 + (x_n - d)^2$.

Thus the path difference between the interfering beams is

$$\Delta = BP - AP = \left\{D^2 + (x_n + d)^2\right\}^{\frac{1}{2}} - \left\{D^2 + (x_n - d)^2\right\}^{\frac{1}{2}}$$

or, $$\Delta+\left\{D^2+(x_n-d)^2\right\}^{\frac{1}{2}}=\left\{D^2+(x_n+d)^2\right\}^{\frac{1}{2}}$$

or, $$\left[\Delta+\left\{D^2+(x_n-d)^2\right\}^{\frac{1}{2}}\right]^2=D^2+(x_n+d)^2$$

or, $$\Delta^2+2\Delta\left[D^2+(x_n-d)^2\right]^{\frac{1}{2}}+D^2(x_n-d)^2=D^2+(x_n+d)^2$$

or, $$\Delta^2+2\Delta\left[D^2+(x_n-d)^2\right]^{\frac{1}{2}}=4x_nd$$

or, $$4x_nd-\Delta^2=2\Delta\left[D^2+(x_n-d)^2\right]^{\frac{1}{2}}$$

Squaring, we get $\left(4x_nd-\Delta^2\right)^2=4\Delta^2\left[D^2+(x_n-d)^2\right]$

or, $$16x_n^2d^2+\Delta^4-8x_nd\Delta^2=4\Delta^2[D^2+(x_n-d)^2]$$

or, $$\frac{16x_n^2d^2+\Delta^4-8x_nd\Delta^2}{4\Delta^2}=D^2+(x_n-d)^2$$

or, $$\frac{4x_n^2d^2}{\Delta^2}+\frac{\Delta^2}{4}-2x_nd=D^2+x_n^2+d^2-2x_nd$$

or, $$x_n^2\left(1-\frac{4d^2}{\Delta^2}\right)+D^2=\frac{\Delta^2}{4}-d^2$$

or, $$\frac{x_n^2}{\dfrac{\Delta^2}{\Delta^2-4d^2}}+D^2=\frac{\Delta^2-4d^2}{4}$$

or, $$\frac{x_n^2}{\dfrac{\Delta^2}{\Delta^2-4d^2}\cdot\dfrac{\Delta^2-4d^2}{4}}+\frac{D^2}{\dfrac{\Delta^2-4d^2}{4}}=1$$

or, $$\frac{x_n^2}{\dfrac{\Delta^2}{4}}-\frac{D^2}{\dfrac{4d^2-\Delta^2}{4}}=1 \tag{4.12}$$

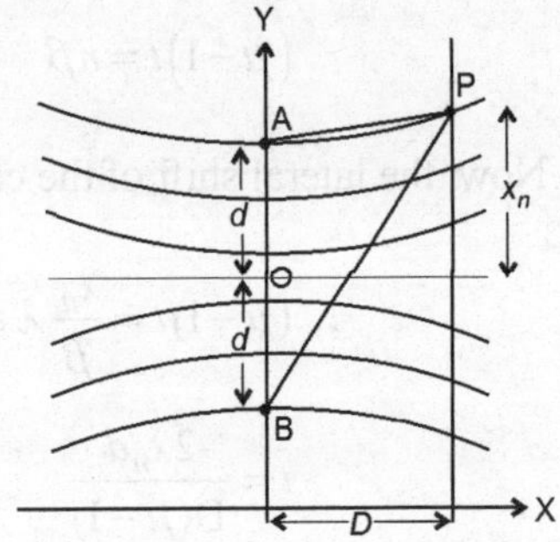

Fig. 4.6: Shape of the fringe

Thus the loci of the points of constant path difference Δ in the x_n-D plane are hyperbola (Fig. 4.6).

4.8 DETERMINATION OF THICKNESS OF A THIN FILM

By thin film we mean a layer of material whose thickness is ranged from fractions of a nanometer to several micrometers. If a thin film of thickness t be introduced into one of the path AP of the interfering waves then the position of the central fringe will be shifted from O to P so that the optical path is again equal *i.e.* AP = BP (Fig. 4.7). Now, the time taken by light in going from A to P and from B to P is equal. Thus,

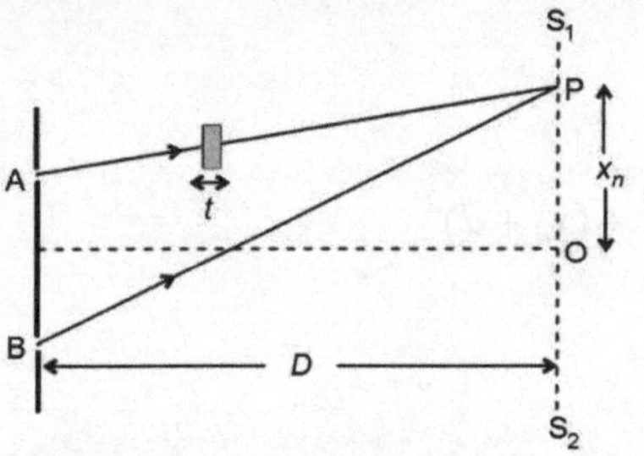

Fig. 4.7: Determination of the thickness of a thin film

$$\frac{BP}{c} = \frac{AP - t}{c} + \frac{t}{v}$$

where v is the velocity of light in the film.

or, $$\frac{BP - AP}{c} = \frac{t}{v} - \frac{t}{c}$$

or, $$BP - AP = \left(\frac{t}{v} - \frac{t}{c}\right)c = t\left(\frac{c}{v} - 1\right)$$

$\therefore$ $$BP - AP = t(\mu - 1) \quad (4.13)$$

If P be the original position occupied by n^{th} order fringe then, $BP - AP = n\beta$

or, $$(\mu - 1)t = n\beta$$

Now, the lateral shift of the central fringe of zero optical path difference is, $x_n = n\beta = n\frac{\lambda D}{2d}$

$\therefore$ $$(\mu - 1)t = \frac{x_n}{\beta}\lambda = \frac{x_n \lambda}{\frac{\lambda D}{2d}} = \frac{2dx_n}{D}$$

$\therefore$ $$t = \frac{2x_n d}{D(\mu - 1)} \quad (4.14)$$

This gives the thickness of the film.

4.9 STOKES' LAW

Stokes interpreted the relative phase of light as found to be reflected at a boundary between materials of varying refractive indices. It states that if waves are reflected from a rarer to denser medium interface then the reflected waves suffer a phase change of π or a path difference of λ/2 compared to the incident waves. Suppose a light wave of amplitude 'a' is incident on the surface and fractions of the amplitude of the incident wave are respectively reflected and refracted above and below the surface (Fig. 4.8).

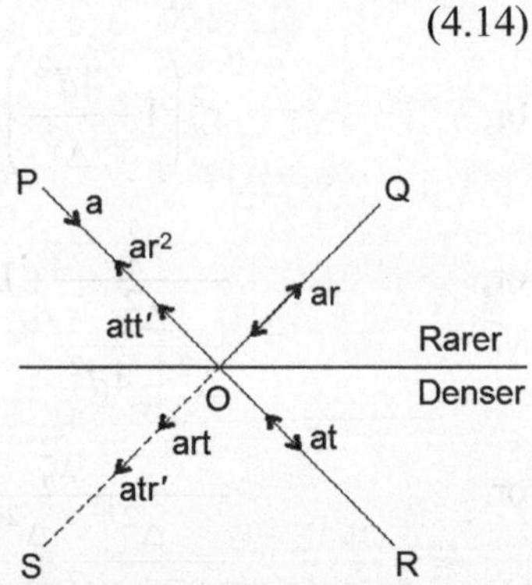

Fig. 4.8: For the interpretation of Stokes' law

Let the amplitude of reflected and refracted waves are respectively 'ar' and 'at'. If there is no absorption of energy, the wave motion is a strictly reversible phenomenon. Now, if the reflected wave along OQ be reversed we get one reflected wave of amplitude 'ar^2' along OP and another refracted wave of amplitude 'art' along OS. Similarly, if the wave along OR is being reversed we get along OS, a reflected wave of amplitude 'atr'' and along OP, a refracted wave of amplitude 'att''. Here, r' and t' are the reflection and transmission coefficient of the denser medium. Since the resultant effect must consists a wave of amplitude 'a' along OP, we have

$$att' + ar^2 = a$$

and $\quad art + atr' = 0$

Thus, we have, $\quad tt' = 1 - r^2 \text{ and } r' = -r \qquad (4.15)$

The last relation shows a phase change of π between reflection in the rarer and the denser medium. Eq. (4.15) represents the Stokes' law.

4.10 INTERFERENCE IN A THIN FILM

4.10.1 Reflected Light

Consider a thin film of thickness t and refractive index μ bounded by two plane surfaces AB and A'B' (Fig. 4.9). Let the ray PO is incident at O on AB. A part of it is reflected along OT and the other part is refracted along OQ.

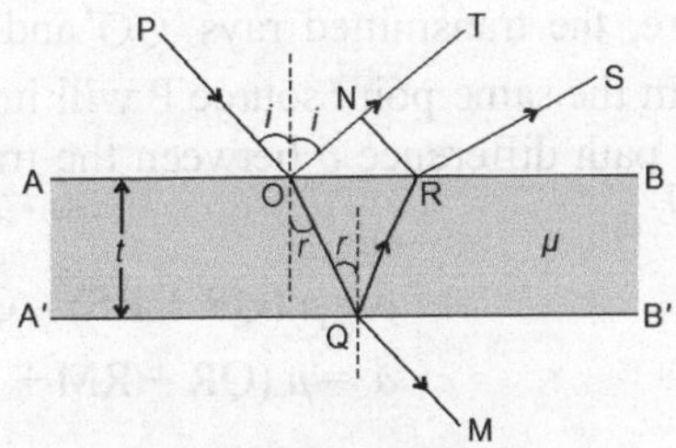

Fig. 4.9: Interference in a thin film for reflected light

Then at Q the refracted ray suffer reflection along QR and transmission along QM. The refracted ray QR then again refracted along RS and the two reflected rays OT and RS interfere with each other.

Now the path difference between these two rays is,

$$\delta = \mu\,(OQ + QR) - ON = 2\mu OQ - ON \qquad (\because OQ = QR)$$

Again, $\quad \cos r = \dfrac{t}{OQ}$

or, $\quad OQ = \dfrac{t}{\cos r}$

and $\quad \sin i = \dfrac{ON}{OR}$

or, $\quad ON = OR \sin i$

$$\therefore \quad \delta = \frac{2\mu t}{\cos r} - OR \sin i = \frac{2\mu t}{\cos r} - 2t \tan r \sin i \qquad \left[\because \tan r = \frac{OR}{2t};\ OR = 2t \tan r\right]$$

$$= \frac{2\mu t}{\cos r}(1 - \sin^2 r) = \frac{2\mu t}{\cos r}\cos^2 r \qquad [\because \sin i = \mu \sin r]$$

or, $$\delta = 2\mu t \cos r \qquad (4.16)$$

Now, an extra path difference of $\dfrac{\lambda}{2}$ is produced along the path OT as it is reflected from the surface of denser medium.

i.e., $$\delta = 2\mu t \cos r \pm \frac{\lambda}{2} \qquad (4.17)$$

For constructive interference, $2\mu t \cos r \pm \dfrac{\lambda}{2} = n\lambda$

or, $$2\mu t\cos r = (2n \mp 1)\frac{\lambda}{2} = \text{odd multiple of } \frac{\lambda}{2} \quad (4.18)$$

For destructive interference, $2\mu t\cos r \pm \frac{\lambda}{2} = (2n \pm 1)\frac{\lambda}{2}$

or, $$2\mu t\cos r = n\lambda = 2n\frac{\lambda}{2} = \text{even multiple of } \frac{\lambda}{2} \quad (4.19)$$

4.10.2 Transmitted Light

Here, the transmitted rays QQ′ and SS′ which are derived from the same point source P will interfere (Fig. 4.10). Now, the path difference δ between the transmitted rays QQ′ and SS′ is,

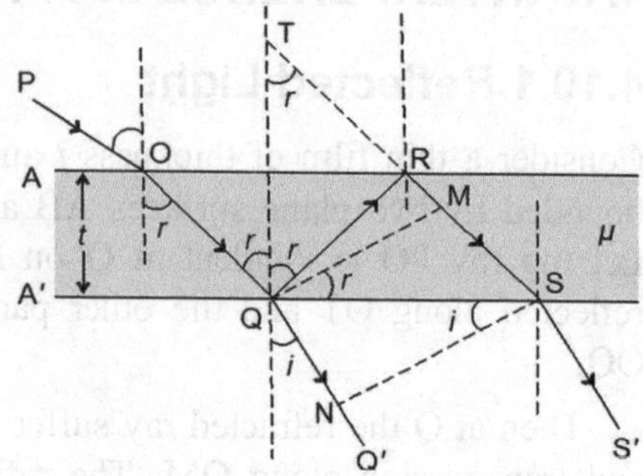

Fig. 4.10: Interference in a thin film for transmitted light

$$\delta = \mu\,(\text{QR} + \text{RS}) - \text{QN}$$

or, $$\delta = \mu\,(\text{QR} + \text{RM} + \text{MS}) - \mu\,\text{MS}$$

$$\left[\because \mu = \frac{\sin i}{\sin r} = \frac{\text{QN/QS}}{\text{MS/QS}} = \frac{\text{QN}}{\text{MS}}\right]$$

or, $$\delta = \mu\,(\text{QR} + \text{RM}) = \mu\,(\text{TR} + \text{RM}) = \mu\text{TM}$$

$$= \mu\text{TQ}\cos r = 2\mu t\cos r \qquad \left[\because \text{TQ} = 2t,\ \cos r = \frac{\text{TM}}{\text{TQ}}\right] \quad (4.20)$$

As the two rays, QQ′ and SS′ come after refraction at rarer medium (air) at points Q and R, there is no additional path difference between the two rays. Therefore, for constructive interference of nth order $[n = 1, 2, 3, \ldots]$

$$2\mu t\cos r = n\lambda \quad (4.21)$$

and for destructive interference of nth order $[n = 1, 2, 3, \ldots.]$

$$2\mu t\cos r = (2n \pm 1)\frac{\lambda}{2} \quad (4.22)$$

Thus, the conditions for bright or dark films for transmitted light are just reverse of that for reflected light.

EXAMPLE 4.7

Example 4.7: A thin film of RI 1.43 is illuminated by white light making incidence at an angle 30°. Find the thickness of the film if the spectroscopic observation of the refracted light reveals dark band corresponding to wavelength 600 nm.

Solution: Given $\lambda = 600$ nm, $\mu = 1.43$, $i = 30^\circ$, $n = 1$ (for minimum thickness)

Now, the thickness of the thin film will be $t = \dfrac{(2n-1)\lambda}{4\mu\cos r}$

or, $t = \dfrac{(2-1)\times 600\times 10^{-9}}{4\times 1.43\times 0.935} = 112.2\,\text{nm}$. $\left[\text{As } \cos r = \sqrt{1-\sin^2 r} = \sqrt{1-\dfrac{\sin^2 i}{\mu^2}} = 0.935\right]$

4.11 NEWTON'S RINGS

Experimental arrangement: Newton's rings are a pattern where interference is developed due to the reflection of light between spherical surface and adjacent flat surface. In the experimental arrangement (Fig. 4.11) for getting Newton's rings a plano-convex lens is placed on an optically flat glass-plate; thus enclosing an air film of varying thickness in between the glass plate and the plano-convex lens.

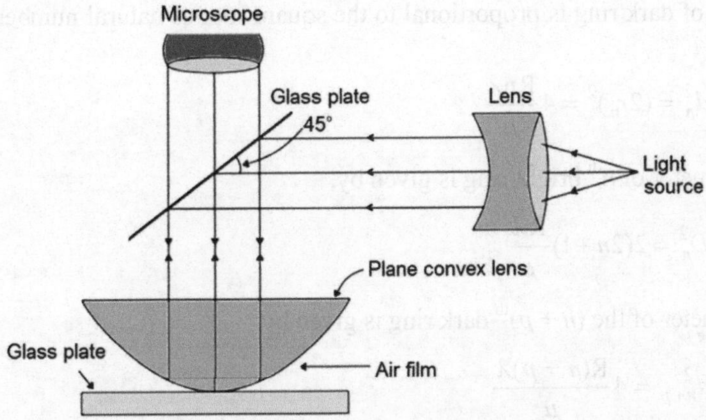

Fig. 4.11: Experimental set up to observe Newton's rings

The film is illuminated by sending light of single wavelength from a broad source. The light is made normal to the surface for Newton's ring apparatus by the help of a plane glass plate held at 45° to the ray from the lens. When the entire system is viewed by a microscope, we see alternate dark and bright concentric rings, which are called Newton's rings. Basically, these rings are produced due to interference between the rays reflected from the top and bottom surfaces of the film enclosed between the lens and glass plate (Fig. 4.12).

Working Formula: The thickness of the film at any point is related to the radius of the circular ring BD (Fig. 4.13) on which it lies and the radius of curvature of the lens R. It the point of observation is dark and radius of nth ring is r_n (say) then,

$$r_n^2 = R^2 - (R-t)^2 = 2Rt - t^2 \tag{4.23}$$

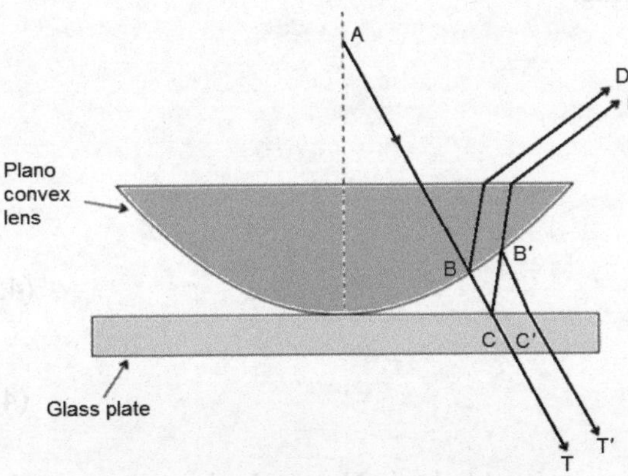

Fig. 4.12: A sectional view of the Newton's rings apparatus

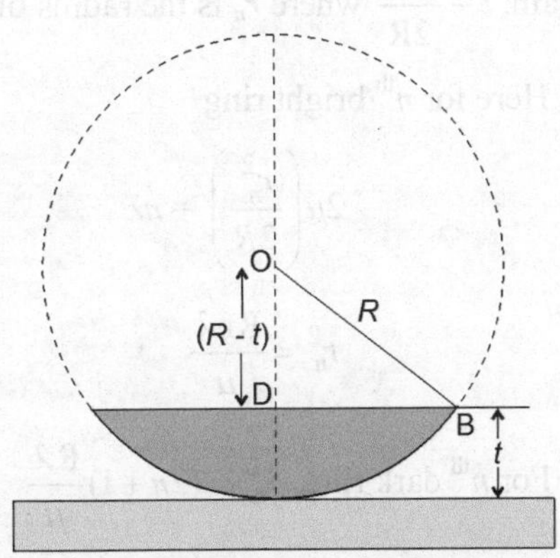

Fig. 4.13: Determination of radii of Newton's rings

As $t \ll R$, $\quad r_n^2 = 2Rt$ (4.24)

For dark fringe, $2\mu t = n\lambda$

$$\text{or,} \quad r_n^2 = \frac{Rn\lambda}{\mu} \tag{4.25}$$

So, the radius of dark ring is proportional to the square root of natural number. The corresponding diameter is

$$d_n^2 = (2r_n)^2 = 4\frac{Rn\lambda}{\mu} \tag{4.26}$$

Similarly, the diameter of nth bright ring is given by,

$$D_n^2 = 2(2n+1)\frac{R\lambda}{\mu} \tag{4.27}$$

Now, the diameter of the $(n + p)^{th}$ dark ring is given by,

$$d_{n+p}^2 = 4\frac{R(n+p)\lambda}{\mu} \tag{4.28}$$

From Eqs. (4.26) and (4.28) we have

$$\lambda = \frac{d_{n+p}^2 - d_n^2}{4pR} \quad \text{[taking } \mu = 1 \text{ (for an air film)]} \tag{4.29}$$

From Eq. (4.29) wavelength λ of the used monochromatic light can be measured.

4.11.1 Newton's Rings for Transmitted Light

For transmitted light central spot is bright.

For n^{th} bright ring, $2\mu t = n\lambda$.

Again, $t = \dfrac{r_n^2}{2R}$ where r_n is the radius of nth bright ring.

Here for n^{th} bright ring

$$2\mu\left(\frac{r_n^2}{2R}\right) = n\lambda$$

$$\text{or,} \quad r_n^2 = \frac{Rn\lambda}{\mu} \tag{4.30}$$

For n^{th} dark ring, $r_n^2 = (2n+1)\dfrac{R\lambda}{\mu}$ (4.31)

When $n = 0$, for bright ring $r_n = 0$. Hence for transmitted light, in Newton's rings, the central ring is bright.

4.11.2 Newton's Rings for Reflected Light

For reflected light the central spot is dark.

For n^{th} bright ring

$$2\mu t = (2n+1)\frac{\lambda}{2}$$

or, $$r_n^2 = \frac{(2n+1)\lambda R}{2\mu} \tag{4.32}$$

For n^{th} dark ring

$$r_n^2 = \frac{n\lambda R}{\mu} \tag{4.33}$$

When $n = 0$, for dark ring $r_n = 0$. Hence for reflected light, in Newton's ring, the central ring is dark.

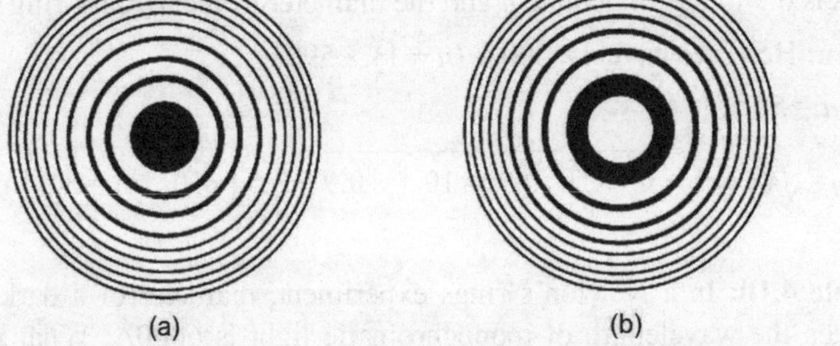

Fig 4.14: Newton's ring for (a) reflected light and (b) transmitted light

4.11.3 Refractive Index of Liquid by Newton's Rings Method

In Newton's rings apparatus, a liquid film can be produced instead of air film by putting few drops of experimental liquid between the plano-convex lens and the glass plate (Fig. 4.15).

If for liquid film, D'_n and D'_{n+p} be the diameter of nth and $(n+p)^{th}$ dark rings respectively, then

$$D'^2_{n+p} - D'^2_n = \frac{4p\lambda R}{\mu} \tag{4.34}$$

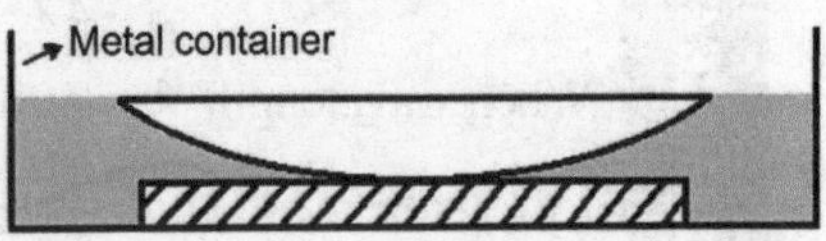

Fig 4.15: Determination of RI of a liquid

where μ is the refractive index of the liquid.

For air film, $\mu = 1$.

So, $$D_{n+p}^2 - D_n^2 = 4p\lambda R \tag{4.35}$$

Thus dividing Eq. (4.35) by Eq. (4.34) we get the refractive index of the liquid as,

$$\mu = \frac{D_{n+p}^2 - D_n^2}{D'^2_{n+p} - D'^2_n} \tag{4.36}$$

EXAMPLE 4.8

Example 4.8: In a Newton's rings experiment, the diameter of 5th dark ring is 0.336 cm and the diameter of the 15th dark ring is 0.590 cm. Find the radius of the plano-convex lens if the wavelength of the light used is 5890 Å.

Solution: Radius of the plano-convex lens will be $R = \dfrac{D_{n+p}^2 - D_n^2}{4p\lambda} = \dfrac{D_{15}^2 - D_5^2}{4 \times 10 \times \lambda}$

Here, $p = 10$, $D_5 = 0.336$ cm, $D_{15} = 0.590$ cm, $\lambda = 5890 \times 10^{-8}$ cm.

$$\therefore \quad R = \frac{(0.590)^2 - (0.336)^2}{4 \times 10 \times 5890 \times 10^{-8}} = 99.83 \text{ cm.}$$

EXAMPLE 4.9

Example 4.9: Newton's rings arrangement is utilized with a source of two wavelengths $\lambda_1 = 600$ nm and $\lambda_2 = 590$ nm. It was observed that the nth dark ring owing to λ_1 coincides with $(n + 1)^{th}$ dark ring by λ_2. If the radius of curvature of the lens is 0.9 m then calculate n and the diameter of the nth dark ring for λ_2.

Solution: Here, we have $n \times 600 = (n + 1) \times 590$

so that $n = 59$

and $D_{59} = \sqrt{4n\lambda_2 R} = \sqrt{4 \times 59 \times 590 \times 10^{-9} \times 0.9} = 3.54 \times 10^{-3}\text{ m} = 3.54$ mm.

EXAMPLE 4.10

Example 4.10: In a Newton's rings experiment, diameter of a dark ring is 0.32 cm, when the wavelength of monochromatic light is 6000Å. What would be the diameter of that ring when the wavelength of light changes to 5000 Å?

Solution: The diameter of the dark ring is given by $D_n^2 = 4n\lambda R$

or, $\quad 4nR = \dfrac{D_n^2}{\lambda}$ where $\lambda = 6000$ Å and $D_n = 0.32$ cm.

$$\therefore \quad nR = \frac{D_n^2}{4\lambda} = \frac{(0.32)^2}{4 \times 6000 \times 10^{-8}} = 426.7$$

When wavelength $\lambda = 5000$Å, using $D_n^2 = 4n\lambda R$ we get,

$$D_n = \sqrt{4n\lambda R} = \sqrt{4 \times 426.7 \times 5000 \times 10^{-8}} = 29.21 \times 10^{-2} \text{ cm}$$

So, $\quad D_n = 0.29$ cm.

EXAMPLE 4.11

Example 4. 11: A film of oil of RI 1.70 is placed between a plane glass plate and an equi-convex lens. The focal length of the lens is 1 m. Determine the radius of the 10th dark ring when light of wavelength 6000 Å is used.

Solution: For equi-convex lens

$$\frac{1}{f} = (\mu - 1)\left(\frac{1}{R} + \frac{1}{R}\right)$$

Here R is the radius of the lens, μ is refractive index of oil, f is the length of the lens.

So, $$\frac{1}{100} = (1.70-1)\left(\frac{2}{R}\right) = 0.70 \times \frac{2}{R}$$

or, $$R = 140 \text{ cm.}$$

Now the radius of the nth dark ring is $r_n = \sqrt{\dfrac{n\lambda R}{\mu}}$

$$\therefore \quad r_{10} = \sqrt{\frac{10 \times 6000 \times 10^{-8} \times 140}{1.70}}$$

$$= \sqrt{\frac{8.4 \times 10^{-2}}{1.70}} = \frac{2.22}{10} = 0.22 \text{ cm} = 2.2 \text{ mm.}$$

EXAMPLE 4.11

4.12 INTERFEROMETERS

Interferometers are tools utilized in different fields of engineering and science. They are named so as they can work by superposing two or more sources of light to produce an interference pattern. This interference pattern can be measured and analyzed by 'Interfere-o-meter' or interferometer. The interference patterns produced by interferometers provide information about the object or phenomenon being investigated. They are generally used for the purpose of very small measurements which cannot be achieved by any other means.

4.12.1 Michelson interferometer

Michelson interferometer is a high precision instrument to make interference fringes by splitting a light beam into two parts. Then it recombines those parts after they traveled separate optical paths (Fig. 4.16).

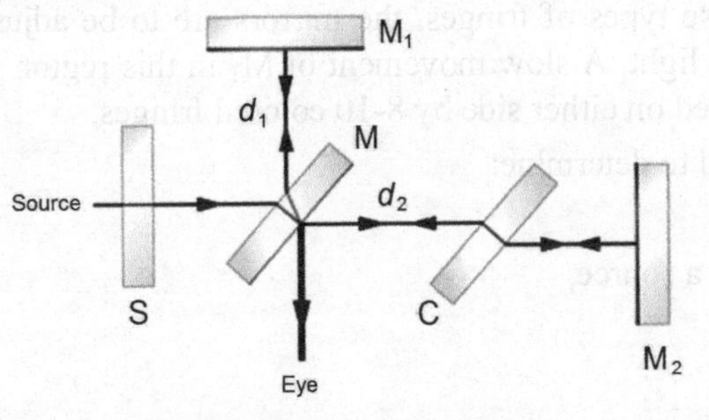

Fig. 4.16: Basic construction of Michelson interferometer

M_1 and M_2 are two optically plane mirrors highly silvered at their front surfaces. M_2 is fixed while M_1 can move accurately with the help of a calibrated screw. Both M_1 and M_1 are capable of slight rotation about their horizontal and vertical axes. M and C are two plane parallel glass plates with equal thickness. The rear side of M is half silvered. Path of a light beam from a single point on extended source S diffuses the light from a monochromatic lamp of fixed wavelength λ.

The beam strikes the half-silvered mirror M, where half is reflected and half passes through the mirror. The reflected light goes to the movable plane mirror M_1, where it is reflected through M to the observer. The transmitted half of the main beam is reflected by the fixed mirror M_2 and then goes toward the observer by M. As both the beams are originated from the same point on the source, they are coherent and hence interfere with each other.

It can be noticed from the figure that one beam passes through M three times and the other only once. To ensure that both beams traverse the same thickness of glass, a compensator plate C of transparent glass is placed in the arm containing M_2. This plate is similar to M (without silvering) and is cut generally from the same piece of glass. With the compensator any phase difference between the two beams is solely owing to the difference in the distances they travel.

Formation of the fringes: If we look along the direction MM_1, we can observe the mirror M_1 together with the image M_2' of the mirror M_2 formed by reflection from the half-silvered surface of M. Thus it can be considered that one of the interfering beams comes by reflection from M_1 and the other appears to come by reflection from M_2'. Depending on the separation (*d*) between M_1 and M_2' and the angle between these surfaces one can obtain fringes of different shapes. These are discussed below:

i) Circular fringes: When the surfaces of the mirrors M_1 and M_2 are perfectly vertical and at right angles to each other, with the half-silvered surface of M making an angle of 45^0 with the incident ray, the image M_2' will be parallel to the mirror M_1 and circular fringes are developed. If *d* be the distance of separation between M_1 and M_2', the corresponding virtual sources will be separated by 2*d*. Now for the angle of inclination θ of the reflected ray with the normal maximum intensity will be observed when

$$2d\cos\theta = m\lambda \tag{4.37}$$

forming concentric circles about the foot of the perpendicular from the eye to the mirrors as the common centre. These types of fringes are known as the fringes of equal inclination. They are situated at infinity and are non-localized.

ii) Localized fringes: When the mirror M_1 and the image M_2' of M_2 are not exactly parallel, they will enclose a wedge shaped film of air between them leading to the development of localized fringes of equal width. Depending on the thickness of the air film, the shape of the fringes may differ (straight or curved).

iii) White light fringes: White light fringes will be observed only when the path difference does not exceed a few wavelengths. To observe these types of fringes, the mirrors are to be adjusted first to get straight fringes with monochromatic light. A slow movement of M_1 in this region will bring fringes having a central dark fringe bordered on either side by 8-10 colored fringes.

Applications: Michelson interferometer can be used to determine:

i) wavelength of a monochromatic light,

ii) small difference of wavelengths of light from a source,

iii) refractive index of a material etc.

4.12.2 Mach–Zehnder Interferometer

It is used to find the variations of the relative phase shift between two collimated beams. This is obtained by splitting light from a single source. The interferometer can be used for measuring phase shifts between two beams by a change in length of one of the paths.

Set-up: A collimated beam is splitted by a half-silvered mirror. Two resulting beams, one is the *sample beam* and the other is the *reference beam*, are reflected. Both the beams then move to a second half-silvered mirror and enter into two detectors. This interferometer is a very good configurable optical instrument in contrast to the Michelson interferometer.

In order to produce fringes in white light, we must be careful to simultaneously equalize the optical paths over all the wavelengths, as otherwise no fringes will be observed.

As seen in Fig. 4.17, a compensating cell made of the same type of glass as the test cell would be placed along the path of the reference beam to match the test cell. Reflecting surfaces of the beam splitters are oriented in such a manner that the reference beams will pass through an equal amount of glass. In this orientation, the test and reference beams experience two front-surface reflections, causing the same number of phase inversions. As a result light traveling an equal optical path length in the test and reference beams produces a white light fringe after constructive interference. Collimated sources result in a non-localized fringe pattern. On the other hand, localized fringes are produced when an extended source is used.

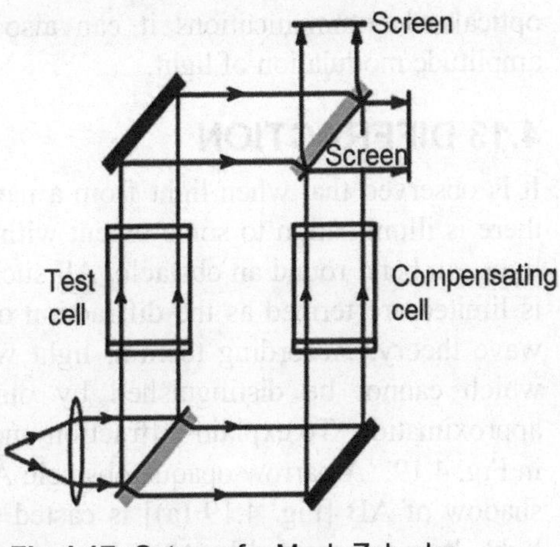

Fig 4.17: Set up of a Mach-Zehnder interferometer

In the figure, we find that the fringes can be adjusted so that they are localized in any desired plane. Most of the time, the fringes are adjusted to lie in the same plane. As a result, the fringes as well as test object can be photographed together. This interferometer is large and its flexibility in locating the fringes has made it for visualizing flow in wind tunnels.

It is used mainly in the fields of plasma physics, aerodynamics and heat. The interferometer is used in electro-optic modulators, electronic devices for fibre optic communication applications. Mach–Zehnder modulators are used in monolithic ICs with high-bandwidth electro-optic amplitude and phase responses over a multiple GHz frequency range. The interferometer can also be used for studying most counterintuitive predictions of quantum mechanics which is called *quantum entanglement*.

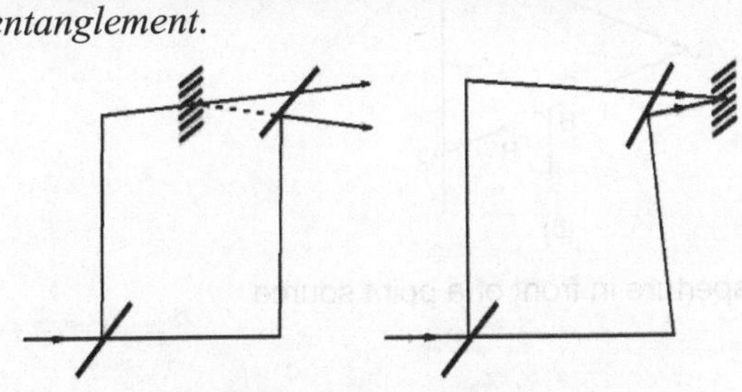

Fig 4.18: Schematic of the working of Mach-Zehnder interferometer

Working: Fresnel equations for a dielectric indicate that there is a phase change upon reflection, when the wave propagates in a lower-refractive index medium from a higher-refractive index medium but never in the opposite case. A phase shift of 180°occurs upon reflection from the front of a mirror, as the medium behind the mirror has a higher RI than the medium the light is moving in air.

No phase shift is there for a rear-surface reflection. Speed of light is lower in media with an index of refraction greater than that of a vacuum, and is given by, $v = c/n$, where c is speed of light in vacuum and n is the RI of the media. This produces an increase in phase shift which is proportional to $(n - 1) \times$ length travelled.

If k represents the constant phase shift, a total of $2k$ phase shift will occur when reflected from the rear side of a mirror. This is owing to the fact that light moving toward the rear side of a mirror will enter the glass plate, causing k phase shift and then reflect from the mirror with no further phase shift.

Applications: The interferometer has a wide range of use. Topics in quantum mechanics, with quantum entanglement, quantum computation, quantum Zeno effect, quantum cryptography, neutron diffraction, bomb tester etc. are some areas where Mach-Zehnder interferometer has a major role. In optical telecommunications it can also be utilized as an electro-optic modulator for phase and amplitude modulation of light.

4.13 DIFFRACTION

It is observed that when light from a narrow linear slit is incident on the sharp edge of an obstacle, there is illumination to some extent within the geometrical shadow of the obstacle. This reveals that light can bend round an obstacle. All such phenomena that are produced when the incident wavefront is limited are termed as the diffraction of light. Fresnel gives an explanation of this on the basis of wave theory. According to him, light wave with small wavelength can produces a small bending which cannot be distinguished by our eye. So, rectilinear propagation of light is only one approximation. To explain diffraction one can take a point source S in front of a screen PQ as shown in Fig. 4.19. A narrow opaque obstacle AB is placed between the source and the screen, such that the shadow of AB [Fig. 4.19 (a)] is casted on the screen, when we consider rectilinear propagation of light. It is represented by A′B′. However, there will be dark and bright fringes near the edge of the shadow. Similarly, when the light is allowed to pass through an aperture AB [Fig. 4.19 (b)] to fall on the screen, then it is observed that the intensity of light is not zero exactly outside the portion A′B′ on the screen, but some dark and bright patches are there, at the edges. So, both these experiments reveal that, while passing through an edge of an obstacle or an aperture, light can bend round it. This phenomenon of bending of light round the sharp edge of an obstacle is what we call diffraction. But this is only possible if the size of the obstacle and the wavelength of light are comparable. For very small wavelengths possessed by light wave, the effect of diffraction is not readily visualized but it can be increased with the increase in wavelength of the incident light wave.

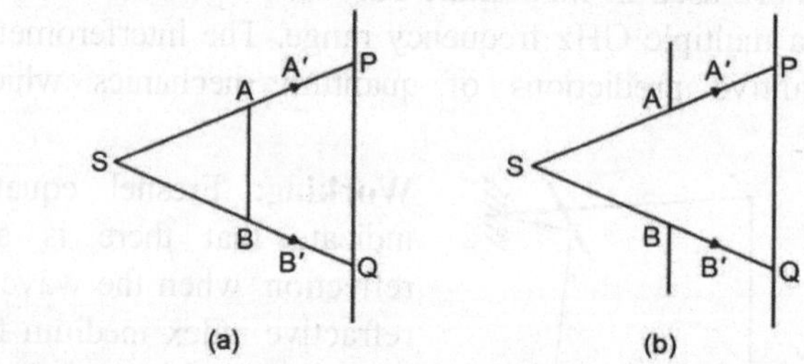

Fig. 4.19: (a) An obstacle and (b) an aperture in front of a point source

4.13.1 Types of Diffraction

Depending on the mutual position of the source and screen, diffraction may be classified into two broad categories.

These are – *Fresnel diffraction* and *Fraunhofer diffraction.*

Fresnel diffraction: In this kind of diffraction, the source and the screen are situated at a finite distance from the obstacle. In our daily life this type of diffraction is more common but it is still difficult to understand physically.

Fraunhofer diffraction: In this kind of diffraction, the source and the screen are situated an infinite distance from the obstacle. It is a limiting case of Fresnel diffraction and is easier to handle physically.

4.13.2 Differences between Fresnel and Fraunhofer diffraction

The basic differences between Fresnel and Fraunhofer diffraction are given below –

Fresnel Diffraction	Fraunhofer diffraction
1. In Fresnel diffraction, either the source or screen or both are situated at a finite distance from the obstacle.	1. In Fraunhofer diffraction, both the source and screen are situated at an infinite distance from the obstacle.
2. In Fresnel diffraction, the incident wavefront is either spherical or cylindrical.	2. In Fraunhofer diffraction, the incident wavefront is generally a plane wavefront.
3. Depending on the number of Fresnel zones, the centre of the diffraction pattern may be bright or dark.	3. In Fraunhofer diffraction, the centre of the diffraction pattern is always bright.

4.14 DIFFERENCES BETWEEN INTERFERENCE AND DIFFRACTION

Both the interference and diffraction phenomena occur due to the superposition of waves and they can occur simultaneously. Some basic differences between *interference and diffraction are as below:*

Interference	Diffraction
i) When two coherent sources originating from a particular point source are propagating in a medium and superpose then interference is produced.	i) Basically this is a kind of interference where the secondary wavelets, originating from the exposed part of the same wavefront, interfere.
ii) The interference fringes may or may not be equally spaced.	ii) The diffraction fringes never spaced equally.
iii) The interference maxima possess equal bright intensity.	iii) The central diffraction maxima possess maximum intensity and it decreases with increase in the order number.
iv) Points of minimum intensity are totally dark	iv) They are partially dark.

4.15 FRESNEL DIFFRACTION IN A CIRCULAR APERTURE

Fresnel approximation in diffraction theory is considered as the diffraction of a focused beam or a plane wave in a circular aperture. It includes effects produced by a finite Fresnel zones, large numerical aperture and the presence of aberrations.

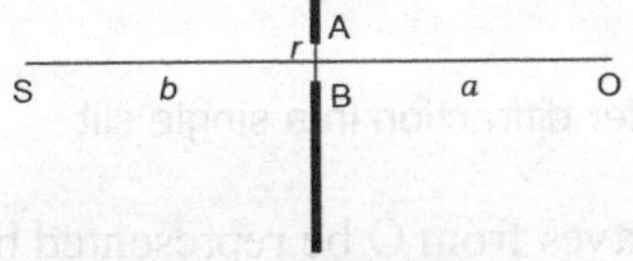

Fig. 4.20: Fresnel diffraction in a circular aperture

When the source S is placed at a distance b from the circular aperture AB then the intensity at a distance a from the aperture at point O may be obtained by dividing the opening into a number of Fresnel zones with respect to O (Fig. 4.20). If the aperture contains even number of zones with respect to O then they mutually cancel each other in pair and the intensity at O will be less. But, if the aperture contains odd number of zones with respect to O then the intensity at O will be more. If r be the radius of the aperture then,

$$\text{OA} = \sqrt{a^2 + r^2} \simeq a + \frac{r^2}{2a} \tag{4.38}$$

Similarly, $$\text{SA} = \sqrt{b^2 + r^2} \simeq b + \frac{r^2}{2b} \tag{4.39}$$

Hence, the path difference between the marginal and the central rays will be,

$$\delta = (\text{OA} + \text{SA}) - (a + b) = \frac{r^2}{2}\left(\frac{1}{a} + \frac{1}{b}\right) \tag{4.40}$$

If, $\delta = \frac{n\lambda}{2}$ then $\frac{r^2}{2}\left(\frac{1}{a} + \frac{1}{b}\right) = \frac{n\lambda}{2}$

or, $$\frac{1}{a} + \frac{1}{b} = \frac{n\lambda}{r^2} \tag{4.41}$$

Thus as a changes, n alternatively changes from odd to even making successively the intensity at O maximum and minimum. For plane incident waves, B= $\propto$.

$\therefore$ $$a = \frac{r^2}{n\lambda} \tag{4.42}$$

This equation shows that for a light of given wavelength λ passing through a circular aperture of fixed radius r, n will change from odd to even as a changes.

4.16 FRAUNHOFER DIFFRACTION IN A SINGLE SLIT

When the source and screen are situated at an infinite distance from the diffracting element, then this type of diffraction is produced. A rectangular aperture with its length large compared to its breadth is termed as a slit. The width of the slit should be at least comparable to the wavelength of light used but it cannot be less than the wavelength.

Let the plane wave front AB of monochromatic light of wavelength λ is propagating normal to a slit S_1S_2 and make incidence on it at an angle of inclination θ (Fig. 4.21). According to Huygens' principle each point of the wavefront in the plane of the slit may be regarded as the source of secondary spherical wavelets. The secondary wavelets travelling normal to the slit are brought to focus by a convex lens L on the screen at C. The wavelets travelling at an angle θ with the normal are brought to focus at Q. Let us calculate the intensity of light at Q.

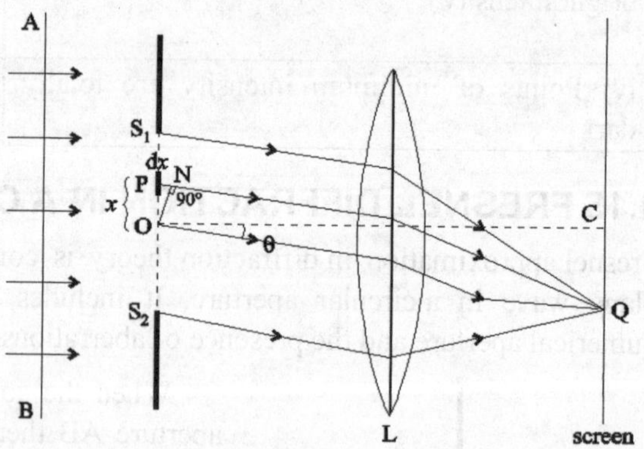

Fig. 4.21: Fraunhofer diffraction in a single slit

Let the complex light disturbance at any instance due to spherical waves from O be represented by $Ae^{i\omega t}$, where A is the amplitude and ω the angular frequency. Now the phase difference between the waves at Q coming from O and P, which is at a distance x from O is given by,

$$\frac{2\pi}{\lambda}\text{ PN} = \frac{2\pi}{\lambda} x \sin\theta = lx$$

where $$l = \frac{2\pi}{\lambda}\sin\theta$$

Now, the complex disturbance at Q due to the secondary waves from P will be proportional to $e^{i(\omega t - lx)}$ and the disturbance at Q due to the diffracting element dx can be written as,

$$dy = CAe^{i(\omega t - lx)}dx \tag{4.43}$$

If a be the width of the slit then the resultant complex disturbance at Q due to the waves coming from all the diffracting elements will be given by,

$$y = \int_{-a/2}^{a/2} CAe^{i(\omega t - lx)}dx = CAe^{i\omega t}\int_{-a/2}^{a/2} e^{-ilx}dx = \frac{CAe^{i\omega t}}{-il}\left[e^{-ilx}\right]_{-a/2}^{a/2}$$

$$= \frac{CAe^{i\omega t}}{il}\left[e^{\frac{ila}{2}} - e^{-\frac{ila}{2}}\right] = \frac{2CA}{l}e^{i\omega t}\sin\frac{la}{2} = CAae^{i\omega t}\frac{\sin\frac{la}{2}}{\frac{la}{2}} \quad \left[\because e^{\frac{ila}{2}} = \cos\frac{la}{2} + i\sin\frac{la}{2}\right]$$

where, $$\frac{la}{2} = \alpha = \frac{2\pi}{\lambda}\sin\theta\frac{a}{2} = \frac{\pi a}{\lambda}\sin\theta \tag{4.44}$$

SCAN ME

for more on single slit diffraction

Therefore, $$y = CAa\frac{\sin\alpha}{\alpha}e^{i\omega t}$$

The resultant intensity I at Q is,

$$I = yy^* = (CAa)^2\frac{\sin^2\alpha}{\alpha^2} = I_0\frac{\sin^2\alpha}{\alpha^2} \tag{4.45}$$

where, $I_0 = (CAa)^2$ is the intensity of the central maximum.

4.16.1 Conditions for maxima and minima

To find the maxima and minima of intensity, let us differentiate Eq. (4.45), with respect to α, so that,

$$\frac{dI}{d\alpha} = \frac{I_0}{\alpha^2}2\sin\alpha\cos\alpha - 2\frac{I_0}{\alpha^3}\sin^2\alpha$$

Now for maximum or minimum value of I, $\frac{dI}{d\alpha} = 0.$

or, $$\frac{I_0}{\alpha^2}2\sin\alpha\cos\alpha - 2\frac{I_0}{\alpha^3}\sin^2\alpha = 0$$

or, $$\frac{2I_0\sin\alpha}{\alpha^3}(\alpha\cos\alpha - \sin\alpha) = 0$$

Thus either, $\sin\alpha = 0$ or, $\alpha\cos\alpha = \sin\alpha$ *i.e.*, $\alpha = \tan\alpha$

Now, $$\frac{d^2I}{d\alpha^2}=\frac{I_0}{\alpha^2}2\cos 2\alpha-\frac{4I_0}{\alpha^3}\sin 2\alpha+\frac{6I_0}{\alpha^4}\sin^2\alpha=\frac{I_0}{\alpha^2}\left[2\left(1-\sin^2\alpha\right)-\frac{4\sin 2\alpha}{\alpha}+\frac{6\sin^2\alpha}{\alpha^2}\right]$$

Thus, $$\left.\frac{d^2I}{d\alpha^2}\right|_{\sin\alpha=0}=\frac{2I_0}{\alpha^2}=\text{positive.}$$

Hence $\sin\alpha=0$ gives the minimum intensity and the condition for minimum is,

$$\alpha=n\pi \text{ with } n=\pm1,\pm2,...$$

or, $$\frac{\pi a\sin\theta}{\lambda}=n\pi$$

or, $$a\sin\theta=n\lambda \tag{4.46}$$

Here $n = 0$ is excluded because for this we get the central maximum as in the limit $\underset{\alpha\to0}{Lt}\frac{\sin\alpha}{\alpha}=1$ is actually giving the maximum.

Again, $$\frac{d^2I}{d\alpha^2}=\frac{I_0}{\alpha^2}\left[2\left(1-\frac{2\tan^2\alpha}{1+\tan^2\alpha}\right)-\frac{8\tan\alpha}{\alpha\left(1+\tan^2\alpha\right)}+\frac{6\tan^2\alpha}{\alpha^2\left(1+\tan^2\alpha\right)}\right]$$

Thus, $$\left.\frac{d^2I}{d\alpha^2}\right|_{tan\alpha=\alpha}=\frac{I_0}{\alpha^2}\left[2\left(1-\frac{2\alpha^2}{1+\alpha^2}\right)-\frac{8\alpha}{\alpha\left(1+\alpha^2\right)}+\frac{6\alpha^2}{\alpha^2\left(1+\alpha^2\right)}\right]=-\frac{2I_0}{1+\alpha^2}$$

Hence $\alpha=\tan\alpha$ gives the maximum intensity and the position for maximum is obtained by solving the equation $\alpha=\tan\alpha$, which is a transcendental equitation and can be solved by graphical method only. An inspection of the graph shows that one value of α is 0 and the other values of α which give the maxima are less than but gradually approaching towards $\pm\frac{3\pi}{2},\pm\frac{5\pi}{2},...$ *etc.* This is illustrated in Fig. 4.22. The intensity of the principal maxima is,

$$I=\underset{\alpha\to0}{\text{Lt}}\frac{I_0\sin^2\alpha}{\alpha^2}=I_0 \tag{4.47}$$

The intensity of the first secondary maxima is,

$$I_1\approx I_0\frac{\sin^2\frac{3\pi}{2}}{\left(\frac{3\pi}{2}\right)^2}=\frac{4I_0}{9\pi^2}=0.045I_0 \tag{4.48}$$

The intensity of the second secondary maxima is,

$$I_2\approx I_0\frac{\sin^2\frac{5\pi}{2}}{\left(\frac{5\pi}{2}\right)^2}=\frac{4I_0}{25\pi^2}=0.016I_0 \tag{4.49}$$

The intensity distribution in the diffraction pattern due to a single slit is shown in Fig. 4.23.

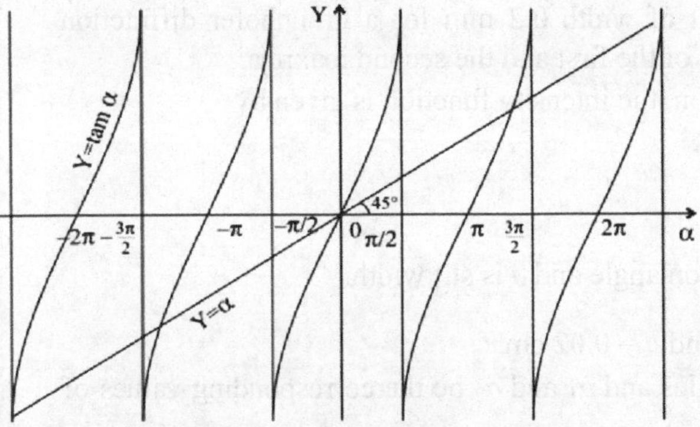

Fig. 4.22: Solution of $\alpha = \tan\alpha$ by transcendental (graphical) method

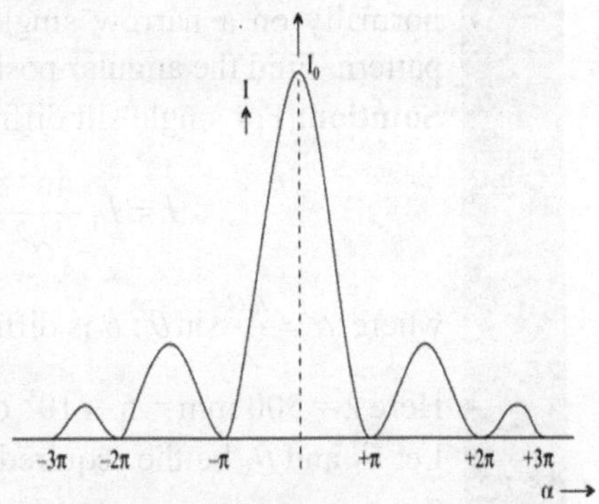

Fig. 4.23: Secondary maxima due to diffraction in a single slit

Example 4.12: A single slit forms a diffraction pattern of Fraunhofer class with white light. The second maximum in the pattern for red light of wavelength 7000 Å coincides with the third maximum of an unknown wavelength. Calculate the unknown wavelength.

Solution: Here $(2n_1+1)\frac{\lambda_1}{2} = (2n_2+1)\frac{\lambda_2}{2}$, where n_1 and n_2 are the order numbers.

Here $n_1 = 2$, $n_2 = 3$ and $\lambda_1 = 7000$ Å.

So, using $\lambda_2 = \frac{5}{7}\lambda_1$ we get

$$\lambda_2 = 5000 \text{ Å}.$$

EXAMPLE 4.12

Example 4.13: A convex lens of 40 cm focal length is used to focus the diffraction pattern of a single slit of 0.3 mm width. Find the linear distance of the first-order dark band from the central band. The wavelength of the light used is 589 nm.

Solution: For n^{th} dark band

$$a \sin\theta = n\lambda,$$

Here, $a = 0.3$ mm $= 0.3\times10^{-3}$ m; $n = 1$; $\lambda = 589\times10^{-9}$ m.

As $\sin\theta \simeq \theta$ for very small θ

so, $$\theta = \frac{\lambda}{a} = \frac{589\times10^{-9}}{0.3\times10^{-3}} \text{ rad}$$

$$= 196.3\times10^{-5} \text{ rad}.$$

$\therefore$ linear distance $x = \theta f = 196.3\times10^{-5}\times40\times10^{-2}$ m

$$= 0.7852 \text{ mm}.$$

EXAMPLE 4.13

Example 4.14: A parallel beam of light of wavelength 500 nm is incident normally on a narrow single slit of width 0.2 mm for a Fraunhofer diffraction pattern. Find the angular position of the first and the second maxima.

Solution: For single slit diffraction, the intensity function is given by

$$I = I_0 \frac{\sin^2 \alpha}{\alpha^2}$$

where $\alpha = \frac{\pi a}{\lambda}\sin\theta$; θ is diffraction angle and a is slit width.

Here $\lambda = 500$ mm $= 5 \times 10^{-5}$ cm and $a = 0.02$ cm.

Let θ_1 and θ_2 be the required angles and α_1 and α_2 be the corresponding values of α.

The first two maxima occur at $\alpha_1 = 1.430\pi$ and $\alpha_2 = 2.462\pi$

Now $\qquad \alpha_1 = \frac{\pi a}{\lambda}\sin\theta_1$ and $\alpha_2 = \frac{\pi a}{\lambda}\sin\theta_2$

$\because \qquad \alpha_1 = \frac{\pi a}{\lambda}\sin\theta_1 = 1.430\pi$

$\therefore \qquad \sin\theta_1 = \frac{1.430\lambda}{a}$

or, $\qquad \theta_1 = \sin^{-1}\left(\frac{1.430\lambda}{a}\right) = 3.57 \times 10^{-3}$ rad.

again $\qquad \alpha_2 = \frac{\pi a}{\lambda}\sin\theta_2 = 2.462\pi$

or, $\qquad \sin\theta_2 = \frac{2.462\lambda}{a}$

or, $\qquad \theta_2 = \sin^{-1}\left(\frac{2.462\lambda}{a}\right) = 6.15 \times 10^{-3}$ rad.

EXAMPLE 4.14

4.17 FRAUNHOFER DIFFRACTION IN A DOUBLE SLIT

The diffraction pattern due to a double slit is caused by the superposition of diffracted light waves by the individual slits. Let a parallel beam of light of wavelength λ be incident normally on a surface containing two slits, each of width a and separated by an opaque space of width b. The distance between any pair of corresponding points of the two slits is $d = a + b$. Fig. 4.24 depicts the Fraunhofer diffraction due to a double slit. The wavelets travelling at an angle θ with the normal are brought to focus at Q. Let us calculate the intensity of light at Q. Let the complex light disturbance at any instant due to the spherical waves from the origin O be represented by $Ae^{i\omega t}$, where A is the amplitude and ω is the angular frequency. Now, the phase difference between the waves at Q coming from O and P which is at a distance x from O is given by,

$$\frac{2\pi}{\lambda}\text{PN} = \frac{2\pi}{\lambda} x \sin\theta = lx \qquad \text{where } l = \frac{2\pi}{\lambda}\sin\theta$$

The disturbance at Q due to secondary waves from P will be proportional to $e^{i(\omega t-lx)}$.

Thus the disturbance at Q due to the diffracting element dx can be written as, $dy = CAe^{i(\omega t-lx)}dx$

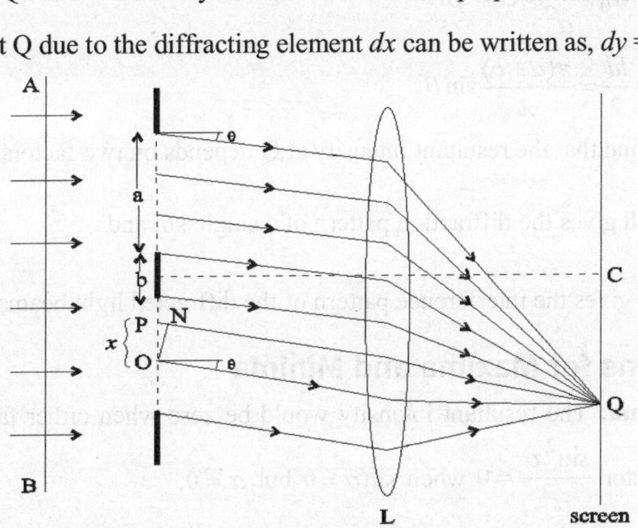

Fig. 4.24: Fraunhofer diffraction due to a double slit

If a be the width of the slit then the resultant complex disturbance at Q due to the waves coming from all the diffracting elements will be given by,

$$y = \int_{-a/2}^{a/2} CAe^{i(\omega t-lx)}dx + \int_{d-a/2}^{d+a/2} CAe^{i(\omega t-lx)}dx = CAe^{i\omega t}\left[\int_{-a/2}^{a/2} e^{-ilx}dx + \int_{d-a/2}^{d+a/2} e^{-ilx}dx\right]$$

$$= \frac{CAe^{i\omega t}}{-il}\left\{\left[e^{-ilx}\right]_{-a/2}^{a/2} + \left[e^{-ilx}\right]_{d-a/2}^{d+a/2}\right\} = \frac{CAe^{i\omega t}}{-il}\left[e^{-\frac{ila}{2}} - e^{\frac{ila}{2}} + e^{-il\left(d+\frac{a}{2}\right)} - e^{il\left(d-\frac{a}{2}\right)}\right]$$

or, $$y = \frac{CAe^{i\omega t}}{il}\left[e^{\frac{ila}{2}} - e^{-\frac{ila}{2}}\right]\left(1+e^{-ild}\right) = \frac{2CA}{l}e^{i\omega t}\sin\frac{la}{2}\left(1+e^{-ild}\right) \quad \left[\because e^{\frac{ila}{2}} = \cos\frac{la}{2} + i\sin\frac{la}{2}\right]$$

or, $$y = CAae^{i\omega t}\frac{\sin\frac{la}{2}}{\frac{la}{2}}\left(1+e^{-ild}\right)$$

Therefore, $$y = CAae^{i\omega t}\frac{\sin\alpha}{\alpha}\left(1+e^{-ild}\right) \text{ where, } \alpha = \frac{la}{2} = \frac{\pi a}{\lambda}\sin\theta$$

Thus the resultant intensity at point Q is,

$$I = yy^* = (CAa)^2\frac{\sin^2\alpha}{\alpha^2}\left(1+e^{-ild}\right)\left(1+e^{ild}\right)$$

$$= (CAa)^2\frac{\sin^2\alpha}{\alpha^2}\left(2+e^{-ild}+e^{ild}\right) = I_0\frac{\sin^2\alpha}{\alpha^2}\left(2+2\cos ld\right) = 4I_0\frac{\sin^2\alpha}{\alpha^2}\cos^2\frac{ld}{2}$$

or, $$I = 4I_0 \frac{\sin^2\alpha}{\alpha^2}\cos^2\beta \tag{4.50}$$

where, $$\beta = \frac{ld}{2} = \frac{\pi(a+b)}{\lambda}\sin\theta$$

From Eq. (4.50) we find that the resultant intensity at Q depends on two factors –

(i) $I_0 \frac{\sin^2\alpha}{\alpha^2}$ which gives the diffraction pattern of a single slit and

(ii) $\cos^2\beta$ which gives the interference pattern of the diffracted light beams from the double slits.

4.17.1 Conditions for Maxima and Minima

Condition for minima: The resultant intensity would be zero when either factor of Eq. (4.50) will become zero. The factor $\frac{\sin^2\alpha}{\alpha^2} = 0$ when $\sin\alpha = 0$ but $\alpha \neq 0$.

i.e., $$\alpha = n\pi \text{ with } n = \pm 1, \pm 2, ...$$

or, $$\frac{\pi a \sin\theta}{\lambda} = n\pi$$

or, $$a\sin\theta = n\lambda \tag{4.51}$$

These minima are known as the diffraction minima.

The positions of minima due to the factor $\cos^2\beta = 0$ are given by, $\cos\beta = 0$.

or, $$\beta = \pm(2p+1)\frac{\pi}{2}$$

or, $$\frac{\pi(a+b)}{\lambda}\sin\theta = \pm(2p+1)\frac{\pi}{2}$$

or, $$(a+b)\sin\theta = \pm(2p+1)\frac{\lambda}{2} \tag{4.52}$$

These minima are known as the interference minima.

Condition for maxima: The positions of maxima due to the factor $\frac{\sin^2\alpha}{\alpha^2}$ are at $\alpha = 0$ and at values of α less than but gradually approaching towards $\pm\frac{3\pi}{2}, \pm\frac{5\pi}{2}, ...$ These are maxima due to diffraction.

The position of maxima due to the factor $\cos^2\beta$ are given by, $\cos^2\beta = 1$.

or, $$\cos\beta = \pm 1 = \cos p\pi$$

or, $$\beta = p\pi$$

or, $$\frac{\pi(a+b)}{\lambda}\sin\theta = p\pi$$

or, $$(a+b)\sin\theta = p\lambda \tag{4.53}$$

This gives the conditions for maxima due to interference.

4.17.2 Missing Order

If the conditions for maxima due to interference pattern and minima due to diffraction pattern are simultaneously satisfied for a given value of θ, then the corresponding interference maxima will be missing due to the presence of diffraction minima. In this case for a diffracting angle $\theta = \theta_1$ we have,

$$\frac{(a+b)\sin\theta_1}{a\sin\theta_1} = \frac{p\lambda}{n\lambda}$$

or, $$\frac{a+b}{a} = \frac{p}{n} \tag{4.54}$$

If $a = b$ then $p = 2n$.

Thus the 2, 4, 6 *etc.* orders of interference maxima are absent due to the presence of 1, 2, 3 *etc.* order of diffraction dark bands and there will be three interference maxima (corresponding to $p = 0, \pm 1, \pm 2$) in the central diffraction maxima.

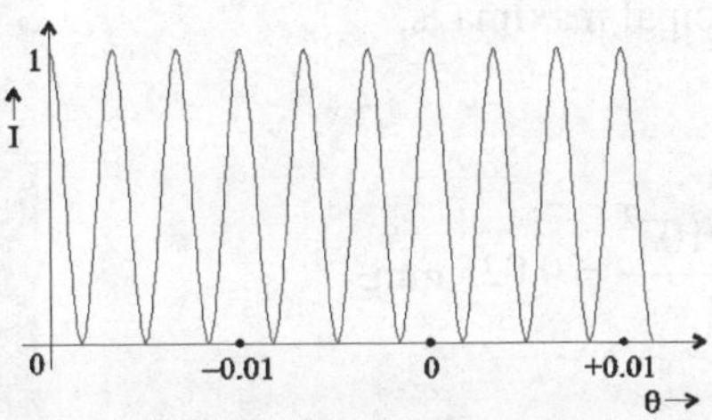

Fig. 4.25: Double slit interference pattern

If $2a = b$ then we have $p = 3n$.

In this case 3, 6, 9 *etc.* orders of interference maxima are absent due to the presence of 1, 2, 3 *etc.* order of diffraction dark bands and there will be five interference maxima (corresponding to $p = 0, \pm 1, \pm 2$) in the central diffraction maxima. The double slit interference pattern is shown in Fig. 4.25 and the intensity distribution (for $2a = b$) of the resultant diffraction pattern is shown in Fig. 4.26.

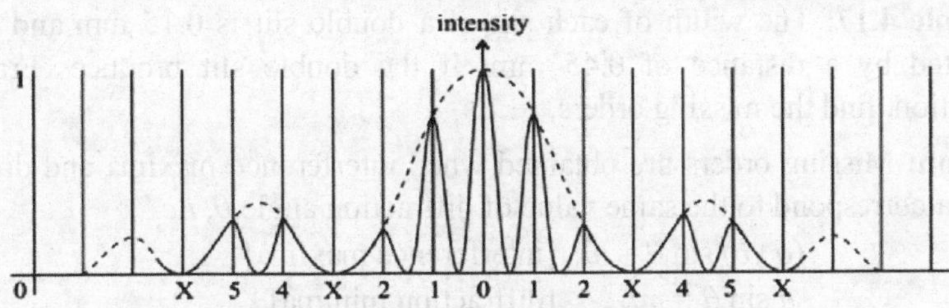

Fig. 4.26: Resultant diffraction pattern

EXAMPLE 4.15

Example 4.15: For Fraunhofer diffraction in a double slit illuminated with light of wavelength 6000 Å, calculate the fringe spacing on the screen 0.8 m away from the slit, if the slit separation is 0.2 mm and the slit width is 0.01 mm.

Solution: The fringe spacing is given by

$$d = \frac{\lambda D}{(a+b)}$$

EXAMPLE 4.15

Here $a = 0.01$ mm $= 0.001$ cm, $b = 0.2$ mm $= 0.02$ cm, $\lambda = 6000$ Å $= 6 \times 10^{-5}$ cm and $D = 0.8$ m $= 80$ cm.

$$\therefore \quad d = \frac{6\times10^{-5}\times80}{0.021} \text{ cm}$$

$$= 0.23 \text{ cm.}$$

So the fringe spacing is 0.23 cm.

EXAMPLE 4.16

Example 4.16: Fraunhofer double slit diffraction pattern is found in the focal plane of a lens of focal length 50 cm. The separation between two maxima adjacent to principal maxima is 5 mm and the fourth order is missing. Find the width of each slits and the distance between their centers. Given $\lambda = 5000$ Å.

Solution: Here fourth order is missing. So,

$$\frac{a+b}{a} = 4 \Rightarrow b = 3a.$$

Condition for first order maxima is $\theta = \dfrac{\lambda}{a+b} = \dfrac{\lambda}{4a}$.

Now, distance between two maxima adjacent to principal maxima is,

$$x = 2f\theta = \frac{2f\lambda}{4a} = \frac{f\lambda}{2a} = 5 \text{ mm.}$$

$$\therefore \quad a = \frac{\lambda f}{2\times5\times10^{-3}} = \frac{5000\times10^{-8}\times50\times10^{-2}}{2\times5\times10^{-3}} = 0.025 \text{ mm.}$$

So, $b = 3a = 0.075$ mm.

EXAMPLE 4.17

Example 4.17: The width of each slit of a double slit is 0.15 mm and they are separated by a distance of 0.45 mm. If the double slit produces Fraunhofer diffraction, find the missing orders.

Solution: Missing orders are obtained when interference maxima and diffraction minima correspond to the same value of diffraction angle θ, *i.e.*

$$(a+b)\sin\theta = n\lambda \quad \text{[interference maxima]}$$

and

$$a\sin\theta = m\lambda \quad \text{[diffraction minima]}$$

So,

$$\frac{a+b}{a} = \frac{n}{m}$$

Here, $a = 0.15$ mm and $b = 0.45$ mm.

$$\therefore \quad \frac{n}{m} = \frac{0.15+0.45}{0.15} = \frac{0.60}{0.15} = 4$$

For values of $m = 1, 2, 3$ etc. $n = 4, 8, 12$ etc.

So, the 4th, 8th, 12th, etc orders of interference maxima will be missing.

4.18 FRAUNHOFER DIFFRACTION AT A PLANE DIFFRACTION GRATING

Diffraction grating is an optical element with a periodic structure that can diffract light into several beams travelling in different diffraction angles. Ideally, a diffraction grating is made up of a set of slits of equal spacings. Fig. 4.27 depicts the Fraunhofer diffraction due to a grating. Let a plane wavefront AB of monochromatic light of wavelength λ propagating normal to a grating S_1S_2 be incident on it. According to Huygens' principle each point of the wavefront in the plane of the slit may be regarded as the source of the secondary spherical wavelets. The secondary wavelets moving normal to the slit are brought to focus by a convex lens on the screen at C. The wavelets travelling at an angle θ with the normal are brought to focus at Q. Let us calculate the intensity of light at Q. Let the complex light disturbance at any instant due to the spherical waves from the origin O be represented by $Ae^{i\omega t}$, where A is the amplitude and ω is the angular frequency.

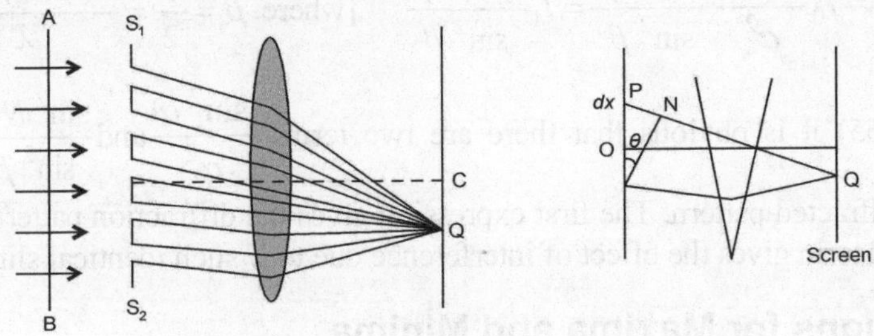

Fig. 4.27: Fraunhofer diffraction due to a grating

Now, the phase different between the waves at Q from O and P which is at a distance x from O is given by, $\frac{2\pi}{\lambda}\text{PN} = \frac{2\pi}{\lambda} x \sin\theta = lx$, where $l = \frac{2\pi}{\lambda}\sin\theta$

The disturbance at Q due to secondary waves from P will be proportional to $e^{i(\omega t - lx)}$. Thus the disturbance at Q due to the diffracting element dx can be written as $dy = CAe^{i(\omega t-lx)}dx$.

If a be the width of the slit then the resultant disturbance at Q due to the waves coming from all the diffracting elements will be given be,

$$y = \int_{-a/2}^{a/2} CAe^{i(\omega t-lx)}dx + \int_{d-a/2}^{d+a/2} CAe^{i(\omega t-lx)}dx + \ldots \int_{(N-1)d-a/2}^{(N-1)d+a/2} CAe^{i(\omega t-lx)}dx$$

$$= CAe^{i\omega t}\left[\int_{-a/2}^{a/2} e^{-ilx}dx + \int_{d-a/2}^{d+a/2} e^{-ilx}dx + \ldots + \int_{(N-1)d-a/2}^{(N-1)d+a/2} e^{-ilx}dx\right]$$

$$= \frac{CAe^{i\omega t}}{-il}\left\{[e^{-ilx}]_{\frac{-a}{2}}^{\frac{a}{2}} + [e^{-ilx}]_{d-\frac{a}{2}}^{d+\frac{a}{2}} + \ldots\ldots + [e^{-ilx}]_{(N-1)d-\frac{a}{2}}^{(N-1)d+\frac{a}{2}}\right\}$$

$$= CAa\frac{\sin\frac{la}{2}}{\frac{la}{2}}\left[1 + e^{-ild} + \ldots + e^{-i(N-1)ld}\right]e^{iwt}$$

$$= CAa\frac{\sin\alpha}{\alpha}\frac{e^{-iNld} - 1}{e^{-ild} - 1}e^{i\omega t} \qquad \left[\text{where } \alpha = \frac{la}{2}\right]$$

Thus the resultant intensity I at Q is,

$$I = yy^* = (CAa)^2 \frac{\sin^2\alpha}{\alpha^2}\frac{(e^{-iNld}-1)(e^{iNld}-1)}{(e^{-ild}-1)(e^{ild}-1)}$$

$$= I_0 \frac{\sin^2\alpha}{\alpha^2}\frac{(2-e^{-iNld}-e^{iNld})}{(2-e^{-ild}-e^{ild})} \quad [\text{where } I_0 = (CAa)^2]$$

$$= I_0 \frac{\sin^2\alpha}{\alpha^2}\frac{(2-2\cos Nld)}{(2-2\cos ld)} = \frac{\sin^2\alpha}{\alpha^2}\frac{(1-\cos Nld)}{(1-\cos ld)} = I_0 \frac{\sin^2\alpha}{\alpha^2}\frac{\sin^2\frac{Nld}{2}}{\sin^2\frac{ld}{2}}$$

$$\therefore \quad I = I_0 \frac{\sin^2\alpha}{\alpha^2}\frac{\sin^2 N\beta}{\sin^2\beta} = I_1 \frac{\sin^2 N\beta}{\sin^2\beta} \quad [\text{where } \beta = \frac{ld}{2} = \frac{\pi(a+b)\sin\theta}{\lambda}] \qquad (4.55)$$

From Eq. (4.55) it is obvious that there are two terms $\frac{\sin^2\alpha}{\alpha^2}$ and $\frac{\sin^2 N\beta}{\sin^2\beta}$ which control the intensity of the diffracted pattern. The first expression gives the diffraction pattern produced by single slit and the second term gives the effect of interference due to N such identical slits.

4.18.1 Conditions for Maxima and Minima

In order to find maxima and minima of the intensity of interference fringes, we must have, $\frac{dI}{d\beta} = 0$.

Now,
$$\frac{dI}{d\beta} = I_1\left[\frac{\sin^2\beta(2N\sin N\beta\cos N\beta) - \sin^2 N\beta(2\sin\beta\cos\beta)}{\sin^4\beta}\right]$$

or,
$$\frac{dI}{d\beta} = 2I_1\left[\frac{N\sin N\beta\cos N\beta}{\sin^2\beta} - \frac{\sin^2 N\beta\cos\beta}{\sin^3\beta}\right]$$

or,
$$\frac{dI}{d\beta} = 2I_1\left(\frac{\sin^2 N\beta}{\sin^2\beta}\right)(N\cot N\beta - \cot\beta) \qquad (4.56)$$

Hence, for maxima or minima, either, $\frac{\sin^2 N\beta}{\sin^2\beta} = 0$ or, $N\cot N\beta - \cot\beta = 0$ *i.e.* $N\cot N\beta = \cot\beta$

Now,
$$\frac{d^2 I}{d\beta^2} = 2I_1\left\{\frac{\sin^2 N\beta}{\sin^2\beta}(-N^2\ \text{cosec}^2 N\beta + \text{cosec}^2\beta) + (N\cot N\beta - \cot\beta) \times \frac{\sin^2\beta(2N\sin N\beta\cos N\beta) - \sin^2 N\beta(2\sin\beta\cos\beta)}{\sin^4\beta}\right\}$$

$$= I_1\left[\frac{2\sin^2 N\beta}{\sin^2\beta}(-N^2 - N^2\cot^2 N\beta + 1 + \cot^2\beta) + \left(\frac{N\cot N\beta - \cot\beta}{\sin^4\beta}\right)(2N\sin^2\beta\sin 2N\beta - 2\sin^2 N\beta\sin 2\beta)\right]$$

(i) When $\sin N\beta = 0$, $\sin \beta \neq 0$ and $N \cot N\beta \neq \cot \beta$ one gets $\dfrac{d^2 I}{d\beta} > 0$ (*i.e.*, positive).

So, $\sin N\beta = 0$ gives the condition of minimum intensity.

(ii) When $N \cot N\beta = \cot \beta$ and $\sin \beta = 0$, one gets $\dfrac{d^2 I}{d\beta} < 0$ (*i.e.*, negative).

So, $N \cot N\beta = \cot \beta$ and $\sin \beta = 0$ gives the condition of maximum intensity.

Principle maxima: When N is very large one obtains intensity maxima at $\sin \beta = 0$ *i.e.*,

$$\beta = \pm m\pi \qquad [m = 0, 1, 2\ldots.etc.]$$

i.e.,
$$\frac{\pi(a+b)\sin\theta}{\lambda} = \pm m\pi \qquad \text{[For normal incidence]}$$

Hence,
$$(a + b) \sin \theta = \pm m\lambda \tag{4.57}$$

These are known as the principal maxima.

For, $\beta = m\pi$, $\dfrac{\sin N\beta}{\sin \beta} = \dfrac{0}{0}$ = indeterminate. But in limit $\beta \to m\pi$ we get

$$\lim_{\beta \to m\pi} \frac{\sin N\beta}{\sin\beta} = \lim_{\beta \to m\pi} \frac{N\cos N\beta}{\cos\beta} = 1 \qquad \text{[Using L. Hospital's rule]}$$

Hence the corresponding resultant intensity is given by

$$I_p = N^2 I_1 \tag{4.58}$$

Thus the intensity of principal maxima increases as the number of slits (N) increases, but due to the presence of the factor I_1, *i.e.*, $\dfrac{\sin^2\alpha}{\alpha^2}$, whose value decreases with the increase of the angle of diffraction (θ), the intensity of principal maxima decreases with the increase of the order of bands.

Secondary minima: From the second condition we get zero intensity, *i.e.* $I = 0$ when $\sin N\beta = 0$ provided $\sin \beta \neq 0$. This gives, $N\beta = \pm p\pi$ [where p is an integer except 0, N, $2N$, $3N$.....*etc*, as in those cases $\sin \beta = 0$ and we get principal maxima]. So,

$$\sin N\beta = 0$$

gives
$$N\beta = \pm p\pi.$$

or,
$$\frac{N\pi(a+b)\sin\theta}{\lambda} = \pm p\pi$$

or,
$$(a + b)\sin\theta = \pm \frac{p\lambda}{N} \tag{4.59}$$

Therefore, $(a + b) \sin \theta = \pm \dfrac{\lambda}{N}, \pm \dfrac{2\lambda}{N} \ldots. \pm \dfrac{(N+1)\lambda}{N}$

Omitting the values of $(a + b) \sin \theta = 0, \dfrac{N\lambda}{N}, \dfrac{2N\lambda}{N} \ldots.$, as these will be giving the principal maxima.

It is obvious that between two principal maxima there are $(N-1)$ equally spaced minima. Hence, there should be $(N-2)$ secondary maxima in between these minima.

Secondary maxima: These maxima are weak in intensity and are given by

$$I_s = I_1 \frac{\sin^2 N\beta}{\sin^2 \beta} = \frac{I_1}{\sin^2 \beta \operatorname{cosec}^2 N\beta} = \frac{I_1}{\sin^2 \beta[1+\cot^2 N\beta]} = \frac{1}{\sin^2 \beta\left[1+\dfrac{\cot^2 \beta}{N^2}\right]}$$

[From the condition $N \cot N\beta = \cot \beta$]

or,
$$I_s = \frac{N^2 I_1}{\sin^2 \beta\left[N^2 + \dfrac{\cos^2 \beta}{\sin^2 \beta}\right]} = \frac{N^2 I_1}{N^2 \sin^2 \beta + \cos^2 \beta} = \frac{N^2 I_1}{(N^2-1)\sin^2 \beta + 1}$$

$$= \frac{I_p}{(N^2-1)\sin^2 \beta + 1}$$

Therefore,
$$\frac{I_s}{I_p} = \frac{1}{1+(N^2-1)\sin^2 \beta} \tag{4.60}$$

Thus greater is the value of N, weaker is the secondary maxima. In practice, N is very large in a grating and so secondary maxima are too weak to be visible in diffraction grating intensity pattern.

In Fig. 4.28 curves are drawn by plotting $\left(\dfrac{\sin \alpha}{\alpha}\right)^2$ against α [Fig. 4.28 (a)], $\dfrac{\sin^2 N\beta}{\sin^2 \beta}$ against β [Fig. 4.28 (b)] and product of them against θ [Fig. 4.28 (c)]. The intensities of secondary maxima fall of between two adjacent principal maxima and are unequally spaced. They also lack symmetry. The lack being greatest adjacent to principal maxima towards which their peaks get shifted.

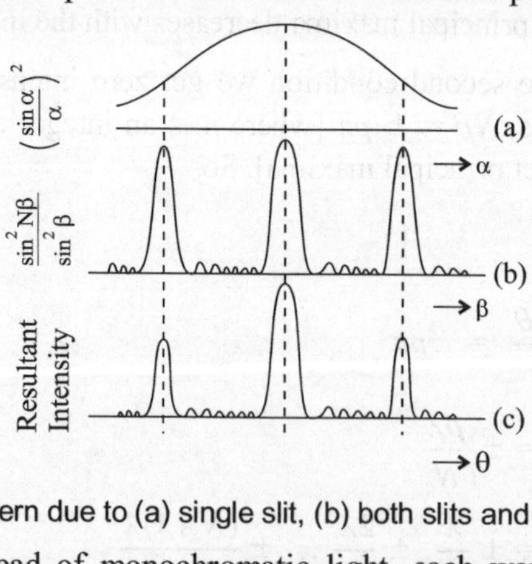

Fig. 4.28: Diffraction pattern due to (a) single slit, (b) both slits and (c) combination of them

If white light is used, instead of monochromatic light, each wavelength will produce its own maxima. So, each maximum will consist of a band of seven colors. Now, the angle of diffraction for red will be greater than that for violet. It means that inner edge of each maximum (*i.e.* edge facing the central maximum) will be violet and the outer edge will be red.

4.18.2 Absent Spectra and Ghost Lines

Condition for absent spectra: For the mth order of principal maximum in the direction θ, we have the condition

$$(a + b) \sin \theta = m \lambda \tag{4.61}$$

If a be such that the pth order diffraction minimum occurs in the same direction θ, then

$$a \sin \theta = p \lambda \tag{4.62}$$

If these two conditions are satisfied simultaneously then m^{th} order principal maximum will be absent from the resultant spectra.

From Eqs. (4.61) and (4.62), we get

$$\frac{a+b}{a} = \frac{m}{p} \tag{4.63}$$

If $b = a$ then from Eq. (4.63), $m = 2p = 2, 4, 6, \ldots$ ($\because p = 1, 2, 3\ldots.$), which implies that the second, fourth, sixth, …, *etc.* order of principal maxima will be absent corresponding to the diffraction minima $p = 1, 2, 3\ldots.$.

Ghost lines: In case of perfect (or ideal) grating the rulings must be equi-spaced, but practically no grating is perfect. The imperfection in rulings of a diffraction grating causes some errors. When these errors are random, the grating gives continuous background illumination. On the other hand, if they are progressive in nature, then one may get sharper spectral lines in planes other than the local plane of the grating. But most frequently the errors repeat themselves periodically because of some defect in the driving mechanism of the machines which make rulings in the diffraction gratings. This results in the generation of false lines along with the principal maxima of a perfect (ideal) grating. These false lines are called *ghost lines.*

4.18.3 Overlapping of Spectral Lines

If the incident light has a large range of wavelengths, then the lines of shorter wavelength and higher order may overlap with the lines of longer wavelength and lower order.

For a grating, having grating element $(a + b)$, the angle of diffraction θ in the mth order spectrum for wavelength λ is given by

$$(a + b) \sin \theta = m \lambda$$

Now, for a given grating $a + b$ = constant. So θ will be the same, *i.e.* overlapping will occur when $m \lambda$ becomes constant.

For instance, red line of 700 nm in the 3rd order will overlap with green line of 525 nm in the 4th order, violet line 420 nm in the 5th order and so on [as $3 \times 700 = 4 \times 525 = 5 \times 420$ = constant].

4.19 ANGULAR DISPERSIVE POWER OF A GRATING

The angular dispersive power of a grating is the rate of change of the angle of diffraction θ with the change in wavelength λ of light.

It is thus denoted by $\frac{d\theta}{d\lambda}$.

The diffraction of the mth order principal maximum for wavelength λ and angle of diffraction θ is given by

$$(a + b) \sin \theta = m \lambda$$

Differentiating we get $(a + b) \cos \theta \, d\theta = m \, d\lambda$

or, $$\frac{d\theta}{d\lambda} = \frac{m}{(a+b)\cos\theta} = \frac{ms}{\cos\theta} \quad (4.64)$$

where $s = \dfrac{1}{(a+b)}$ = number of lines per unit length of the grating. Eq. (4.64) gives the expression for the angular dispersive power of a grating. Here $d\theta$ is the angular separation between two lines with difference in wavelengths as $d\lambda$.

From Eq. (4.64) the following conclusions can be drawn:

(i) The dispersive power is directly proportional to the order of spectrum (m).

(ii) The dispersive power is inversely proportional to the grating element $(a + b)$.

(iii) The dispersive power is inversely proportional to cos θ, *i.e.*, larger the value of θ, greater is the dispersive power.

EXAMPLE 4.18

Example 4.18: *A parallel beam of light of wavelength 5890 Å falls normally on a plane transmission grating having 4250 lines/cm. Find the angle of diffraction for maximum intensity in first order.*

Solution: We known that $(a + b) \sin \theta = n\lambda$

or, $$\sin\theta = \frac{n\lambda}{(a+b)}$$

Here, $n = 1$ and $\dfrac{1}{(a+b)} = 4250$ lines/cm and $\lambda = 5890$ Å $= 5890 \times 10^{-8}$ cm.

Therefore, $\sin\theta = 5890 \times 10^{-8} \times 4250 = 0.25$

or, $\theta = 14.5°$.

EXAMPLE 4.19

Example 4.19: Light of wavelength 6000 Å is incident normally on a plane transmission grating having 5000 lines/cm. Find the difference in deviations in the first and second order spectra.

Solution: Given $\lambda = 6000$Å $= 6 \times 10^{-5}$ cm, $\dfrac{1}{(a+b)} = 5000$.

For the first-order spectrum $(a + b) \sin \theta_1 = \lambda$ [Here $n = 1$]

or, $$\sin\theta_1 = \frac{1}{(a+b)} = 6\times 10^{-5} \times 5000 = 0.3$$

or, $\theta_1 = 17.5°$.

For the second-order spectrum $(a + b) \sin \theta_2 = 2\lambda$ [Here $n = 2$]

or, $$\sin \theta_2 = \frac{2\lambda}{a+b} = 2 \times 6 \times 10^{-5} \times 5000$$

$$= 0.6$$

or, $$\theta_2 = 36.9°.$$

So, the difference in deviations is $(\theta_2 - \theta_1) = 21.4°$.

EXAMPLE 4.19

4.20 RAYLEIGH CRITERION FOR RESOLUTION

For the resolution of two nearby objects, Lord Rayleigh proposed a criterion. According to him "two nearby point objects are said to be just resolved by a suitable optical system, if the position of the central maximum of diffraction pattern of one coincides with the first minimum of the diffraction pattern of the other". It is also applicable for the resolution of spectral lines of equal intensity formed by a prism or a grating. In case of spectral lines, the criterion says, the angular separation between the principal maxima of two spectral lines in a given order should be equal to half angular width of either maximum. The smallest separation between two point objects where they appear just separated is known as the limit of resolution of an optical instrument and the reciprocal of the limit of resolution is called resolving power.

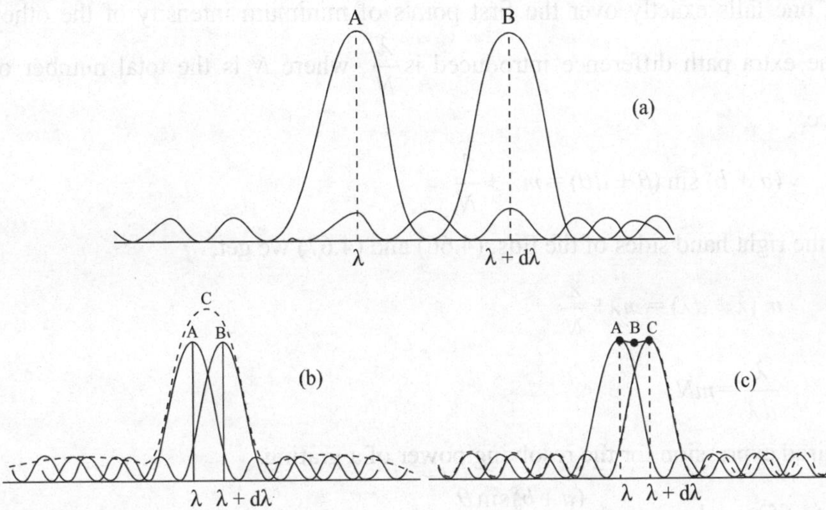

Fig 4.29: Rayleigh criterion for resolution

To illustrate the above criterion, consider the diffraction pattern of the two spectral lines of wavelengths λ and $(\lambda + d\lambda)$ and let A and B are the central maximum due to them respectively [Fig. 4.29 (a)]. If the difference in the angle of diffraction is high, the two images are seen as separate.

In Fig. 4.29 (b) the central maxima corresponding to wavelengths λ and $(\lambda + d\lambda)$ are very close and therefore the two images overlap and cannot be distinguished as separate images. In Fig. 4.29 (c), the position of central maxima of A corresponding to wavelength λ, coincide with the position of first minima corresponding to wavelength $(\lambda + d\lambda)$. Similarly, the central maximum of B coincides with the first minima of A.

The resultant intensity curve reveals a dip at the point C in the middle of the central maxima of A and B. It is estimated that the intensity at C is about 20% less than that at A or B. The position of central maxima of A and B coincides with the first minima of diffraction pattern of each other and it satisfies the Rayleigh criterion. The spectral lines are, therefore, said to be just resolved.

4.21 RESOLVING POWER OF A GRATING

The resolving power of a plane diffraction grating is its ability to just distinguish (resolve) two nearby spectral lines having close wavelengths. If the wavelength of two nearby spectral lines be λ and $(\lambda + d\lambda)$, the resolving power of the grating is defined mathematically as $\frac{\lambda}{d\lambda}$, being the smallest wavelength difference for which the spectral lines can be just resolved at wavelength λ.

The mth order principal maximum of a spectral line of wavelength λ of a grating for normal incidence is

$$(a + b) \sin \theta = m\lambda \tag{4.65}$$

where $(a + b)$ is the grating element and θ is the diffraction angle is corresponding to m^{th} order.

The condition (4.65) for wavelength $(\lambda + d\lambda)$ will be,

$$(a + b) \sin (\theta + d\theta) = m (\lambda + d\lambda) \tag{4.66}$$

According to Rayleigh criterion, these two spectral lines will be just resolved, when the principal maximum of one falls exactly over the first points of minimum intensity of the other line. This is possible if the extra path difference introduced is $\frac{\lambda}{N}$, where N is the total number of lines of the grating surface.

So, $$(a + b) \sin (\theta + d\theta) = m\lambda + \frac{\lambda}{N} \tag{4.67}$$

Equating the right hand sides of the Eqs. (4.66) and (4.67) we get,

$$m (\lambda + d\lambda) = m\lambda + \frac{\lambda}{N}$$

or, $$\frac{\lambda}{d\lambda} = mN \tag{4.68}$$

which is required expression for the resolving power of a grating.

From Eq. (4.65), we have, $m = \frac{(a+b)\sin\theta}{\lambda}$

So, $$\frac{\lambda}{d\lambda} = mN = \frac{N(a+b)\sin\theta}{\lambda} \tag{4.69}$$

or, $$\frac{\lambda}{d\lambda} = W\frac{\sin\theta}{\lambda} \tag{4.70}$$

where $W = N(a + b)$, the total width of the ruled surface of the grating.

For maximum resolving power $\sin\theta = 1$. Hence maximum resolving power $= \frac{W}{\lambda}$.

The relation $\frac{\lambda}{d\lambda} = mN$ shows that resolving power increases with

(i) the order of the spectrum,

(ii) increases with total number of lines in the effective part of the grating and

(iii) it does not depend on grating element.

Example 4.20: Calculate the observed range of orders by a grating having 4000 lines/cm if the illuminated wavelength ranges from 4000 Å to 7500 Å.

Solution: The maximum order visible with grating is $n_{max} = \frac{a+b}{\lambda}$.

Given, $(a + b) = \frac{1}{4000}$

Hence $n_{max} = \frac{1}{4000\lambda}$

For $\lambda = 4000\text{Å} = 4 \times 10^{-5}$ cm, $n_{max} = \frac{1}{4000 \times 4 \times 10^{-5}} = 6.25$

and for $\lambda = 7500\text{Å} = 7.5 \times 10^{-5}$ cm, $n_{max} = \frac{1}{4000 \times 7.5 \times 10^{-5}} = 3.3$

Thus, the observed orders will range between 3 to 6.

EXAMPLE 4.20

Example 4.21: What is the minimum number of lines of a grating which resolve the third order spectrum of two lines having wavelengths of 5890 Å and 5896 Å.

Solution: Given $\lambda_1 = 5890$ Å, $\lambda_2 = 5896$ Å, $d\lambda = 6$ Å,

The average wavelength $\lambda = \frac{\lambda_1 + \lambda_2}{2} = 5893$ Å and $n = 3$.

So, the resolving power

$$(\text{R.P.}) = \frac{\lambda}{d\lambda} = \frac{5893 \times 10^{-8}}{6 \times 10^{-8}} = 982.$$

Further, R.P. $= nN$ [N is the number lines on the grating]

or, $nN = 982$

or, $3N = 982$

or, $N = \frac{982}{3}$

$= 327.33 \approx 327$

So, the minimum number of lines in the grating will be 327.

EXAMPLE 4.21

Example 4.22: Sodium light of wavelength 589 nm and 589.6 nm are made incident normally on a grating having 500 lines/cm. Determine the angular dispersion of the lines in the spectrum of first order. By a simple calculation can you justify these lines are resolved or not?

Solution: Here wavelength of the given rays λ_1 = 589 nm and λ_2 = 589.6 nm, number of rulings/cm in the given grating is N = 500 lines/cm.

We have $\lambda = \dfrac{\sin\theta_n}{nN}$, where n is the order.

For $\lambda = \lambda_1$ = 589 nm, let, $\theta_n = \theta_1'$ and $n = 1$.

so we get $\lambda_1 = \dfrac{\sin\theta_1'}{nN}$

or, $\sin\theta_1' = 0.02945 = \sin 1.688°$

or, $\theta_1' = 1.688°$

For $\lambda = \lambda_2$ = 589.6 nm, let $\theta_n = \theta_1''$ and $n = 1$.

so we get $\lambda_2 = \dfrac{\sin\theta_1''}{nN}$

or, $\sin\theta_1'' = 0.02948 = \sin 1.689°$

or, $\theta_1'' = 1.689°$

$\therefore$ angular dispersion is given by $\theta_d = \theta_1'' - \theta_1'$

$= 1.689° - 1.688° = 0.001°$.

Let N_n be the number of rulings per cm just required to resolve n^{th} order spectrum.

$\therefore$ Resolving power $RP = nN_n$

To resolve 1st order the expression reduces to $RP = N_1$

Again, resolving power is given by

$$RP = \frac{\lambda}{d\lambda}$$

$$\therefore \quad N_1 = RP = \frac{\lambda}{d\lambda}; \text{ where } \lambda = \frac{\lambda_1 + \lambda_2}{2} = 589.3\,\text{nm}.$$

$$\text{or,} \quad N_1 = \frac{589.3}{0.6} = 982.1 \text{ rulings/cm}.$$

The given grating has N = 500 rulings/cm

$\because \quad N < N_1$

So, the lines of 589 nm and 589.6 nm cannot be resolved by the given grating.

EXAMPLE 4.22

Example 4.23: Examine if the two spectral lines of $\lambda = 589$ nm and $\lambda = 589.6$ nm can be clearly resolved in (i) first order, (ii) second order by diffraction grating of 2 cm width and 425 lines/cm.

Solution: No. of slits on the grating = width of the total no. of lines × no. of lines/cm.

or, $$N = 2 \times 425 = 850$$

Again, $$\text{RP} = \frac{\lambda}{d\lambda} = \frac{(589.6 + 589)/2}{589.6 - 589} = 982 = nN.$$

For, $n = 1$; $N = 982 > 850$ (no. of slits on the grating).

So, the lines will not be resolved.

For, $n = 2$; $N = 982/2 = 491 < 850$ (no. of slits on the grating)

So, the lines will be resolved in the 2nd order.

EXAMPLE 4.23

Example 4.24: A 2 cm wide diffraction grating is just able to resolve sodium *D*-lines (having wavelengths 589 nm and 589·6 nm) in 2nd order. Find the number of rulings per mm.

Solution: No. of slits on the grating

= width of the total no. of lines × no. of lines/cm = 200 *s*

where *s* is no. of lines/mm

Again, $$\text{RP} = \frac{\lambda}{d\lambda} = \frac{(589.6 + 589)/2}{589.6 - 589} = 982 = nN = 2N$$

$\therefore$ $$N = 982.16/2 = 491 = 200s$$

or, $$s = \frac{491}{200} = 2.455 \text{ mm.}$$

EXAMPLE 4.24

Example 4.25: Calculate the least width that a grating must have to resolve two components of the sodium D-line on the second order. The grating is having 800 lines/cm. The wavelength for D_1 and D_2 lines of sodium are 5893 Å and 5896 Å respectively.

Solution: No. of slits on the grating = width of the total no. of lines × no. of lines/cm = 800 *x*

where *x* is the width of the total no. of lines.

Again, $$\text{RP} = \frac{\lambda}{d\lambda} = \frac{(5896 + 5893)/2}{5896 - 5893} = 1965 = nN = 2N.$$

or, $$N = 982.5 = 800\,x$$

or, $$x = \frac{982.5}{800} = 1.228 \text{ cm.}$$

EXAMPLE 4.25

4.22 DIFFERENCES BETWEEN GRATING AND PRISM SPECTRA

Grating Spectrum	Prism Spectrum
(i) It is formed by diffraction.	(i) It is formed by dispersion.
(ii) The grating spectra are fainter as the order of the spectrum increases.	(ii) The prism spectrum is bright.
(iii) The grating forms spectra of many orders.	(iii) The prism forms one spectrum.
(iv) The grating spectra do not depend on the material of the grating.	(iv) The prism spectrum depends on the material of the prism.
(v) The grating spectra are pure.	(v) In most cases prism spectrum is not pure.
(vi) The resolving power of a grating is large.	(vi) The resolving power of a prism is small.

4.23 FRAUNHOFER DIFFRACTION IN CIRCULAR APERTURE

Let a parallel light beam of wavelength λ is incident normally on a circular aperture of radius a. Each point of the incident wavefront in the plane of the aperture may be considered as the source of a secondary disturbance as according to the Huygens' theory. Let us now consider the rays diffracted at an angle θ with the normal to the aperture (Fig. 4.30) and calculate the resultant intensity at point Q.

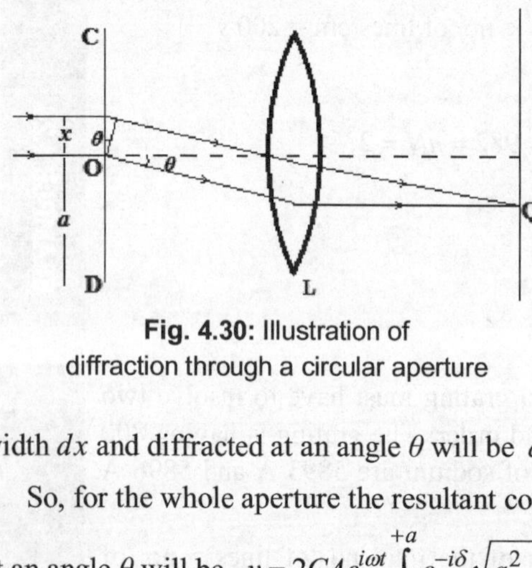

Fig. 4.30: Illustration of diffraction through a circular aperture

Here, the phase difference between the rays diffracted at an angle θ from O and from the point P on the horizontal diameter at a distance x from O is $\delta = \frac{2\pi}{\lambda} x \sin\theta$.

Now, the area of the strip of width dx at a distance x from O is $= 2\sqrt{a^2 - x^2}\,dx$

Hence, the complex disturbance at the given point due to the rays coming from the strip AB of width dx and diffracted at an angle θ will be $dy = 2CAe^{i(\omega t - \delta)}\sqrt{a^2 - x^2}\,dx.$

So, for the whole aperture the resultant complex disturbance at a given point for the rays diffracted at an angle θ will be, $y = 2CAe^{i\omega t}\int_{-a}^{+a} e^{-i\delta}\sqrt{a^2 - x^2}\,dx$

$$= 2CAe^{i\omega t}\int_{-a}^{+a} e^{-i\frac{2\pi}{\lambda}x\sin\theta}\sqrt{a^2 - x^2}\,dx$$

The real amplitude of the disturbance is thus

$$R = 2CA\int_{-a}^{+a}\sqrt{a^2 - x^2}\cos\frac{2\pi x\sin\theta}{\lambda}\,dx$$

Let $x = au$ and $\frac{2\pi a \sin\theta}{\lambda} = m$

Hence $dx = adu$

$$\therefore \quad R = 2CAa^2 \int_{-1}^{+1} \sqrt{1-u^2}\cos mu\,du$$

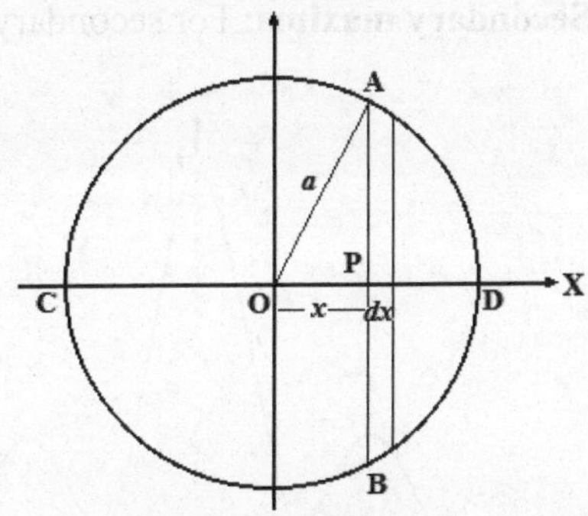

Fig 4.31: Projected view of the circular aperture

Now, $\int_{-1}^{+1} \sqrt{1-u^2}\cos mu\,du$ is the Bessel's function of the first order and the value of the function is $J_1(m)\frac{\pi}{m}$.

$$\therefore \quad R = 2CAa^2 J_1(m)\frac{\pi}{m} \tag{4.71}$$

So, for the whole aperture the resultant intensity at a given point for the rays diffracted at an angle θ will be,

$$I = R^2 = \left(CA\pi a^2\right)^2 \left[\frac{2J_1(m)}{m}\right]^2 = I_0 \left[\frac{2J_1(m)}{m}\right]^2 \tag{4.72}$$

4.23.1 Conditions for maxima and minima

As θ increases, m also increases but $\frac{J_1(m)}{m}$ alternately increases and decreases.

Principle maxima: When $m = 0$, $\frac{J_1(m)}{m}$ becomes indeterminate but in the limit $m \to 0$ we have,

$$\lim_{m\to 0} \frac{J_1(m)}{m} = \frac{1}{2}$$

Hence $I = I_0$ (4.73)

This gives the condition for the principal maxima. As $m \to 0$, $\theta \to 0$. So, the maximum intensity occurs at the geometrical image of the source.

Secondary minima: Secondary minima will occur at the points where $J_1(m) = 0$ and $m \neq 0$.

For the first zero of the function $J_1(m)$ we have $m = 1.22\pi$.

$$\therefore \quad \frac{2\pi a \sin\theta}{\lambda} = 1.22\pi$$

$$\text{or,} \quad \sin\theta = \frac{1.22\lambda}{2a} = \frac{1.22\lambda}{D} \tag{4.74}$$

For the second and third zero of the function $J_1(m)$ we have $m = 2.233\pi$ and $m = 3.238\pi$.

Corresponding angle of diffraction will be, $\sin\theta' = \frac{2.233\lambda}{D}$ and $\sin\theta'' = \frac{3.328\lambda}{D}$ (4.75)

Secondary maxima: For secondary maxima $\frac{d}{dm}\left[\frac{J_1(m)}{m}\right]=0.$

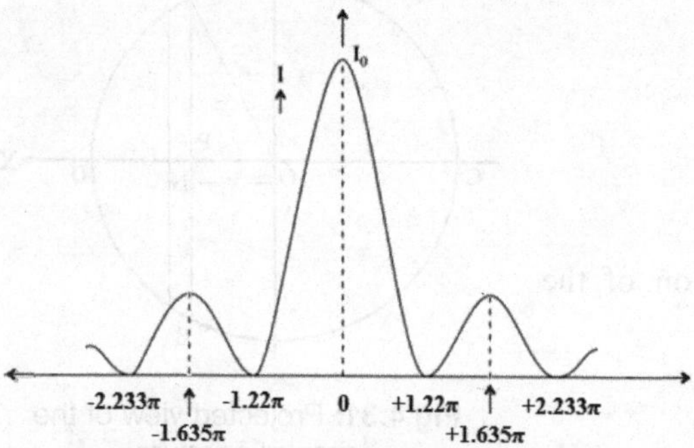

Fig. 4.32: Intensity distribution due to a circular aperture

This will give the first and the second secondary maxima for $m=1.635\pi$ and $m=2.679\pi$ respectively. These values of m will give the intensity of the secondary maxima as

$$I_1=\frac{I_0}{56} \tag{4.76}$$

and

$$I_2=\frac{I_0}{119} \tag{4.77}$$

The intensity curve of the diffraction pattern produced for a circular aperture is shown in Fig. 4.32.

4.24 RESOLVING POWER OF MICROSCOPE

The primary function of microscope is not to magnify an object but to reveal those finer details in the object which are invisible to unaided eye. The limit of resolution of a microscope is measured by the least permissible linear distance between two point objects so that the two images just appear to be resolved. The reciprocal of the resolution limit is called the resolving power of the microscope.

Let us consider two point objects separated by a distance x and they are very close to the objective and normal to the axis of the microscope. If θ is the semi-vertical angle of the cone of the rays by the objective of the microscope then it can be shown that the two images just appear to be resolved if,

$$2x\sin\theta=1.22\lambda \tag{4.78}$$

or,

$$x=\frac{1.22\lambda}{2\sin\theta} \tag{4.79}$$

which is the required limit of resolution.

The high resolving power microscopes are generally oil immersion type in which the space between the objective of the microscope and the object is filled with an oil of refractive index μ. Then,

$$x=\frac{1.22\lambda}{2\mu\sin\theta} \tag{4.80}$$

where $\mu\sin\theta$ is known as numerical aperture (NA) of the objective.

Resolving power is defined as,

$$\text{RP}=\frac{1}{x}=\frac{2\mu\sin\theta}{1.22\lambda} \tag{4.81}$$

By using ultraviolet light and quartz lenses the resolving power of the microscope can also be increased. Such a microscope is called *ultra microscope.*

At present, highest resolving power is obtained in electron microscope where electrons are used instead of white light.

Example 4.26: Estimate the resolving limit of a microscope, the objective of which gathers light over a cone of semi-angle 30° for visible light (5800 Å).

Solution: The resolving limit of a microscope is $x = \dfrac{1.22\lambda}{2\sin\theta}$.

Here, $\lambda = 5800$ Å $= 5.8 \times 10^{-5}$ cm, $\theta = 30°$.

So, $$x = \frac{1.22 \times 5.8 \times 10^{-5}}{2\sin 30°} = \frac{1.22 \times 5.8 \times 10^{-5}}{2 \times 0.5} = 7.08 \times 10^{-5} \text{ cm}.$$

EXAMPLE 4.26

Example 4.27: A microscope is used to resolve two point sources separated by a 4×10^{-5} cm distance. If the wavelength of light be 5461 Å, calculate the numerical aperture of the objective.

Solution: Numerical aperture of the objective will be

$$\text{NA} = \frac{1.22\lambda}{2d} = \frac{1.22 \times 5461 \times 10^{-8}}{2 \times 4 \times 10^{-5}} = 0.832.$$

EXAMPLE 4.27

Example 4.28: An oil immersion microscope resolves the rulings of a grating having 3900 lines/ mm when light of wavelength 400 nm is employed. Determine the numerical aperture of the lens.

Solution: Resolving power RP $= \dfrac{1.22\lambda}{\text{NA}}$, where NA is numerical aperture of the lens. So, $$\text{NA} = \frac{1.22\lambda}{\text{RP}}.$$

Now, resolving power (RP) = grating element $= \dfrac{1}{3900}$ mm $= \dfrac{10^{-3}}{3900}$ m and $\lambda = 400$ nm $= 400 \times 10^{-9}$ m.

$$\therefore \quad \text{NA} = \frac{1.22\lambda}{\text{RP}} = 0.95.$$

EXAMPLE 4.28

Example 4.29: Two stars situated at a distance of 9.5×10^{12} km are sending light of wavelength 600 nm, to a telescope of diameter 20 cm. Find the distance of a separation of the stars for which they are just resolved.

Solution: The diameter of the telescope $D = 1.22\dfrac{\lambda L}{y}$.

Here, L = distance of telescope from stars = 9.5×10^{12} km = 9.5×10^{15} m, y = distance of separation of the stars, λ = wavelength of light = 600 mm = 600×10^{-9} m and D = telescope diameter = 20 cm = 20×10^{-2} m.

$$\therefore \quad y = 1.22\frac{\lambda L}{D} = 3.477 \times 10^{7} \text{ km}.$$

EXAMPLE 4.29

UNIT SUMMARY

- ***Coherent sources***

  Two sources of light having a constant phase difference between them.

  Temporal Coherence

  Phase difference of any two points does not change in a fixed time interval.

  Spatial Coherence

  Correlations between the two light fields at two different space points.

- ***Huygens' theory of wave propagation***

  Each point on a primary wavefront acts as a secondary disturbance.

  Amplitude of the secondary wavelet is highest in the forward direction.

  It is zero in the backward direction.

- ***Interference of light***

  Formation of maxima and minima due to superposition of monochromatic waves of light from two closely spaced coherent sources.

  Condition for sustained interference pattern

  i) Two sources of light must be coherent and very narrow.

  ii) Their wavelengths are equal and amplitudes are equal or nearly equal.

  iii) For polarized light, the state of polarization should be the same.

  iv) The path difference should be even multiple of $\lambda/2$ for constructive interference and odd multiple of $\lambda/2$ for destructive interference.

- ***Young's double-slit experiment***

  $$I = A^2 = 4a^2 \cos^2 \frac{\phi}{2}$$

  $$I_{\max} = 4a^2$$

  $$I_{\min} = 0$$

  $$I_{av} = 2a^2$$

- ***Fringe width***

  $$\beta = \frac{\lambda D}{2d}$$

- ***Shape of the interference fringes***

  $$\frac{x_n^2}{\frac{\Delta^2}{4}} - \frac{D^2}{\frac{4d^2 - \Delta^2}{4}} = 1$$

- ***Determination of thickness of a thin film***

  $$t = \frac{2x_n d}{D(\mu - 1)}$$

- ***Stokes' law***

  $$tt' = 1 - r^2$$

  and $r' = -r$

- ***Interference in a thin film***

  Reflected light

  Constructive interference: $2\mu t \cos r = (2n \mp 1)\dfrac{\lambda}{2}$ = odd multiple of $\dfrac{\lambda}{2}$

  Destructive interference: $2\mu t \cos r = n\lambda = 2n\dfrac{\lambda}{2}$ = even multiple of $\dfrac{\lambda}{2}$

  Transmitted light

  Constructive interference: $2\mu t \cos r = n\lambda$

  Destructive interference: $2\mu t \cos r = (2n \pm 1)\dfrac{\lambda}{2}$

- ***Newton's rings***

  $\lambda = \dfrac{d_{n+p}^2 - d_n^2}{4pR}$ [for an air film]

  Newton's rings in transmitted light

  For n^{th} bright ring: $r_n^2 = \dfrac{Rn\lambda}{\mu}$

  For n^{th} dark ring: $r_n^2 = (2n+1)\dfrac{R\lambda}{\mu}$

  Newton's ring in reflected light

  For n^{th} bright ring: $r_n^2 = \dfrac{(2n+1)\lambda R}{2\mu}$

  For n^{th} dark ring: $r_n^2 = \dfrac{n\lambda R}{\mu}$

  Refractive index of a liquid: $\mu = \dfrac{D_{n+p}^2 - D_n^2}{D_{n+p}'^2 - D_n'^2}$

- ***Interferometers***

  Michelson interferometer

  $2d\cos\theta = m\lambda$

  Mach–Zehnder interferometer

  Phase shift increase is proportional to $(n - 1)$ time the length travelled.

- ***Diffraction***

  Fresnel Diffraction

  The source and the screen are situated at a finite distance from the obstacle.

  Fraunhofer diffraction

  The source and the screen are situated at an infinite distance from the obstacle.

- ***Fresnel diffraction in a circular aperture***

  $\dfrac{1}{a} + \dfrac{1}{b} = \dfrac{n\lambda}{r^2}$

- ***Fraunhofer diffraction in a single slit***

$$I = I_0 \frac{\sin^2 \alpha}{\alpha^2}$$

Condition for maxima and minima

For minimum intensity: $a \sin\theta = n\lambda$ when $\alpha = n\pi$; $n = \pm 1, \pm 2, ...$

Principal maxima: $I = I_0$ when $\alpha \to 0$

First secondary maxima: $I_1 = \frac{4I_0}{9\pi^2}$ when α approaches $\pm\frac{3\pi}{2}$

Second secondary maxima: $I_2 = \frac{4I_0}{25\pi^2}$ when α approaches $\pm\frac{5\pi}{2}$

- ***Fraunhofer diffraction in a double slit***

$$I = 4I_0 \frac{\sin^2 \alpha}{\alpha^2} \cos^2 \beta$$

Condition for maxima and minima

Diffraction minima condition: $a \sin\theta = n\lambda$

Interference minima condition: $(a+b)\sin\theta = \pm(2p+1)\frac{\lambda}{2}$

Interference maxima condition: $(a+b)\sin\theta = p\lambda$

Missing order

$$\frac{a+b}{a} = \frac{p}{n}$$

If $a = b$ then $p = 2n$. If $2a = b$ then $p = 3n$.

- ***Fraunhofer diffraction due to a plane diffraction grating***

$$I = I_0 \frac{\sin^2 \alpha}{\alpha^2} \frac{\sin^2 N\beta}{\sin^2 \beta} = I_1 \frac{\sin^2 N\beta}{\sin^2 \beta}$$

Condition for maxima and minima

Principal maxima: $I_p = N^2 I_1$ when $(a + b) \sin\theta = \pm m\lambda$

Secondary maxima: $I_s = \frac{I_p}{(N^2 - 1)\sin^2 \beta + 1}$ when $N \cot N\beta = \cot\beta$

Secondary minima: $(a + b) \sin\theta = \pm\frac{\lambda}{N}, \pm\frac{2\lambda}{N} .... \pm\frac{(N+1)\lambda}{N}$

when $N\beta = \pm p\pi$; where p is an integer except 0, N, $2N$, $3N$.....*etc*

- ***Angular dispersive power of a grating***

$$\frac{d\theta}{d\lambda} = \frac{m}{(a+b)\cos\theta} = \frac{ms}{\cos\theta}$$

- ***Resolving power of a grating***

$$\frac{\lambda}{d\lambda} = mN = \frac{N(a+b)\sin\theta}{\lambda} = W\frac{\sin\theta}{\lambda}$$

- ***Fraunhofer diffraction in a circular aperture***

$$I = I_0\left[\frac{2J_1(m)}{m}\right]^2$$

Condition for maxima and minima

Principal maxima: $I = I_0$ when $m \to 0$

First secondary maxima: $\sin\theta = \dfrac{1.22\lambda}{D}$ for $m = 1.22\pi$

Second secondary maxima: $\sin\theta' = \dfrac{2.233\lambda}{D}$ for $m = 2.233\pi$

Third secondary maxima: $\sin\theta'' = \dfrac{3.328\lambda}{D}$ for $m = 3.238\pi$

First secondary maxima: $I_1 = \dfrac{I_0}{56}$ for $m = 1.635\pi$

Second secondary maxima: $I_2 = \dfrac{I_0}{119}$ for $m = 2.679\pi$

- ***Resolving power of microscope***

$$\text{RP} = \frac{1}{x} = \frac{2\mu\sin\theta}{1.22\lambda}$$

EXERCISES

Multiple Choice Questions

4.1 The radii of the Newton's rings (R_n) depend on wavelength of the incident light (λ) and the radius of curvature of the convex surface of the plano-convex lens (r) as follows:

(a) increases with λ and decreases with r (b) increases with r and decreases with λ
(c) increases both with λ and r (d) decreases both with λ and r

4.2 The relation between the path difference (Δl) and phase difference ($\Delta\phi$) is

(a) $\Delta\phi = \dfrac{2\pi}{\lambda}\Delta l$ (b) $\Delta\phi = \dfrac{\pi}{\lambda}\Delta l$ (c) $\Delta\lambda = \dfrac{2\pi}{\lambda}\Delta\phi$ (d) $\Delta\lambda = \dfrac{\pi}{\lambda}\Delta\phi$

4.3 Double slit interference pattern is the limiting case of double slit diffraction pattern when the

(a) distance between the slit, tends to zero
(b) distance of the source and the slit tend to infinity
(c) slit widths tend to zero
(d) distance between the slits tends to infinity

4.4 Newton's rings experiment is based on

(a) division of amplitude (b) division of wavefront
(c) division of frequency (d) division of phase angle

4.5 A source of light emits lights of frequencies between ν and $\nu + \Delta\nu$. If coherence time of the emergent light beam is T_c, then

(a) $T_c \propto \Delta\nu$ (b) $T_c \propto \frac{1}{\Delta\nu}$ (c) $T_c \propto \nu$ (d) $T_c \propto \frac{1}{\nu}$

4.6 In an interference pattern produced by the identical coherent sources of monochromatic light, the intensity at the sight of central maximum is I. If the intensity at the same spot when either of the two slits is closed is I_0, we must have,

(a) $I = I_0$ (b) $I = 2I_0$
(c) $I = 4I_0$ (d) I and I_0 are not related

4.7 Two sources are said to be coherent when the waves produced by them have

(a) same wavelength (b) same wavelength and same phase
(c) same wavelength and constant phase difference
(d) same amplitude and constant phase difference

4.8 In an arrangement for viewing Newton's rings, if the lens which rests on a glass plate were moved upwards by one wavelength (of the viewing light), which of the following will be observed

(a) the central spot becomes bright (b) no change of fringe pattern is observed
(c) the rings shift towards the center (d) the rings move out from the center.

4.9 In Newton's rings experiment the diameters of the dark rings are proportional to

(a) odd natural numbers (b) square root of the natural numbers
(c) square root of the odd natural numbers (d) square of the natural numbers

4.10 How does reducing the slit separation, affect the appearance of the fringes in the double slit interference?

(a) the fringe width increases and the fringes are brighter
(b) the fringe width increases but the brightness remains unchanged
(c) the fringe width decreases and the fringes are brighter
(d) the fringe width is unchanged but the fringes are less bright

4.11 Young's double slit experiment is based on

(a) division of wavefront (b) division of amplitude
(c) division of both amplitude and wavefront (d) none of these

4.12 When interference takes place at some region, light energy is

(a) created (b) destroyed (c) redistributed (d) none of these

4.13 The central spot of a Newton's rings experiment with a monochromatic light is

(a) bright (b) dark (c) white (d) none of these

4.14 In Young's double slit experiment using two identical slits, the intensity of the maxima at the centre of the screen is I. What will be the intensity at the centre of the screen if one of the slits is closed?

(a) I (b) 2I (c) $\frac{I}{4}$ (d) $\frac{I}{2}$

4.15 In Young's double slit experiment the two coherent sources are produced by

(a) division of wavefront (b) division of amplitude

(c) all of these (d) none of these

4.16 An interference pattern is observed by Young's double slit experiment. If the separation between the coherent sources is halved and the distance of screen from the coherent sources is doubled, the new fringe width

(a) becomes double (b) becomes one fourth
(c) remains same (d) becomes four times

4.17 If the intensities of two coherent light waves are in the ratio 25 : 1, then the ratio of maximum to minimum intensity in the region of superposition of two waves is

(a) 4 : 9 (b) 9 : 4 (c) 11 : 12 (d) 12 : 11

4.18 Two sinusoidal waves of same frequency and having amplitudes A_1 and A_2 respectively superpose coherently. The ratio of the maximum intensity and the minimum intensity of the resultant wave is

(a) $\frac{A_1^2 + A_2^2}{A_1^2 - A_2^2}$ (b) $\frac{A_1^2 - A_2^2}{A_1^2 + A_2^2}$ (c) $\frac{A_1^2}{A_2^2}$ (d) $\frac{A_1 + A_2}{A_1 - A_2}$

4.19 For constructive interference of two light waves, the path difference should be

(a) odd multiple of $\lambda/2$ (b) even multiple of $\lambda/2$
(c) fraction of $\lambda/2$ (d) none of these

4.20 For a pair of sources to be coherent, they should emit light waves of

(a) constant amplitude (b) constant frequency
(c) constant wavelength (d) constant phase difference

4.21 In Young's double slit experiment, if the distance between the coherent sources is d, the distance between the source and screen is D; then the fringe width will be

(a) $\frac{d}{\lambda D}$ (b) $\frac{dD}{\lambda}$ (c) $\frac{\lambda D}{d}$ (d) none of these

4.22 If Young's double slit experiment is carried out in water, then the fringe width will

(a) increases (b) decreases
(c) remains unchanged (d) none of these

4.23 The ratio of the fringe width in Young's double slit experiment when the separation between the slits is doubled and the distance between the slits and screen is halved is

(a) 1 : 1 (b) 1 : 4 (c) 4 : 1 (d) 2 : 1

4.24 Two waves having the intensities in the ratio 9 : 1 produce interference. The ratio of maximum to minimum intensity is equal to

(a) 1 : 81 (b) 9 : 1 (c) 4 : 1 (d) 2 : 1

4.25 The fact that light is a wave is evidenced by

(a) black body radiation (b) photoelectric effect
(c) interference phenomenon (d) none of these

4.26 In interference phenomena the law of conservation of energy is

(a) not violated (b) always violated
(c) sometimes violated (d) violated depending on frequency

4.27 During interference of light waves energy is
(a) created (b) destroyed (c) redistributed (d) none of these

4.28 In Fraunhofer diffraction, the incident wavefront is
(a) plane (b) spherical (c) cylindrical (d) of arbitrary shape

4.29 The separation (Δ) between diffraction lines for the pattern formed by a transmission grating depends on the number of rulings per inch (n) and the wavelength of the incident light (λ) as
(a) Δ increases with n but decreases with λ (b) Δ decreases with n but increases with λ
(c) Δ increases both with n and λ (d) Δ decreases both with n and λ

4.30 The intensity of principal maximum in a Fraunhofer diffraction spectrum produced by a grating with N number of lines is proportional to
(a) $\frac{1}{N}$ (b) N (c) N^2 (d) $\frac{1}{N^2}$

4.31 Double slit interference pattern is the limiting case of double slit diffraction pattern when the
(a) distance between the slits tends to zero
(b) distance between the source and the slits tends to infinity
(c) slit width tends to zero (d) distance between the slits tends to infinity

4.32 In a plane transmission grating light
(a) diffracts to produce the resultant pattern
(b) diffracts and interferes to produce the resultant pattern
(c) interferes to produce the resultant pattern (d) none of these

4.33 In a single slit Fraunhofer diffraction, when the slit width is increased the width of the central maximum
(a) increases (b) decreases
(c) remains unchanged
(d) may increase or decrease and that depends upon the initial value of the slit width

4.34 The resolving power of a grating in the nth order is
(a) $\frac{n}{N}$ (b) nN (c) $\frac{N}{n}$ (d) $\frac{N}{n^2}$

4.35 Diffraction occurs when the width of the narrow slit is
(a) comparable to the wavelength of the light used
(b) equal to the wavelength of the light used
(c) much greater than the wavelength of the light used
(d) much smaller than the wavelength of the light used

4.36 A diffraction pattern is obtained with a plane transmission grating using a beam of red light. What happens after red light is replaced by blue light?
(a) the lines will shift towards lower angles (b) the times will shift towards higher angles
(c) the lines will disappear (d) no visible changes will be noticed

4.37 Missing orders are found in case of double slit diffraction patterns due to
(a) unequal value of two slit widths
(b) superposition of diffraction minima and interference maxima

(c) superposition of diffraction maxima and interference minima
(d) oblique incidence of light

4.38 Resolving power of microscope objective placed in air with light of wavelength λ_1 is the same when immersed in oil of refractive index μ and wavelength λ_2. If the semi-vertical angle to be the same in these two cases then
(a) $\lambda_1 = m\lambda_2$ (b) $\lambda_1 = m\lambda_1$ (c) $\lambda_1\lambda_2 = m(\lambda_1 + \lambda_2)$ (d) $\lambda_1 + \lambda_2 = m\lambda_1\lambda_2$

4.39 Dispersive power of grating increases with increase of
(a) number of lines per cm (b) order number
(c) number of lines and order number (d) intensity of incident light

4.40 In a single slit Fraunhofer diffraction, if the slit width is increased, the width of central maximum will
(a) increases (b) remains same (c) decreases (d) uncertain

4.41 For a point source of light the nature of wavefront is
(a) cylindrical (b) spherical (c) straight line (d) plane

4.42 If W be the width of a grating, λ be the wavelength of the incident light and θ be the diffraction angle then resolving power $\left(\frac{\lambda}{d\lambda}\right)$ is
(a) $W\lambda\sin\theta$ (b) $\frac{W\sin\theta}{\lambda}$ (c) $\frac{\sin\theta}{W\lambda}$ (d) $\frac{W\lambda}{\sin\theta}$

4.43 In Fraunhofer diffraction the incident wavefront is
(a) plane (b) spherical (c) cylindrical (d) ellipsoidal

4.44 When white light is incident on a plane diffraction grating the central maxima will be
(a) dark (b) white (c) blue (d) red

4.56 Ghost line in grating spectra appear due to
(a) regular arrangement of slits (b) irregular arrangement of slits
(c) interference from many slits (d) polarization from many slits

4.46 When the slit width remains constant and the slit separation is increased, the number of interference maxima around the central maxima in a single slit diffraction pattern will
(a) remain the same (b) increase (c) decrease (d) none of these

4.47 The ratio of intensity of the first secondary maximum to the central maximum in a single-slit diffraction pattern is
(a) 1 : 2 (b) 1 : 12 (c) 1 : 22 (d) none of these

4.48 Resolving power of a grating will increase when the order number is
(a) more (b) less (c) constant (d) none of these

4.49 If opaque space width is equal to the slit width, then the number of interference maxima around the central maximum will be
(a) 1 (b) 2 (c) 3 (d) 4

4.50 The maximum number of orders possible (if $N = 5000$ lines/cm and $\lambda = 5893$ Å) is
(a) 1 (b) 2 (c) 3 (d) 4

4.51 If in a single slit experiment, the slit width is reduced, the fringe width will become
(a) sharper (b) wider (c) brighter (d) darker

4.52 Diffraction pattern in a single slit diffraction minima are
(a) equally spaced (b) unequally spaced (c) randomly spaced (d) none of these

4.53 The number of lines/cm in a diffraction grating is given by
(a) $(a-b)$ (b) $\frac{1}{(a+b)}$ (c) $\frac{1}{ab}$ (d) none of these

4.54 The double slit diffraction pattern will reduces to that of a single slit when the width of the opaque space (b) is related to slit width (a) as
(a) $a=0$ (b) $a=b$ (c) $a=2b$ (d) $a=4b$

4.55 When a soap bubble is illuminated with white light multiple colors are observed due to
(a) reflection (b) diffraction (c) interference (d) polarization

Answers of Multiple Choice Questions

4.1 (c), 4.2 (a), 4.3 (c), 4.4 (a), 4.5 (b), 4.6 (c), 4.7 (c), 4.8 (b), 4.9 (b), 4.10(b), 4.11 (a), 4.12 (c), 4.13(b), 4.14(c), 4.15 (a), 4.16 (d), 4.17 (b), 4.18 (d), 4.19 (b), 4.20 (d), 4.21 (c), 4.22 (b), 4.23 (c), 4.24 (c), 4.25 (c), 4.26 (a), 4.27 (c), 4.28 (a), 4.29 (c), 4.30 (c), 4.31 (c), 4.32 (c), 4.33 (a), 4.34 (b), 4.35 (a), 4.36 (a), 4.37 (b), 4.38 (b), 4.39 (c), 4.40 (c), 4.41 (b), 4.42 (b), 4.43 (a), 4.44 (b), 4.45 (b), 4.46 (b), 4.47 (c), 4.48 (a), 4.49 (c), 4.50 (c), 4.51 (b), 4.52 (a), 4.53 (b), 4.54 (a), 4.55 (c)

Short and Long Answer Type Questions

Category I

4.1 Justify that energy remains conserved in interference phenomenon.

4.2 Find the difference between temporal coherence and spatial coherence.

4.3 If distance between the two coherent sources of light with wavelength λ is d and D be the source-screen distance then show that the fringe width separation is $\beta=\frac{\lambda D}{2d}$.

4.4 Can you obtain interference with wide slits? Why?

4.5 Can light from two different sources produce interference? Why?

4.6 A mica sheet of thickness t and refractive index μ is introduced in the path of one the interfering beams in Young's double slit experiment. Find out the shift of the nth bright fringes in terms μ and t.

4.7 How coherent sources are produced in Young's double-slit experiment? Explain the mechanism.

4.8 Can you observe interference pattern if two coherent sources are separated by a distance less than a wavelength of light?

4.9 Prove that the diameters of the dark Newton's rings are proportional to the square root of the natural numbers and the diameters of the bright Newton's rings are proportional to the square root of the odd natural numbers.

4.10 Prove that the width of Newton's rings decreases with order number of the rings.
4.11 Mention the essential conditions for sustained interference pattern.
4.12 Find the difference between coherent and non-coherent sources of light.
4.13 In how many classes diffraction of light can be classified?
4.14 Explain the difference between diffraction spectra and interference spectra.
4.15 Compare Fresnel and Fraunhofer diffractions.
4.16 Prove that the intensities of secondary maxima formed by Fraunhofer diffraction at a single slit are of decreasing order.
4.17 Find the intensity distribution for the Fraunhofer diffraction pattern through a single slit. Hence find the position of maxima and minima.
4.18 Write down the expression for the intensity of light due to Fraunhofer diffraction in a transmission grating and hence find the condition for secondary minima in the interference pattern.
4.19 Draw the intensity distribution pattern of diffraction grating.
4.20 Show that the intensity of the 1st secondary maxima formed by a single slit Fraunhofer diffraction process is nearly 4·5% of the principal maxima.
4.21 Explain Fraunhofer diffraction by a single slit with necessary theory. Point out also the graphical representation of intensity distribution.
4.22 Explain the meaning of absent spectra in the diffraction pattern of grating.
4.23 Explain how the diffraction patterns due to a single slit depend on the slit width.
4.24 Discuss Fraunhofer diffraction due to a double slit.
4.25 Explain the terms absent spectra and ghost lines.
4.26 Derive an expression for intensity distribution for Fraunhofer diffraction in double slit. Draw the intensity distribution curve and explain it.
4.27 Explain how resolving power of a grating can be increased.
4.28 Find the missing orders in a double slit diffraction pattern.
4.29 Find an expression for the resolving power of a grating having N slits.
4.30 Explain the role of resolving power for an optical instrument.
4.31 Obtain an expression for the resolving power of a microscope. How it can be increased?
4.32 Explain the action of a diffraction grating. How it can be used to measure wavelength of light?
4.33 Show that the resolving power is given by the product of the total number of rulings and the order number of the spectrum.
4.34 State and explain Rayleigh's criterion of resolution.
4.35 What happens to fringe width if the Young's double slit set up is immersed in water? Explain.
4.36 Can two real sources of light act as a coherent source?
4.37 Mention the applications of interference in engineering?
4.38 With a neat diagram explain the working of a Michelson interferometer. Also give some of its applications.
4.39 With a neat diagram explain the working of a Mach-Zehnder interferometer. Also give some of its applications.
4.40 Explain the formation of Fresnel's zone due to Fresnel diffraction in a circular aperture.
4.41 Explain the conditions of maxima and minima due to Fraunhofer diffraction in a circular aperture.

Category II

4.42 (a) Distinguish between interference and diffraction of light. Is energy conserved in interference phenomenon? Explain.

(b) In a Newton's ring arrangement, the diameter of the 5th dark ring is 0·3 cm and the diameter of 25th dark ring is 0·5 cm. If the radius of curvature of the plano convex lens is 100 cm, find the wavelength of the light used.

(c) What are the conditions to get sustained interference pattern?

4.43 (a) Deduce an expression for the intensity of light at a point due to superposition of waves coming from two light sources. Hence, find the conditions of destructive and constructive interference.

(b) In an interference experiment, d is the distance between two coherent sources of light with wavelength λ and D is the source-screen distance. Show that the separation between the two consecutive dark bands is given by $\beta = \frac{\lambda D}{2d}$.

4.44 Establish the condition of constructive and destructive interference for light coming from two coherent sources.

4.45 (a) Explain the phenomenon of interference of light. Describe and explain Young's experiment demonstrating interference of light.

(b) Give the theory of Newton's rings experiment to determine the wavelength of monochromatic light and show how the refractive index of a liquid can be determined?

(c) Describe how Newton's ring apparatus can be employed to measure (i) the radius of curvature of a long focal length spectacle glass and (b) the refractive index of a liquid.

4.46 (a) Explain the nature of change in the fringe in Newton's ring experiment when

(i) some oil is placed between the glass plate and the plano-convex lens, and

(ii) the plano-convex lens is gradually moved away from the glass plate.

(b) The diameter of the nth Newton's ring changes from 1·2 cm to 1 cm, when the air space between the lens and the plate is replaced by a transparent liquid. Find the refractive index of the liquid.

4.47 (a) Deduce an expression for calculating very small thickness of mica sheet with the help of Young's double slit experiment.

(b) A piece of glass of refractive index 1.5 is introduced in one of the interfering beams in Young's experiment. The central bright fringe is observed to shift to the position previously occupied by the 4th bright fringe at a distance 2 mm from the central maximum. If the fringe width and wavelength of light used are 1.1 mm and 559 nm respectively, find the thickness of the plate.

(c) Two slits in Young's double slit experiment are 1.3 mm apart from each other and the screen is placed 1.6 m away from the slits. Calculate the distance of the fifth bright fringe. The wavelength of light used is 600 nm.

(d) In what ways the interference in Young's experiment and interference in Newton's rings experiment differ?

4.48 (a) Two independent sources of light of same wavelength cannot produce interference. Justify.

(b) Plot the intensity distribution curve of Young's double-slit interference experiment and label it.

(c) Can you measure the refractive index of a liquid by Newton's ring experiment? Explain.

(d) In Young's experiment, the width of the fringe obtained with light of wavelength 6000Å is 2.0 mm. What will be the fringe width if the entire apparatus is immersed in a liquid of refractive index 1.33?

(e) Newton's rings experiment is performed with reflected light of wavelength 5500 Å using a plano-convex lens and a plane glass plate. What would be the observation when the glass plate is moved away from the lens along the axis of the lens by 10^{-5} m?

4.49 (a) What are missing orders in case of double slit diffraction pattern?

(b) Starting from the intensity distribution on the screen due to single slit diffraction, deduce the conditions for maxima and minima. Hence show that maximum light is contained in central bright fringe.

(c) Why don't we observe diffraction of light waves whereas we do for sound waves under ordinary circumstances? Explain mathematically. What will happen if we keep on reducing the slit widths of double slit diffraction set up? Give mathematical results to support your answer.

(d) Calculate the distance at which a millimetre scale would appear blurred to a human eye. (Take the diameter of the pupil to be 2 mm and $\lambda = 600$ nm).

(e) Examine if the two spectral lines of $\lambda = 559$ nm and $\lambda = 559.6$ nm can be clearly resolved in (i) first order, (ii) second order by diffraction grating of 2 cm width and 425 lines/cm.

4.50 Prove that the intensity of secondary maxima formed for Fraunhofer diffraction at a single slit are of decreasing order.

4.51 (a) Find the intensity distribution for the Fraunhofer diffraction pattern through a single slit.

(b) Hence find the position of maxima and minima.

4.52 (a) Calculate the least width of a grating to resolve two components of the sodium D-line for the second order. Assume that the grating is having 500 lines/cm. The wavelength for D_1 and D_2 lines of sodium are 5593 Å and 5596 Å respectively.

(b) Deduce the missing order for a double slit Fraunhofer diffraction pattern if the slit width are 0.16 mm and they are 0.5 mm apart.

(c) Two stars situated at a distance of 9.5×10^{12} km are sending light of wavelength 500 nm to a telescope of diameter 25 cm. Find the distance of a separation of the stars for which they are just resolved.

4.53 (a) Derive the intensity distribution of light for diffraction of Fraunhofer class due to a single slit. Sketch the intensity distribution.

(b) A single slit forms a diffraction pattern of Fraunhofer class with white light. The second maximum in the pattern for red light of wavelength 5000 Å coincides with the third maximum of an unknown wavelength. Calculate the unknown wavelength.

4.54 (a) Write down the expression of intensity (explaining the symbols) distribution for Fraunhofer diffraction due to a double slit. Draw the distribution graphically.

(b) Find the condition of diffraction minima and interference maxima from the above.

(c) In a double slit experiment if the slit width is 2 mm and the separation between the slits is 4 mm then find the missing orders. Derive the formula that you use.

(d) Define resolving power of a grating. What factors does it depend upon?

4.55 (a) Distinguish between Fresnel and Fraunhofer class of diffraction.

(b) Draw and explain the intensity distribution curve due to single slit diffraction phenomenon.

(c) State and explain Rayleigh criterion for resolution of spectral lines.

(d) An oil immersion microscope just resolves the rulings of a grating having 3900 lines/ mm when light of wavelength 400 nm is employed. Find the numerical aperture of the lens.

4.56 (a) What is missing order in case of double slit diffraction pattern?

(b) A 2 cm wide diffraction grating is just able to resolve sodium *D*-lines (having wavelengths 559 nm and 559·6 nm) in 2nd order. Find the number of rulings per mm.

(c) Obtain an expression for resultant intensity and hence find the conditions for maxima and minima in a single slit Fraunhofer diffraction process.

4.57 (a) Prove that the intensity of secondary maxima formed for Fraunhofer diffraction at a single slit are of decreasing order.

(b) Explain missing order in an *N*-slit diffraction.

(c) State Rayleigh criterion. Write a short note on resolving power of a grating.

4.58 (a) A microscope is used to resolve two point sources separated by a distance 4×10^{-5} cm. If the wavelength of light be 546 nm, calculate the numerical aperture of the objective.

(b) Write down the expression for the intensity due to Fraunhofer diffraction in a single slit. How this expression will be modified in case of Fraunhofer diffraction in double slit? Explain. Find the conditions for interference maxima for Fraunhofer diffraction in double slit.

Numerical Problems

4.1 In Young's double slit interference the distance between the coherent sources is 1.15 nm. Calculate the fringe width that would be observed on a screen placed at a distance of 55 cm from the source. The wavelength of light used is 5593 Å.

4.2 In a Newton's ring experiment the diameter of 5th dark ring and 15th dark ring are 0.37 cm and 0.59 cm. Find the radius of the plano-convex lens if the wavelength of light used is 559 nm.

4.3 A mica sheet of RI 1.558 is introduced in one of the interfering beams in Young's experiment and the central fringe gets shifted by 0.2 cm. The distance between the sources is 0.1 cm and the screen is placed at a distance 25 cm from them. Determine the thickness of the mica sheet.

4.4 Newton's rings are observed in reflected light of wavelength $\lambda = 590$ nm. The diameter of the 10th dark ring is 5 mm. Find the radius of curvature of the lens and the thickness of the air film.

4.5 Two waves having intensities in the ratio 4 :1 produce interference. Find the ratio of maximum to minimum intensity.

4.6 Fringes are produced with monochromatic light of wavelength 659 nm. A thin film of glass of refractive index 1.52 is placed in the path of one of the interfering rays. The central bright fringe is found to move to a position occupied by the 5th bright band from the centre. Calculate the thickness of the glass film.

4.7 In a Newton's ring experiment, if the diameter of the 10th ring changes from 1.4 cm to 1.25 cm when a liquid is introduced between the lens and the plate, find the RI of the plate. [*Ans*: 1.215]

4.8 Two straight and narrow parallel slits, 5 mm apart, are illuminated by monochromatic light of wavelength 6×10^{-5} cm. If fringes are obtained on a 0.5 m distant screen from the slits, calculate the fringe width. [*Ans*: 0.06 mm]

4.9 In Young's double silt experiment, if slits are separated by 0.1 cm and the screen is placed 0.5 m away and the slits are illuminated by a monochromatic light of wavelength 4500 Å, calculate the fringe width. [*Ans*: 24×10^{-5} cm]

4.10 What is the maximum number of lines of a grating which will resolve the 3rd order spectrum of two lines having wavelength 5590 Å and 5596 Å? [*Ans*: 325]

4.11 In a Newton's ring experiment, the diameters of the mth and $(m + 5)^{th}$ dark rings are 0.5 cm and 1.2 cm respectively. Find the wavelength of the light used if the radius of curvature of the lower surface of the lens is 10 m. [*Ans*: 4000 Å]

4.12 In a Newton's ring experiment, if the radius of curvature of the lower surface of the lens is 1.45 m, find the diameter of the 3rd dark ring produced by a light of wavelength 5590 Å. [*Ans*: 3.2 mm]

4.13 Find the thickness of soap film (RI = 1.33) producing constructive interference of 2nd order of reflected light of wavelength 5000 Å.

4.14 A convex lens of focal length 40 cm is employed to focus the Fraunhofer diffraction pattern of a single slit of 0.3 mm width. Find the linear distance of the first order dark band from the central band. The wavelength of the light used is 559 nm.

4.15 Calculate the least width that a grating must have, to resolve two components of the sodium-D lines for the second order. The grating is having 500 lines/cm. The wavelength for D_1 and D_2 lines of sodium are 5593 Å and 5596 Å respectively. [*Ans*: 1.23 cm]

4.16 In a plane transmission grating the angle of diffraction for second order maxima for wavelength 5000 Å is 30°. Calculate the number of lines in one centimeter of the grating surface.

4.17 A light of wavelength 500 mm is incident normally on a slit of variable width and large length. A detector is placed at a large distance on the other side at an angle of 20^0 with normal. For what width of the slit would the receiver show zero intensity?

4.18 A plane transmission grating having 1500 line/cm is being used under normal incidence of light.

(i) What is the longest wavelength of light for which a spectrum can be seen?

(ii) What is the highest order spectrum that can be seen for the light of 559.3 nm wavelength?

(iii) The spectral line of 559.3 nm in the second order spectrum overlaps with another spectral line in the next order. Find wavelength of the other spectral line.

(iv) Find the expression of the dispersive power for a given order.

(v) If 90% of the width of the grating is covered, how the width of the spectral lines are changed?

4.19 Deduce the missing orders for a double slit Fraunhofer diffraction pattern if the slit width is 0.16 mm and they are 0.8 mm apart. [*Ans*: 6th, 12th, 18th ...]

4.20 In a plane transmission grating the angle of diffraction of the 2nd order maxima for a wavelength 5×10^{-5} cm is 300. Calculate the number of lines/cm of the grating surface. [*Ans*: 5000 lines/cm]

4.21 Find the highest visible orders in a grating spectra, if the wavelength of the incident radiation is 5593 Å and the number of lines on the grating is 5000 lines/cm? [*Ans*: 3]

4.22 A light of wavelength 5593×10^{-8} cm falls normally on a slit of width 0.5 mm. Calculate the total angular width of the central maximum. [*Ans*: 2.4×10^{-3} rad]

4.23 A single slit of width 1 mm is illuminated by light of wavelength 559 nm. Find the observed angular spread of the central maxima of diffraction pattern. [*Ans*: 0.065°]

4.24 A slit of width 12000 Å is illuminated by a light of wavelength 6000 Å. Calculate the angular width of the central maxima in the diffraction pattern. [*Ans*: 25.5°]

4.25 Light of wavelength 6000 Å is incident normally on a plane transmission grating having 5000 lines/cm. Find the difference in deviations in the first and second order spectra. [*Ans*: 21.4°]

4.26 Sodium light is made to fall normally on a grating having 500 lines/cm. Calculate the least width of the grating so that it can resolve the two components of sodium light having wavelength 559 nm and 559.6 nm in the 2nd order. [*Ans*: 6.15 mm]

4.27 Find the limit of resolution of a microscope, if numerical aperture is 0.61 and the wavelength of light used is 5000 Å. [*Ans*: 3500Å]

4.28 In a double slit experiment, if the slit width is 2 mm and the separation between the slits is 4 *mm then find the missing orders.* [*Ans*: 3rd, 6th, 9th ...]

PRACTICALS

1. Young's double-slit experiment for interference of light

Aim

To understand the phenomenon of interference through interference at a double slit

Apparatus

Double slit, light source, screen

Theory

If there are two sources of waves, the waves could be made to interfere. As light is an electromagnetic wave, it exhibit interference effects under appropriate conditions. In Young's experiment, sunlight was allowed to pass through a pinhole on a board. The emerging beam from the first board fell on two pinholes on a second board. The light coming out from those two pinholes then fell on a screen where a pattern of bright and dark spots can be observed. This pattern is called fringes and can only be explained through interference, a wave phenomenon.

Procedure

We can study double slit interference with the help of the experimental arrangement, shown in Fig. (i).

This is an arrangement analogous to Young's experiment. Light from a monochromatic source is allowed to fall on a slit S_0. The light emanating from there is incident on two other slits S_1 and S_2 which are equidistant from S_0. A pattern of *interference fringes* will be produced on the screen. All the slits in this experiment are assumed to be so narrow that they can be considered secondary point sources for Huygens' wavelets. Let the slits S_1 and S_2 are separated by a distance d (<1 mm) and the distance between the screen and the slits is D ($\approx$ 1 m). Fringes so produced by interfering Huygens' wavelets from slits S_1 and S_2 are observed on the screen. In this double slit experiment, we have to use monochromatic light of single wavelength.

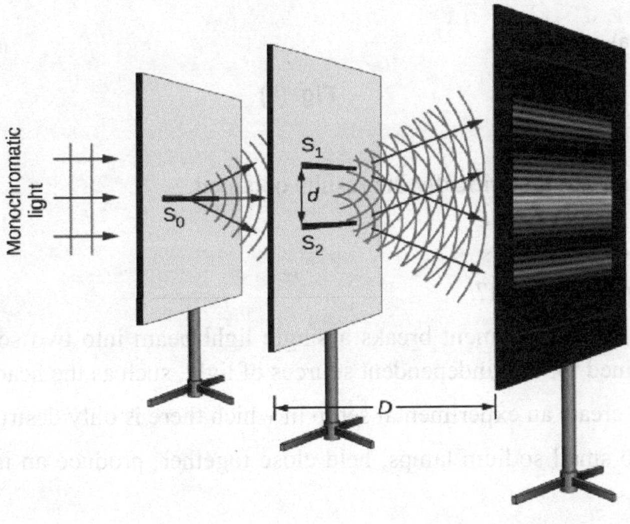

Fig. (i)

Result

Table 1: Determination of the fringe width and wavelength of the light used

Slit separation d (mm)	Number of fringes (N)	Distance between N fringes (cm)	Fringe spacing Δy (cm)	L (m)	λ (nm)

Fig. (ii) (a) and Fig. (ii) (b) respectively shows the pure constructive and destructive interference of two waves having the same wavelength and amplitude. When the amplitudes of waves are added, we get a pure constructive interference. Under this situation identical waves are in phase. On the other hand, we get pure destructive interference when identical waves are exactly out of phase, or shifted by half a wavelength.

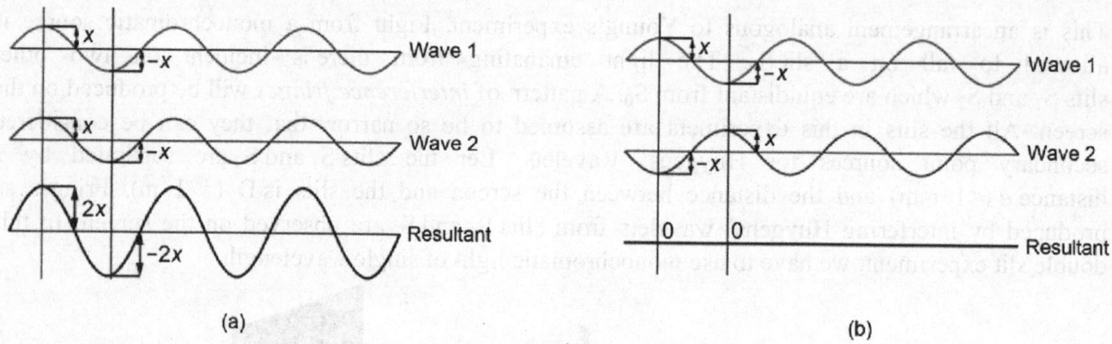

Fig. (ii)

Precaution

Analyze the graphs and discuss critically the results obtained.

Viva-Voce Questions

1. What is interference of light?
2. Young's double-slit experiment breaks a single light beam into two sources. Would the same pattern be obtained for two independent sources of light, such as the headlights of a distant car?
3. Is it possible to create an experimental setup in which there is only destructive interference?
4. Why won't two small sodium lamps, held close together, produce an interference pattern on a distant screen?
5. What if the sodium lamps were replaced by two laser pointers held close together?

2. Combined interference and diffraction experiment

Aim

To observe interference and diffraction properties of light

Apparatus

Double slit, light source, screen.

Theory

Light is an electro-magnetic wave. The electric field component (*i.e.*, amplitude) of one beam can be added or subtracted from that of the other, which is termed as interference of light. Again, the electric field interferes with itself when propagating through an aperture, which is known as diffraction.

Interference Experiment: When a coherent source of light illuminates two slits, the two spherical waves emerging from the slits interfere with each other to produce a symmetrical pattern of varying intensity. This can be found on a screen as an alternating pattern of bright and dark fringes. The bright interference fringe occur when,

$$h \sin\theta = \lambda\, m_i \quad \text{where } m_i = 0, \pm 1, \pm 2, \ldots \qquad (i)$$

where h is the separation between the two slits.

As the angle that the fringes subtend with the optical axis is too small, we may consider

$$\tan\theta \approx \theta$$

The distance on the screen (x) from a given interference maxima to the central maximum is then given by,

$$x = \frac{m_i \lambda t}{h}$$

where x is measured in mm, λ is in μm, h is in mm and t is in m.

Diffraction Experiment: The fringes produced by diffraction of light at a single slit are a simple demonstration for the wave nature of light. When wavelets of light after propagating from the slit produce destructive interference, dark fringes (minima) occur among the wavelets. This occurs when,

$$d\sin\theta = \lambda\, m_d \quad \text{where } m_d = \pm 1, \pm 2, \ldots \tag{ii}$$

In Eq. (*ii*), d is the slit width, θ represents the angle the dark fringe subtends from the axis, m_d is the fringe order and λ is the wavelength of the light source. Geometry for the formation of the fringes is shown in Fig. (i).

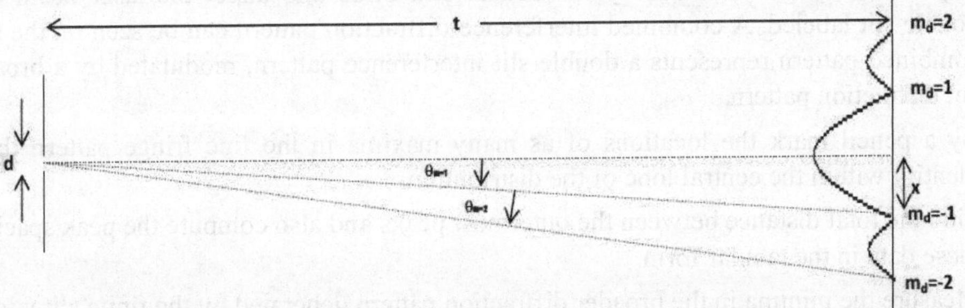

Fig. (i)

The distances from the central peak to the periodic minima is obtained from,

$$x = \frac{m_d \lambda t}{d} \quad \text{where } m_d = \pm 1, \pm 2, \ldots$$

For a circular aperture, the formula can be modified as,

$$x = \frac{1.22\lambda t}{d}$$

where d represents the diameter of the aperture.

Combined Interference and Diffraction: The double-slit apertures are comprised of two slits which have a finite width and due to this the double slit interference patterns contain effects of diffraction owing to each slit aperture. The combined effect is a fine double slit pattern whose overall intensity is modulated by the diffraction pattern of a single slit. Fig. (ii) represents the combined pattern. The peaks caused by interference are labeled as $m_i = 0, \pm 1, \ldots$ and the minima caused by diffraction are labeled as $m_d = \pm 1, \pm 2, \ldots$

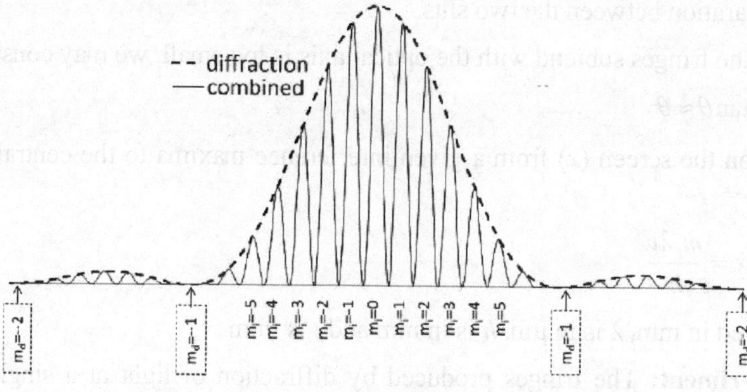

Fig. (ii)

Procedure

For getting Diffraction and Interference Parameters

1. Move the paper up in order to get a blank imaging area or if required put a new piece of paper on the wall.
2. Replace the single slit slide by the double slit. Slide and direct the laser beam through the double slit labeled. A combined interference-diffraction pattern can be seen on the screen. The combined pattern represents a double-slit interference pattern, modulated by a broader single-slit diffraction pattern.
3. By a pencil mark the locations of as many maxima in the fine fringe pattern that you can identify within the central lobe of the distribution.
4. Find the total distance between the outermost peaks and also compute the peak spacing. Record these data in the tabular form.
5. Measure the minima in the broader diffraction pattern generated by the finite slit width.
6. From each of the average spacing, determine the wavelength of light of the He-Ne laser.
7. Label the drawing with the slit width, slit separation and slit to screen distance.
8. Next move or replace your image screen as required and repeat your observations using the double slit slides labeled

For Diffraction

1. Put the slide with the single slit in the slide holder and then keep it on the optical bench.
2. Mount the He-Ne laser on the optical bench at a distance of 20 cm from the slide.
3. Tape a piece of paper to the wall for an image screen; it is positioned in a manner so that the laser light will fall near the top.
4. Direct the laser beam through the single slit marked as, say, 0.16 mm (w = 0.16 mm). You will find a diffraction pattern on the wall. Measure the distance from slit to screen. Maintain this distance throughout the entire experiment. If for some reason you move the apparatus, then you need to re-measure t.
5. By a pencil mark the location of the central peak and then mark the locations of the minima in

the diffraction pattern on either side of the peak up to the highest order that you can properly observe (*i.e.*, $m_i = +/- 10$).

6. Find the distance between the last minima on either side of the central peak. Then divide this distance by two times the order of the last minima to get the average fringe spacing. Using this value determine the wavelength of the laser.
7. Label your drawing with the slit width and also slit to screen distance.
8. Now, move the paper higher up on the wall to get a blank area for the next demonstration. Repeat the steps.

To estimate the diameter of the aperture

1. Now mount a screen on the track to bring it to about 50 cm from the aperture holder and determine the distance *t*.
2. Replace the double slit slide by the slide containing the circular apertures.
3. Shine the beam through the larger circular aperture and see the airy disk diffraction pattern. If you find any trouble to see the first ring in the airy disk, then bring the screen closer or turn down the room lights further. Measure the distance from the central peak to the first minimum which is *x* in the formula for the circular aperture.
4. Knowing the wavelength and aperture to screen distance, measure the diameter of the aperture.
5. Repeat the steps for the smaller circular aperture

Table 1: Determination of the diffraction and interference parameters

Slits	t (cm)	D (mm)	Average diffraction minima spacing	θ_1 (Diffraction)	λ (nm)	h (cm)	Avg. diffraction maxima spacing	θ_1 (Interference)	λ (nm)

Table 2: To find the wavelength

Demo	Distance to wall t (cm)	Slit width d (mm)	Average fringe width (mm)	θ (degree) for $m_d = 1$	λ (nm) for $m_d = 1$

$$\text{Percentage error in the measurement of } \lambda = 100\left|\frac{\lambda - \lambda_{\text{measured}}}{\lambda}\right|$$

Table 3: Determination of the diameter of the aperture

Aperture	*t* (cm)	λ (nm)	Actual aperture diameter (cm)	Measured aperture diameter (*d*) (cm)	% error

Precautions

1. The photo detector should be as far from the slit as possible.
2. Scale of vernier should be rotated slowly and carefully.

Viva-Voce Questions

1. What is meant by interference of light?
2. What is diffraction of light? Was there any change in the spacing of the diffraction minima, the width of the maxima or the intensity of the bright fringes?
3. What is the difference between interference and diffraction of light?
4. What is meant by the interference fringes?
5. For a given slit width, what happened when you changed the slit separation?
6. Did the effects of decreasing slit separation differ between the narrower and wider slits?
7. What happened to the fringes when you decreased the slit width?
8. He-Ne laser emits a wavelength of 632.8 nm. How accurately were you able to compute the wavelength?

3. To establish whether a light source is polychromatic or not using a CD as a reflection grating

Aim

To get a quick and effective method to establish whether a light source is polychromatic or not

Apparatus

Digital audio CD or CD-ROM, light sources, to be selected from the following or individually chosen: daylight bulb, sodium vapour lamp, street lamps, incandescent bulb, light-emitting diode

Theory

A combination of the silvering and the regular grooves on a compact disc (CD) makes it very quick and effective reflection gratings. Polychromatic light is found to reflect in seven rainbow colours while monochromatic light cannot be split and so remains as single colour. In CD we have seen the rainbow patterns they produce. The purpose of this experiment is to use this effect for gathering information about light sources utilized in daily life. It can be done in the laboratory or outside.

Procedure

1. Observe the reflection that occurs due to a white light bulb on the CD.
2. Draw what you find in the intensity and position of the colours. Record the corresponding angle at which you held the CD.
3. Maintaining the geometry of the set up as similar as possible, observe for other light sources for polychromatic or monochromatic.
4. These light sources can be chosen: a candle, sodium street lights, a 'daylight' light bulb, a coloured light bulb, light shining through a coloured transparent layer like dyed cellophane, etc.
5. Try to distinguish between, say, monochromatic yellow light and visual yellow (*i.e.*, light that shows yellow but is a mixture of red and green).

Observations

The following two properties can be observed:

1. Some very common light sources are monochromatic.
2. Compact discs can make useful reflection gratings.

Special Notes: This simple experiment could be taken further by the students to show the pattern formed in true monochromatic light. The particular pattern so obtained can be used as a basis for discussion. One very useful area of observation may be to include the difference between natural daylight bulbs and white bulbs for distinguishing between true monochromatic sources and visual monochromes.

KNOW MORE

Wave optics, is that branch of optics which considers interference, diffraction, polarization and other phenomena where the ray approximation of geometric optics is no longer valid.

Activity

In order to establish his theory, Huygens' considered wavefront emphasizing a wave crest advancing. For this he imagined each point along the wave crest to be source point for small, circular, expanding wavelets. These wavelets expand with the speed of the wave. The surface tangent drawn to these wavelets finds the contour of the advancing wave. When light was considered as a wave, the superposition principle is valid to find the constructive and destructive interferences for light waves. Interference in light waves is not very easy to find as the wavelengths are too short. Further for getting the constructive interference, two waves must have the two contributing crests and the two troughs arriving at the same time. On the other hand, for producing a destructive interference, a crest from one wave and a trough from the other must arrive at a given point at the same time. Thomas Young was the scientist who first demonstrated interference from light waves with a double slit.

Interesting facts

If we examine a compact disc under white light, noting the colors observed and locations of the colors, we can observe some interesting facts. Let us determine if the spectra are formed by diffraction from circular lines centered at the middle of the disc. If by diffraction the spectra are formed, then we have to note their spacing or the type of spacing. Next, explore the spectra of a few light sources, such as a candle flame, halogen light, incandescent bulb and fluorescent light. Measuring the spacing of the rows of pits in the compact disc, estimate the maximum spacing that will permit the given number of megabytes of information to be stored. In optics, there is a standard way to measure diffraction effects. In radio, this approximation is utilized to calculate some effects related to optical effects. It models much interference, diffraction and polarization effects but not the dependence of diffraction on polarization. Since it is a high frequency approximation, this is more accurate in optics than for radio.

Analogy

Physical optics is the name of an approximation commonly used in optics, electrical engineering and applied physics. It is an intermediate method between geometric optics, which is not concerned with wave effects, and full wave electromagnetism which is mainly theory based.

This approximation consists of using ray optics for estimating the field on a surface and then integrating that field over the surface for the purpose of measuring the transmitted or scattered field.

History

Since superposition is a property of waves, what Huygens' developed was essentially the first wave theory of light. But the fact is that he imagined light as a series of pulses, like hand claps, rather than as a sinusoidal wave. The history is too interesting from another consideration. Isaac Newton loved much the atomic theory of matter and so he searched for evidence that light was also made of tiny particles.

The paths of his light particles would correspond to rays but the only significant difference between a ray model and a particle model of light would occur if one could isolate individual particles. Newton never did this but he outlined his model as a particle model. It is therefore, more accurate to say that he was one of the builders of the ray model.

Timelines

Ancient India (5th to 6th century BC): The philosophical schools of Samkhya and Vaisheshika, formulated theories on light.

In the second century: Claudius Ptolemy, in his *Optics* discussed reflection and refraction. He measured the angles of refraction between air, water, and glass.

Dignaga in the 5th century and Dharmakirti in the 7th century: The Indian Buddhists developed a type of momentary flashes of light or energy.

The English Franciscan, Roger Bacon (1214–1294): He was influenced by Grosseteste's writings on the importance of light. In his optical writings he cited many translated optical works.

Hypothesis of Light of 1675: Newton posited the existence of the ether for transmitting forces between particles and ultimately in 1704, Newton published *Opticks*, where he expounded his corpuscular theory of light.

1665: The diffraction properties of light were characterized by Francesco Maria Grimaldi.

1678: Huygens' explained that every point where a luminous disturbance meets turns into a source of the spherical wave itself.

1801: Thomas Young through an experiment strongly inferred the wave-like nature of light. He justified that some type of interaction would occur when two light waves meet.

1815 and 1818: Augustin-Jean Fresnel made some studies and calculations on diffraction of light which largely supported the wave theory of light.

Applications (Real Life / Industrial)

If we look at the reds, blues and greens in a sunlit soap bubble, we can find how straw-colored soapy water could produce them which can only be explained by the wave character of light. It is equally true for the colors found in an oil slick or in the light reflected from a compact disc. Many other interesting phenomena, such as the dispersion of white light into a rainbow of colors when allowed to pass through a narrow slit cannot be explained fully by geometric optics. In all these phenomena, light interacts with small objects and exhibits its wave characteristics. This branch of optics that considers the behavior of light when it exhibits wave characteristics is known as the wave optics having a significant role in both real life and industry.

Case Study (Environmental / Sustainability / Social / Ethical Issues)

By taking help of electron microscopes, images of individual atoms can be taken, but a visible-light microscope cannot be used to do so. Stereo speakers create the illusion of music which comes from a band arranged in the living room but the stereo illusion does not work with bass notes. Similarly, computer chip manufacturers invest huge amount of money in equipment to etch chips with X-rays instead of visible light. The explanation to all of these questions has to do with the subject of wave optics.

The interaction of light waves with matter and its practical applications to optical devices like mirrors have a major role. Simultaneously, we need to understand the phenomena associated with the wave optics and the devices that can only be developed using the wave model of light.

Inquisitiveness and Curiosity Topics

Light wave has a wavelength about a million times shorter than the water waves. This is for the reason that the diffraction effects scale uniformly. If we enlarge or reduce the whole diffraction situation by the same factor, including both the wavelengths and the sizes of the obstacles the wave encounters, the result is still a valid solution. It is true that water waves and light waves differ in many ways, not just in scale only.

The principle of least time is a very general term and does not apply just to refraction and reflection. It can even be used to prove that light rays travel in a straight line through empty space. This very general approach to wave motion was used by Richard Feynman, one of the pioneers who in the 1950's reconciled quantum mechanics with relativity and a readable explanation of the same was given by him in a book titled as 'The Strange Theory of Light and Matter' in which he wrote for the general people.

REFERENCES AND SUGGESTED READINGS

1. S. A. Akhmanov and S. Yu. Nikitin, *Physical Optics*, Oxford University Press, 1997.

2. R. A. Serway and J. W. Jewett, *Physics for Scientists and Engineers*, 6th Edition, Brooks/Cole, 2004.
3. P. Ya. Ufimtsev, *Fundamentals of the Physical Theory of Diffraction*, John Wiley and Sons, 2007.
4. *Huygens' Principle*, Math Pages, Retrieved 2017-10-03.
5. C. Huygens, *Traité de la Lumière* (drafted 1678; published in Leyden by Van der Aa, 1690), translated by S. P. Thompson as Treatise on Light (London: Macmillan, 1912; Project Gutenberg edition, 2005.
6. O. S. Heavens and R. W. Ditchburn, *Insight into Optics*, Chichester: Wiley and Sons, 1987.
7. M. Born and E. Wolf, *Principles of Optics*, Cambridge University Press, 1999.
8. C. Balanis, *Advanced Engineering Electromagnetics*, John Wiley and Sons, 2012.
9. M. V. Klein and T. E.Furtak, *Optics*, 2nd Edition, New York: John Wiley and Sons, 1986.
10. J. Goodman, *Introduction to Fourier Optics*, 3rd Edition, Roberts and Co Publishers, 2005.
11. J. Baggott, *The Quantum Story*, Oxford Press, 2011.
12. B. B. Baker and E. T. Copson, *The Mathematical Theory of Huygens' Principle*, Oxford, 1939, 1950; AMS Chelsea, 1987.
13. J. A. Stratton, *Electromagnetic Theory*, McGraw-Hill, 1941.
14. https://nptel.ac.in/courses/115/105/115105083/
15. https://nptel.ac.in/courses/115/106/115106119/
16. http://srjcstaff.santarosa.edu/~lwillia2/41/41ch22_s17.pdf
17. https://www2.physics.ox.ac.uk/sites/default/files/2011-10-19/optics_lectures_2012_pdf_11269.pdf

5 Lasers

UNIT SPECIFICS

In this unit we have considered the following aspects on LASER (Light Amplification by Stimulated Emission of Radiation):

- Einstein's theory of matter radiation interaction;
- Einstein's A and B coefficients;
- Amplification of light by population inversion;
- Different types of gas lasers (He-Ne, CO_2);
- Different types of solid-state lasers (ruby, Neodymium);
- Dye lasers;
- Properties of laser beams: mono-chromaticity, coherence, directionality and brightness;
- Applications of lasers in science, engineering and medicine.

The practical applications of the topics are discussed for generating further curiosity and creativity as well as improving problem solving capacity.

Adding a large number of multiple choice questions and also of short and long answer types following lower and higher order of Bloom's taxonomy we have attempted to improve the understanding of the subject. Assignments through a number of numerical problems are provided for exercises. A list of references and suggested readings have been included in the unit so that more curious students can go through them for getting further information on various topics. We have also provided some QR codes in different sections which can be scanned for gathering supportive knowledge.

We have designed a practical based on the topics covered in the unit. Moreover, there is a "Know More" section for giving the supplementary information. This section clearly focuses on the early activity when the topics started to develop. Examples of some interesting facts, analogy, salient observations and findings, timelines up to the recent developments have been largely covered in the unit. Applications of the subject for day-to-day real life and also some industrial applications have been mentioned. Some case studies related to environmental, sustainability, social and ethical issues are considered and finally inquisitiveness and curiosity topics relevant to the unit are included.

RATIONALE

Laser physics is a branch of optics which describes the theory and practice of lasers. It is concerned with the propagation of laser beam with various laser applications as well as with related fields like nonlinear optics and quantum optics. Laser is produced when the electrons in atoms in special glasses, crystals or gases absorb energy from an electrical current or another laser. These excited electrons move from a lower-energy orbit to an orbit of higher-energy around the atom's nucleus.

This unit will help students to get a clear idea of laser technology starting from the basics. Einstein's theory of matter radiation interaction and the amplification of light by population inversion are some of the aspects which can provide to develop the subject in a greater detail and to apply the same using different type of lasers like gas laser, solid-state laser etc. for practical purposes, particularly on engineering aspects.

PRE-REQUISITE

Physics: Modern Physics (Class XII)

UNIT OUTCOMES

List of learning outcomes is as follows:

U5-O1: Differentiate between the laser and ordinary light

U5-O2: Illustrate the concept of spontaneous and stimulated emission of radiation in different types of lasing actions

U5-O3: Demonstrate optical pumping and population inversion concepts in the working of a laser

U5-O4: Apply the fundamental idea of lasing action in different types of lasers

U5-O5: Apply the principle of laser in science and engineering problems

Unit-5 Outcome	EXPECTED MAPPING WITH COURSE OUTCOMES *(1- Weak Correlation; 2- Medium Correlation; 3- Strong Correlation)*					
	CO-1	CO-2	CO-3	CO-4	CO-5	CO-6
U5-O1	1	-	-	-	1	1
U5-O2	-	-	-	-	2	1
U5-O3	-	-	-	-	3	1
U5-O4	-	-	-	-	3	2
U5-O5	-	-	-	-	3	3

5.1 INTRODUCTION

In this unit lasers and their applications are briefly discussed. LASER stands for '*Light Amplification by Stimulated Emission of Radiation*'. The process can take place in atoms, ions and molecules of solids, liquids and gases to produce a pure intense beam of light. The frequency of the emitted radiation extends from the ultraviolet (UV) to radio frequency (RF) region of electromagnetic spectrum. The output power of radiation of a laser varies from a few mW to MW. Lasers can emit radiation in continuous or pulsed mode. Laser is a source that can produce highly concentrated and parallel beam of intense coherent light.

5.2 LASER AND ORDINARY LIGHT

Both laser and ordinary light are electromagnetic radiations but laser is somehow distinct from that of an ordinary light. These are:

(i) Monochromatic: Ordinary light is a combination of different wavelengths or frequencies. On the other hand, for a laser beam, all the photons emitted possess the same energy and wavelength. Hence, the laser beams have a particular wavelength or color and simultaneously it possesses a particular frequency.

(ii) High directivity: We consider two points at distances r_1 and r_2 respectively coming from the windows of a laser. Let the measured diameters of the laser spot at those points are d_1 and d_2 respectively. Then the angle of divergence between those two points can be mathematically expressed as,

$$\theta = \frac{d_2 - d_1}{2(r_2 - r_1)} \tag{5.1}$$

and it is measured to be a few mili-radian/m.

(iii) High degree of coherence: Coherence is related to definite phase relationship between two sources at different points of space and time and is categorized as:

a) **Temporal coherence:** If two waves always maintain fixed amplitude and phase difference while passing through a fixed point in space then they are temporally coherent for a specified time (τ_c) and the length of wave pulse corresponding to that time is the coherence length given by

$$l_c = c\tau_c \tag{5.2}$$

b) **Spatial coherence:** It is the phase relationship between two waves present at two space points at the same instant. It depends on the dimension or size of the source.

(iv) High brightness: A 1 mW He-Ne laser is brighter (100 times) than a mid-day Sun and so one can directly look at a glowing 100 W bulb but not at a 1 mW He-Ne laser.

5.3 EINSTEIN'S A AND B COEFFICIENTS

Laser was invented by Einstein in 1917. He clearly pointed out that in the simplest form of two energy levels laser system, the distribution of atoms vary due to absorption or emission of radiation. Einstein implemented three empirical coefficients for quantifying the change of population of the two levels [See Annexure-III for further details]. Absorption of photon can excite an atom from ground to an excited state. But the atom cannot remain in the excited state for a long time and return to the ground state or to some intermediate (metastable) state by emission process.

5.3.1 Absorption of Radiation

In an active medium the atoms have a large number of quantized energy states. In absence of any external influence they are usually in the ground state with energy E_1. Due to irradiance of photon of frequency υ they can be excited to some energy level E_2 such that the frequency of transition is,

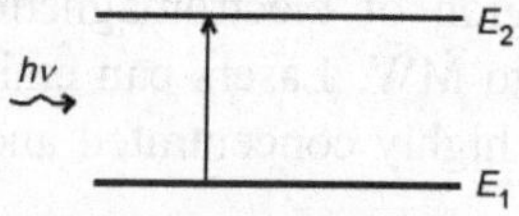

Fig. 5.1 (a): Absorption by radiation

$$\upsilon = \frac{E_2 - E_1}{h} \quad \text{(where } h \text{ is Planck's constant)} \tag{5.3}$$

This is known as absorption or excitation of atoms. This type of absorption is known as the stimulated absorption [Fig. 5.1 (a)] of radiation.

5.3.2 Spontaneous Emission of Radiation

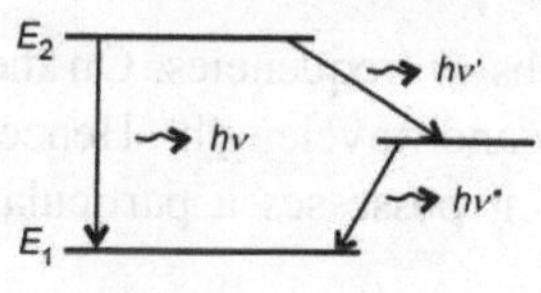

Fig. 5.1 (b): Spontaneous emission of radiation

An atom cannot stay in an excited state for long time and comes to its ground state by emitting a photon of frequency υ within a very short interval of time ($\sim 10^{-8}$ s). Sometimes, there may be two steps where at the first step the excited atoms come to a metastable state by an emission of frequency υ' and then to the ground state by another emission of frequency υ''. This process of radiation is called as the spontaneous emission [Fig. 5.1 (b)] of radiation.

5.3.3 Stimulated Emission of Radiation

When a photon of frequency υ interacts with an atom which is already in the excited state E_2, it fails to excite the atoms which were already excited. Eventually it de-excites the atom emitting an additional photon of same frequency υ and brings it back to the ground state. This kind of transition is known as induced or stimulated emission of radiation [Fig. 5.1(c)]. If we consider an assembly of atoms in an active medium some atoms are in the excited state and they will excite at a certain rate according to statistical probability theory.

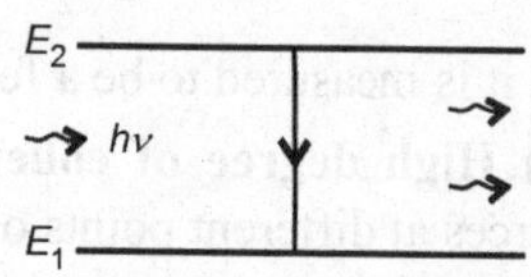

Fig. 5.1 (c): Stimulated emission of radiation

Let, the temperature of the assembly be T with the radiation frequency υ and energy density $u(\upsilon)$. Let, N_1 and N_2 are the number of atoms per unit volume in the active medium at the energy levels E_1 and E_2 at any instant of time. The rate of transition from state 1 to state 2 due to absorption process depends on the energy density of radiation and the number of atoms making transition from ground level to the excited level is given by,

$$N_1 B_{12}\, u(\upsilon)$$

where B_{12} is a constant known as Einstein's coefficient of absorption.

Now, the probability of spontaneous emission is independent of the energy density and depends only on the number of atoms in the excited state. Thus the number of atoms which will make spontaneous emission from state 2 to state 1 is given by,

$$N_2 A_{21}$$

where A_{21} is the constant of proportionality known as Einstein's coefficient of spontaneous emission.

Finally, the probability of stimulated emission depends on number of atoms N_2 in the excited state and also on the energy density of radiation.

Thus the number of atoms which will make transition from state 2 to state 1 is given by,

$$N_2 B_{21} u(\upsilon)$$

where B_{21} is known as Einstein's coefficient of stimulated emission.

At equilibrium, the rate of emission and absorption must be equal. Thus we have,

$$N_2 A_{21} + N_2 B_{21} u(\upsilon) = N_1 B_{12}\, u(\upsilon)$$

or,
$$u(\upsilon)\,[N_1 B_{12} - N_2 B_{21}] = N_2 A_{21}$$

or,
$$u(\upsilon) = \frac{N_2 A_{21}}{N_1 B_{12} - N_2 B_{21}} = \frac{A_{21}}{B_{21}} \frac{1}{\frac{N_1}{N_2}\frac{B_{12}}{B_{21}} - 1} = \frac{A_{21}}{B_{21}} \frac{1}{\frac{N_1}{N_2}\left(e^{\frac{h\upsilon}{kT}} - 1\right)} \tag{5.4}$$

Since according to Maxwell–Boltzmann distribution law, the relationship between the populations of two levels at a temperature T may be written as

$$N_2 = N_1 e^{-\frac{h\upsilon}{kT}}$$

or,
$$\frac{N_1}{N_2} = e^{\frac{h\upsilon}{kT}}$$

Now, according to Planck's law of radiation the energy density of radiation is given by,

$$u(\upsilon) = \frac{8\pi h\upsilon^3}{c^3} \frac{1}{\left(e^{\frac{h\upsilon}{kT}} - 1\right)} \tag{5.5}$$

So, comparing Eq. (5.4) with Eq. (5.5) we may write, $\frac{A_{21}}{B_{21}} = \frac{8\pi h\upsilon^3}{c^3}$

or,
$$A_{21} = \frac{8\pi h\upsilon^3}{c^3} B_{21} \tag{5.6}$$

and
$$\frac{B_{12}}{B_{21}} = 1$$

or,
$$B_{12} = B_{21} \tag{5.7}$$

Thus, the probability of stimulated absorption is equal to the probability of stimulated emission and the probability of spontaneous emission increases as υ^3. Spontaneous emission probability becomes rapid when the difference in energy between the levels 1 and 2 is large.

EXAMPLE 5.1

Example 5.1: Find the probability of stimulated emission in a laser action which emits radiation of 5000Å wavelength if the probability of spontaneous emission is 10%.

Solution: Given $A_{21} = 0.1$ and $\lambda = 5000\text{Å} = 5000 \times 10^{-10}$ m.

So using, $\frac{A_{21}}{B_{21}} = \frac{8\pi h\upsilon^3}{C^3} = \frac{8\pi h}{\lambda^3}$

EXAMPLE 5.1

we get, $B_{21} = \frac{A_{21}\lambda^3}{8\pi h}$

$$= \frac{0.1\times(5000\times10^{-10})^3}{8\times3.14\times6.625\times10^{-34}} = 7.5\times10^{11}.$$

Thus, the probability of stimulated emission is 7.5×10^{11}.

5.4 WORKING OF LASER

The working of laser is based on the phenomenon of spontaneous and stimulated emission. During the lasing process the active medium should be such that it should have at least one additional third state (metastable state) besides the ground state and excited state (Fig. 5.2).

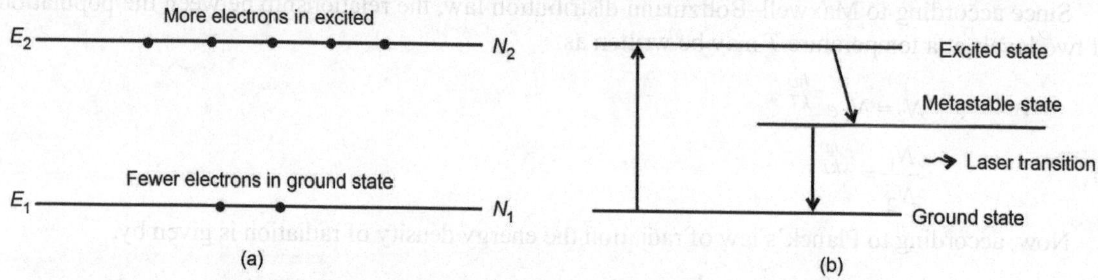

Fig. 5.2: Illustration of (a) Lasing action, (b) Population inversion

The lifetime of an atom in excited state is $\sim10^{-8}$ s which is less than the life time in metastable state. So, an atom can stay longer time in metastable state than its excited state. When such atoms are excited with a light of suitable wavelength from lower energy state (E_1), the atoms jump to the excited state (E_2) by absorbing incident photons. Due to shorter life-time of the excited state the atoms jump to the relatively longer lived metastable state. So, there exist a large number of atoms in the metastable state than that available in the ground state and this phenomenon of having higher number of atoms in the longer lived metastable state than the ground state is called *population inversion* [Fig. 5.2 (b)].

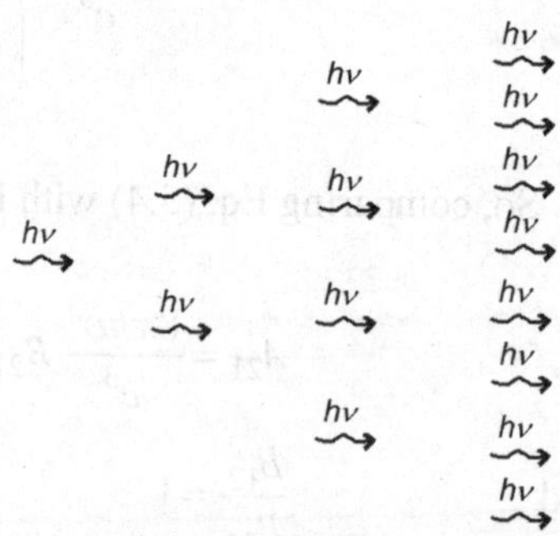

Fig. 5.3: Light amplification by stimulated emission of radiation

The process through which population inversion can be achieved is known as *pumping*. When population inversion is achieved, a photon obtained by spontaneous emission strike an atom in the metastable state and the excited atom of the metastable state is thus stimulated to emit photon of same energy, as that of the incident photon and hence a large number of coherent photons are produced (Fig. 5.3) by repeated stimulated emission. In this way light amplification takes place due to multiplication of photons. As these photons are having the same frequency and same phase, a strong coherent beam of light is generated.

5.5 LIGHT AMPLIFICATION AND PUMPING

Adding energy through a process called pumping is a situation in which equilibrium is not established. Pumping may be electrical, optical or chemical. The purpose is to raise sufficient number of atoms to the upper energy level. This mechanism is termed as population inversion. When the population inversion becomes positive, light is amplified.

5.5.1 Population Inversion

According to Maxwell-Boltzmann (M-B) statistics the numbers of atoms present in a particular energy state is,

$$N_2 = N_1\, e^{-\frac{h\upsilon}{kT}} \tag{5.8}$$

where N_1 and N_2 are respectively the number of atoms present per unit volume in the lower and higher energy states, k is the Boltzmann constant and T the absolute temperature.

As, $E_2 > E_1;\ N_1 > N_2$

i.e., the number of atoms in the lower energy state is larger than that in the higher energy state. But to achieve strong laser beam, the number of atoms in the higher energy state should be larger than the number of atoms in the ground state. This phenomenon is termed as the population inversion.

Example 5.2: A relative population (Boltzmann ratio) of 1/e is the representative of the ratio of populations in two energy states at room temperatures (T = 27^{0}C). Determine the wavelength of the radiation emitted at this temperature.

Solution: According to the problem,

$$e^{-\frac{hc}{\lambda kT}} = \frac{1}{e}$$

or, $$hc = \lambda kT$$

or, $$\lambda = \frac{hc}{kT} = \frac{6.626\times10^{-34}\times3\times10^{8}}{1.23\times10^{-23}\times300} = 5387\times10^{-8}\,\text{m}.$$

EXAMPLE 5.2

5.5.2 Pumping

After the attainment of population inversion, the influence leading to the emission of stimulated photons for the transition of atoms from the metastable state to the ground state is the lasing action and the process through which the atoms from lower energy state are excited to create population inversion to higher energy state in the active medium is called pumping.

Various modes of pumping: Pumping can be produced by light source (optical pumping) by incorporating pumping cavities, flash lamp pumping, arc lamp pumping, external laser pumping or other optical pumping methods, electrical discharge (electrical pumping), gas dynamic pumping or by chemical pumping or nuclear reaction.

5.6 OPTICAL RESONANCE

Optical resonator: It is used to sustain the lasing action for a long time through oscillations of the light waves. It is a device which can direct the photons back and forth through the laser medium. The number of photons thus generated

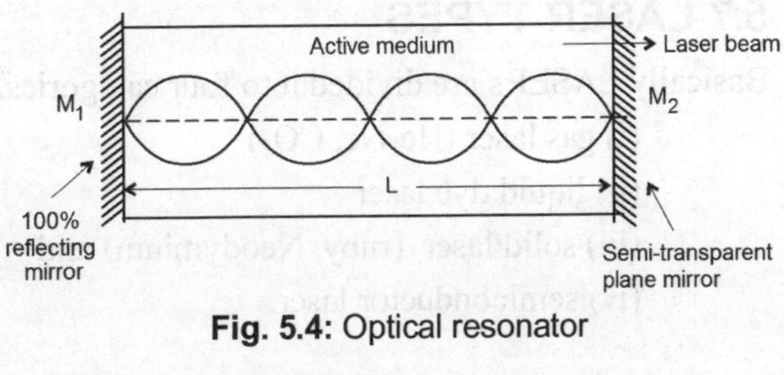

Fig. 5.4: Optical resonator

gets multiplied by stimulated emission to achieve a coherent laser beam. An active medium is placed between the mirrors M_1 and M_2 [Fig. 5.4]. The mirror M_2 is semi-transparent so that more than 90% of incident light is reflected from it and a small fraction is transmitted through it as the laser beam.

If L be the length of the resonator cavity, then the change in phase in one round of trip of the wave of wavelength λ between the mirrors will be,

$$\Delta = \frac{2\pi}{\lambda}.2L + 2\delta \quad (5.9)$$

where δ is the phase change due to one reflection in any one of the mirrors.

For constructive interference we must have $\Delta = 2n\pi$ (where $n = 1, 2, 3\ldots$), so that standing waves are formed inside the cavity.

$$\therefore \quad \frac{2\pi}{\lambda}.2L + 2\delta = 2n\pi$$

or, $$\frac{2\pi L}{\lambda} + \delta = n\pi$$

or, $$\lambda = \frac{2\pi L}{n\pi - \delta}$$

or, $$\frac{c}{\upsilon} = \frac{2\pi L}{n\pi - \delta}$$

or, $$\upsilon = \frac{c}{2\pi L}(n\pi - \delta) \quad (5.10)$$

SCAN ME

to see the schematic understanding of Laser

This is the relationship between the length of the resonator and the frequency of the laser. So, standing waves are formed within the two mirrors, the reflectors form the nodes. Also the wavelength of the laser must be smaller than the length (L) of the cavity. So, the separation in frequency between two consecutive modes of vibration is,

$$\Delta\upsilon = \upsilon_{n+L} - \upsilon_n = \frac{c}{2L} \quad (5.11)$$

Threshold condition: After population inversion is achieved in the cavity, light amplification takes place within the resonator by the active medium. But during this time the light radiation suffers some losses and to sustain lasing action this loss must be compensated by induced transition in the active medium. So, a minimum strength of population inversion is required and this is called the threshold population inversion. The most important losses are the losses due to absorption by the active medium, scattering and diffraction by the mirror *etc.*

5.7 LASER TYPES

Basically LASERs are divided into four categories.

(i) gas laser (He-Ne, CO_2)

(ii) liquid dye laser

(iii) solid laser (ruby, Neodymium) and

(iv) semiconductor laser.

Here we will describe some of these lasers along with their properties.

5.7.1 Ruby LASER

It is one important solid state laser having the following specifications:

Crystal: Ruby

Shape: Ends are flat; one end completely silvered and the other end partially silvered (forming a resonant cavity).

Chemistry: Ruby is crystalline Al $_2O_3$ doped with Cr. The triply ionized Cr^{3+} ions replace some Al^{3+} ions giving a transparent crystal.

Color: Pink or red depending on the concentration of Cr^{3+}, *e.g.*, Maiman used pink ruby with a concentration of about 0.5% of Cr.

Energy levels of Cr^{3+}: States E_1 and E_2 have very short life time ($\leq 10^{-9}$ s) while metastable state M has a large life time (~ ms). Transitions to E_1 and E_2 are caused by radiation of wavelength 6600 Å and 4000 Å respectively. The chromium ion make a very fast non-irradiative transition from the excited to the metastable state M and since the state has a large lifetime, the number atoms in this state keeps on increasing, this eventually leads to a state of population inversion between state M and G. Once a state of inversion is achieved, lasing action is triggered spontaneously by emitted photons. The energy level diagram is shown in Fig. 5.5.

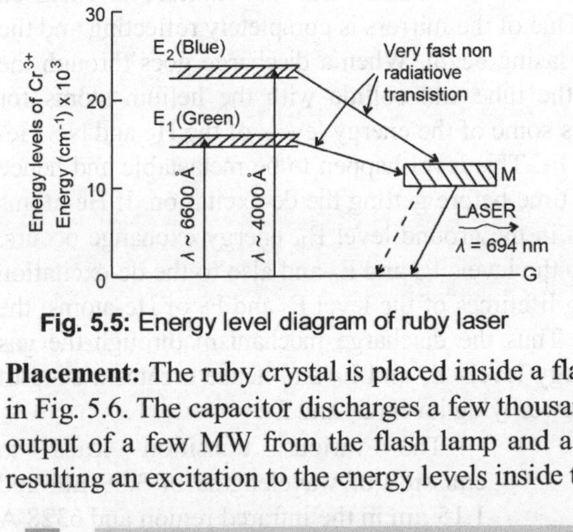

Fig. 5.5: Energy level diagram of ruby laser

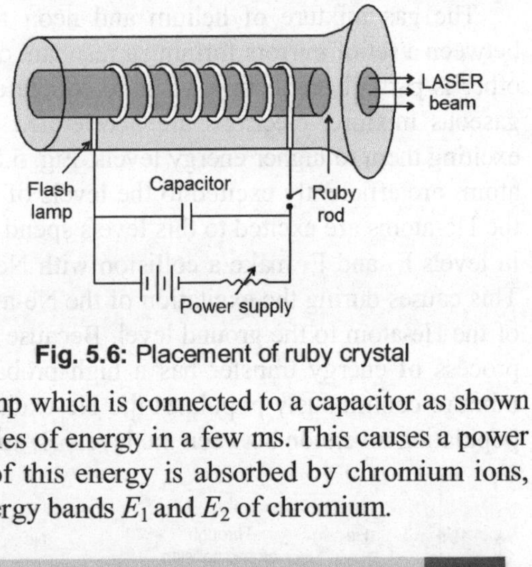

Fig. 5.6: Placement of ruby crystal

Placement: The ruby crystal is placed inside a flash lamp which is connected to a capacitor as shown in Fig. 5.6. The capacitor discharges a few thousand joules of energy in a few ms. This causes a power output of a few MW from the flash lamp and a part of this energy is absorbed by chromium ions, resulting an excitation to the energy levels inside the energy bands E_1 and E_2 of chromium.

Example 5.3: If the wavelength of radiation of ruby laser is 6943Å what is the energy of a photon emitted?

Solution: The wavelength of radiation of ruby laser is 6943Å.

So, the energy of a photon emitted will be

$$E = \frac{hc}{\lambda} = \frac{6.626 \times 10^{-34} \times 3 \times 10^{8}}{6943 \times 10^{-10} \times 1.6 \times 10^{-19}} \text{eV}$$

$$= 1.787 \text{ eV.}$$

EXAMPLE 5.3

EXAMPLE 5.4

Example 5.4: Find the wavelength spread of a 1ns pulse from a ruby LASER with a wavelength of 630 nm.

Solution: Wavelength spread

$$d\lambda = \frac{\lambda^2}{c\tau_c} = \frac{\left(630\times10^{-9}\right)^2}{3\times10^8\times1\times10^{-9}}$$

$$=1.323\times10^{-12}\,\text{m}=1.323\,\text{pm}.$$

5.7.2 Helium-Neon LASER

It was the first gas laser to be operated successfully. Ali Javan and his co-workers who were working in the Bell Telephone laboratories in USA fabricated this laser in 1961. A He-Ne laser consists of a long narrow discharge tube of diameter 1 cm and length 80 cm as shown in Fig. 5.7. Its high energetic medium consists of a mixture of 10 : 1 ratio of He and Ne at a total pressure of about 1 torr.

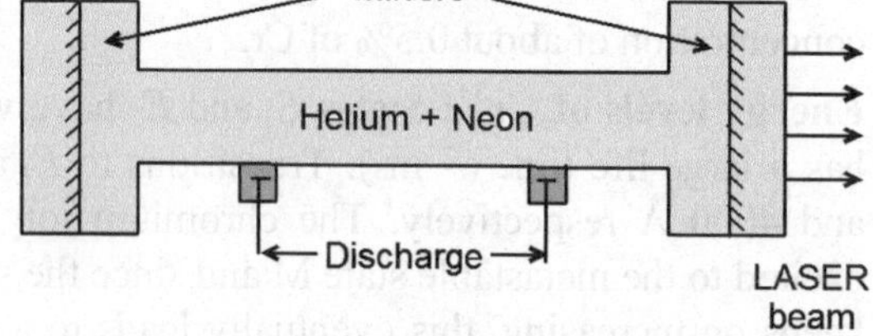

Fig. 5.7: Schematic diagram of a He-Ne LASER

The gas mixture of helium and neon forms the lasing medium and this mixture is enclosed between a set of mirrors forming a resonant cavity. One of the mirrors is completely reflecting and the other is partially reflecting, so as to couple out the lasing beam. When a discharge goes through the gaseous mixture, electrons are accelerated down the tube and collide with the helium atoms for exciting them to higher energy levels. Fig. 5.8 shows some of the energy levels of the He and Ne. He-atoms are efficiently excited to the levels of F_2 and F_3. This level happen to be metastable and hence the He-atoms are excited to this levels spend a long time before getting the de-excitation. If He-atoms in levels F_2 and F_3 make a collision with Ne-atoms in the ground level E_1, energy exchange occurs. This causes during the excitation of the Ne-atoms to the levels E_4 and E_6 and also to the de-excitation of the He-atom to the ground level. Because of long lifetimes of the level F_2 and F_3 of He-atoms, the process of energy transfer has a high probability. Thus the discharge mechanism through the gas mixture continuously populates the Ne excited energy levels E_4 and E_6 and hence creates a state of population inversion between the levels E_4 and the lower lying levels E_5 and E_3.

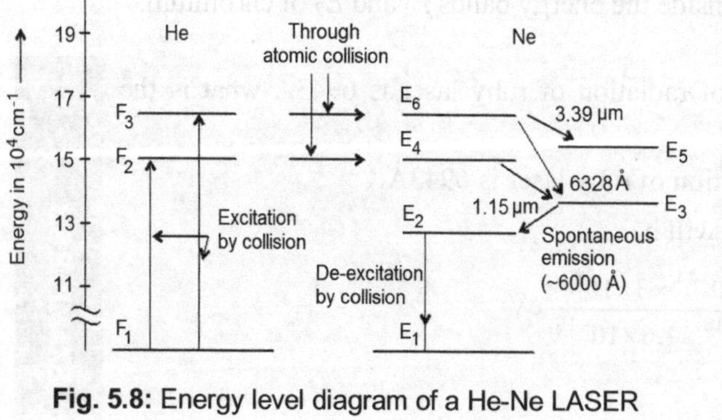

Fig. 5.8: Energy level diagram of a He-Ne LASER

The various transition leads to emission of wavelengths of 3.39 μm and 1.15 μm in the infrared region and 6328 Å in the visible range. The last wavelength corresponds to the well known red light from He-Ne laser. The excited Ne atoms drop down from level E_3 to E_2 by spontaneously emitting a photon around λ = 6000 Å. When a narrow tube is used the Ne atoms in the energy level E_2 collide with the walls of the tube and gets de-excited to the level E_1.

Advantages

(i) Light from gas laser compared to solid laser is considerably more directional and monochromatic.

(ii) Gas lasers are capable of supplying a continuous laser beam without elaborate cooling arrangement.

Uses: He-Ne laser can be used in

(i) holography,
(ii) bar code reading,
(iii) laser printing,
(iv) laser modulation telemetry,
(v) interferometry *etc.*

Example 5.5: In a He-Ne laser transition from 3s to 2p-level, a laser transition of wavelength 632.8 nm is produced. If the energy of the 2p-level is 15.2×10^{-19} J, how much pumping energy is required, if there is no energy loss in the He-Ne laser?

Solution: Here the energy of the emitted laser beam is

$$E = \frac{hc}{\lambda} = \frac{(6.625\times10^{-34})\times(3\times10^{8})}{632.8\times10^{-9}} = 3.1\times10^{-19}\,\text{J} = 1.935\text{ eV}.$$

Again energy of the 2p-level is $15.2\times10^{-19}\,\text{J} = 9.488\text{eV}$.
Hence, the energy of the pumping source is = (9.488 + 1.935) eV = 11.42 eV.

EXAMPLE 5.5

Example 5.6: Find out the number of photons required to be emitted per second to give output power 2 mW corresponding to wavelength 632.8 nm.

Solution: Energy of a single photon,

$$E = \frac{hc}{\lambda} = \frac{6.626\times10^{-34}\times3\times10^{8}}{632.8\times10^{-9}} = 3.14\times10^{-15}\,\text{J}.$$

So, no. of photons required to be emitted per second to give output power 2 mW is

$$n = \frac{2\times10^{-3}}{3.14\times10^{-15}} = 6.37\times10^{11}.$$

EXAMPLE 5.6

Example 5.7: Calculate the mode of separation of a He-Ne laser system with the two plane mirrors forming the resonant cavity separated by 0.6 m. Also find the coherence length (in vacuum) of laser beam produced.

Solution: The frequency difference between two consecutive modes is

$$\Delta\nu = \frac{c}{2d}$$

where c = velocity of light in vacuum = 3×10^{8} m/s, d = 0.6 m.

EXAMPLE 5.7

EXAMPLE 5.7

So, $\Delta\nu = \dfrac{3\times10^8}{2\times0.6} = 250\times10^6\ \text{Hz} = 250\ \text{MHz}.$

Thus, coherence time $\tau_c = \dfrac{1}{\Delta\upsilon} = \dfrac{1}{250\times10^6}$,

and coherence length $l_c = c\tau_c = \dfrac{3\times10^8}{250\times10^6} = 1.2\ \text{m}.$

EXAMPLE 5.8

Example 5.8: In He-Ne laser transition from 3*s* to 2*p* level, a laser beam of wavelength 632.8 nm is produced. If the 2*p* level has energy equal to 15.2×10^{-19} J, calculate the required pumping energy assuming no loss of energy.

Solution: The pumping takes place between the energy levels E_0 and E_{3s} and the transition takes place between E_{3p} and E_{2p} levels for which

$$\Delta E = E_{3s} - E_{2p} = \frac{hc}{\lambda}$$

or, $$\Delta E = \frac{6.627\times10^{-34}\times3\times10^8}{632.8\times10^{-9}}$$

or, $$\Delta E = 3.14\times10^{-19}\ \text{J}.$$

So, the required pumping energy is given by,

$$E = E_{3s} - E_0 = E_{2p} + \Delta E$$

or, $$E = 15.20\times10^{-19} + 3.14\times10^{-19}\ \text{J}$$

or, $$E = 18.34\times10^{-19}\ \text{J}.$$

5.7.3 CO_2 Molecular Gas Laser

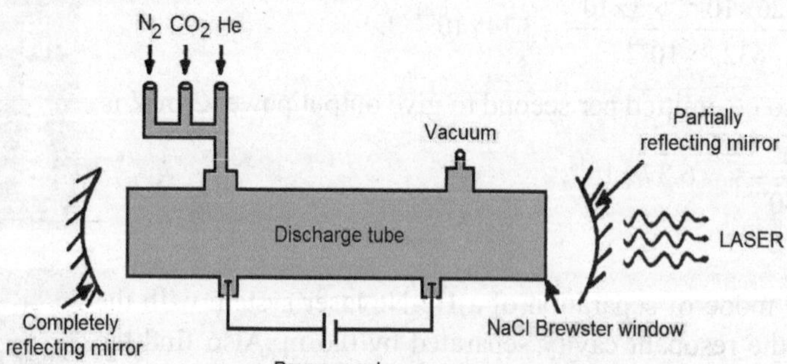

Fig. 5.9: Schematic diagram of a CO_2 Laser

In any molecular gas laser, laser action is obtained by transitions between vibrational and rotational energy levels of molecules. The basic construction is very simple and the obtained output of this laser is continuous.

In the CO_2 molecular gas laser, transition occurs between the vibrational states of carbon dioxide molecules.

Pillai developed the CO_2 laser, the first molecular gas laser which is a four level laser system. It is highly efficient and operates at 10.6 μm in the far infra red (IR) region.

Construction: It consists of a quartz tube of 5 m length and 2.5 cm diameter. This discharge tube is filled with gaseous mixture of CO_2 (active medium), helium (He) and nitrogen (N_2) with proper partial pressures. The terminals of the discharge tubes are connected to a source of power supply. The two ends of the discharge tube are fitted with NaCl Brewster windows. As a result, the generated laser light is polarized. Two concave mirrors are used for the purpose; out of which one is fully reflecting and the other is partially reflecting. These two mirrors produce the optical resonator.

Working Principle: The active medium of this gas laser is a gas mixture of CO_2, N_2 and He. The laser transition occurs between the vibrational states of CO_2 molecules. When electric discharge starts in the gas, the electrons collide with nitrogen molecules raising it to excited states. This process is expressed as, $N_2 + e^* = N_2^* + e$ (5.12)

where, N_2 represents the nitrogen molecule in ground state, e* represents the electron with kinetic energy, N_2^* represents the nitrogen molecule in the excited state and e is same electron with smaller energy. In the excited state N_2 molecules collide with CO_2 atoms in ground state causing an excitation to higher electronic, vibrational and rotational levels. This mechanism is mathematically expressed as,

$$N_2^* + CO_2 = CO_2^* + N_2 \quad (5.13)$$

where, N_2^* represents the nitrogen molecule in excited state, N_2 represents nitrogen molecule in ground state, CO_2^* is the carbon dioxide atoms in excited state and CO_2 represents the carbon dioxide atoms in ground.

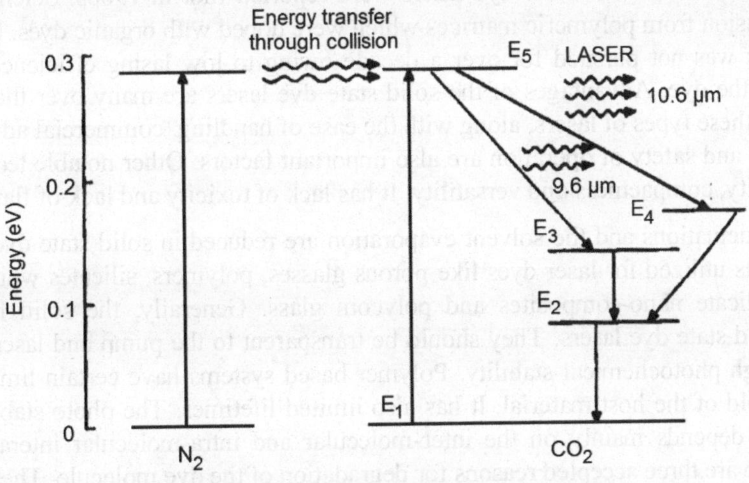

Fig. 5.10: Energy level diagram of CO_2 Laser

As the excited level of N_2 is very close to the E_5 level of CO_2 atom, population in E_5 level increases. As soon as population inversion is reached, any of the spontaneously emitted photon will trigger laser action in the tube. There are two categories of laser transition possible:

(i) Transition E_5 to E_4: This produces a laser beam corresponding to a wavelength 10.6 μm,

(ii) Transition E_5 to E_3: This produces a laser beam corresponding to a wavelength 9.6 μm.

Generally, 10.6 μm transition is more intense than the transition at 9.6 μm. The output power from this laser is 10 kW. The CO_2 laser produces a beam which is invisible. It can propagate through the atmosphere and can be used as a welding or cutting device in industry. CO_2 lasers are used sometimes for the purpose of target designation. It can be used to induce fluorescence in trace materials, with the object of detection, where only the beam has to propagate short distances through the atmospheric medium.

5.7.4 Dye Lasers

Organic dye laser systems have importance due to their wide tunability, capability for producing ultra-short light pulses and versatility.

Properties of organic dyes: Organic dyes are a class of coloured substances useful for their ability to impart colour to other liquid, solid or gaseous substances which are characterized by their strong absorption bands between the ultraviolet and the near infrared region. The chemical structure is an alternating structure of single and double carbon to carbon bond (conjugate bonds) as shown below:

$$C = C - C = C - C$$

Practical pumping arrangements: There are few pumping methods currently used in dye lasers. Different methods are applied for various laser setups. In most cases the pumping beam gets totally absorbed in the dye and does not proceed much beyond the dye chamber. This absorption is a result of Beer's law:

$$I = I_0 e^{-aNd} \tag{5.14}$$

where, I is transmitted light intensity, I_0 is incident light intensity, a is molar extinction coefficient in cm^{-1}, N is concentration in mole/litre and d is thickness of the absorbing layer in cm.

Solid State Dye Laser: Solid-state dye lasers were reported first in 1960s. Scientists demonstrated stimulated emission from polymeric matrices which were doped with organic dyes. But work on solid-state dye lasers was not pursued for over a decade owing to low lasing efficiencies and fast photo degradation of the dye. Advantages of the solid state dye lasers are many over the traditional liquid dye lasers. For these types of lasers, along with the ease of handling, commercial advantage of the low production cost and safety of operation are also important factors. Other notable technical advantages are manageability, compactness and versatility. It has lack of toxicity and lack of flammability.

The flow fluctuations and the solvent evaporation are reduced in solid state dye lasers. There are several materials utilized for laser dyes like porous glasses, polymers, silicates which are organically modified or silicate nano-composites and polycom glass. Generally, the solid host materials are suitable for solid state dye lasers. They should be transparent to the pump and laser wavelengths and should have high photochemical stability. Polymer based systems have certain limitations like small damage threshold of the host material. It has also limited lifetimes. The photo stability of dye doped solid materials depends mainly on the inter-molecular and intra-molecular interactions of the dye molecule. There are three accepted reasons for degradation of the dye molecule. These are:

(i) photo-deactivation from the excited state caused by chemical oxidation

(ii) formation of dimers that absorbs irradiation without any fluorescence and

(iii) thermal destruction.

Application of dye laser: Commercial dye-laser manufacturing got a significant boost in different domains like absorption spectroscopy and saturation spectroscopy, with a very narrow bandwidth resolution (down to 0.01 wave number). Another area that contributed to atomic and molecular dynamics was the use of the ability of the non-linear frequency mixing combined with the ability to tune the laser over a wide bandwidth of frequencies. Wide range of frequencies can be used in many fields. In physical chemistry they are used to investigate the processes of photo-ionization and photo-dissociation, and in this way exhibit many aspects on the molecular and atomic dynamics. These methods are applied in biology and many medical applications.

5.7.5 Nd:YAG Laser

Neodymium-doped Yttrium Aluminum Garnet (Nd:YAG) laser is a type of solid state laser where Nd:YAG is used as a laser medium. Nd:YAG laser is a four-level laser system. All the four energy levels are involved in laser action for operating in both pulsed and continuous mode. It can produce laser light commonly in the near-infrared region of the spectrum corresponding to 1064 nm. It emits laser light also at few wavelengths, *e.g.*, 1440 nm, 1320 nm, 1120 nm and 940 nm.

Construction: Nd:YAG laser has three basic elements: (i) an energy source, (ii) active medium, and (iii) optical resonator [Fig. 5.11].

i) **Energy source:** The energy source (or pump source) supplies energy to the active medium to get population inversion. In Nd:YAG laser, light energy sources, *e.g.*, flashtube or laser diodes are used as energy source for supplying energy to the active medium. Nowadays, laser diodes are usually chosen over flashtubes due to its high efficiency and low cost.

ii) **Active medium:** The laser medium of Nd:YAG laser is made by synthetic crystalline material [Yttrium Aluminum Garnet (YAG)]. This is doped with a chemical element [neodymium (Nd)] where the lower energy state electrons of neodymium ions are excited to higher energy state making the lasing action in the active medium.

iii) **Optical resonator:** The Nd:YAG crystal is kept between two mirrors which are optically coated or silvered (one partially and the another completely).

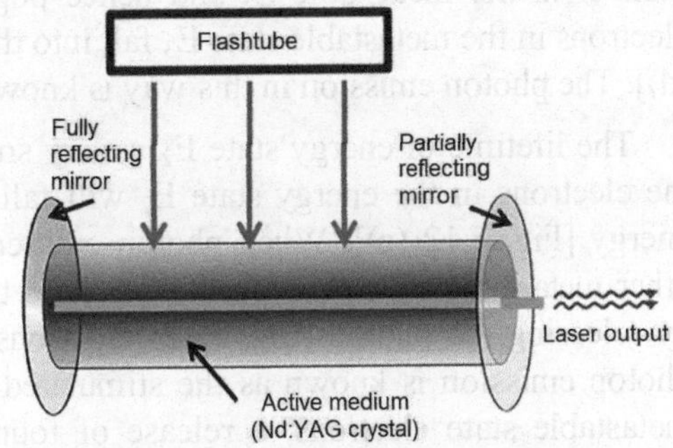

Fig. 5.11: Schematic diagram of a Nd:YAG Laser

As the fully silvered mirror completely reflects the light it is called as fully reflecting mirror. A major part of light is reflected by the partially silvered mirror but a small part of light is permitted through it to form the laser beam. This mirror is called partially reflecting mirror.

Working of Nd:YAG laser: Nd:YAG laser is a four-level laser system which indicates that all the four energy levels are involved in laser action. Light energy, like laser diode or flashtube, is used to supply energy to the active medium.

For the case of Nd:YAG laser, the lower energy state electrons in the neodymium ions are excited to higher energy state to get population inversion. We consider a Nd:YAG crystal active medium with N number of electrons consisting of four energy levels E_1, E_2, E_3, and E_4 such that the number of electrons in the energy states E_1, E_2, E_3 and E_4 are N_1, N_2, N_3, and N_4 respectively.

Let the energy levels are such that

$$E_1 < E_2 < E_3 < E_4,$$

where the energy level E_1 represents the ground state, E_2 is the next higher energy state or excited state, E_3 is the metastable state and E_4 is the pump state. Let initially, the population is such that $N_1 > N_2 > N_3 > N_4$ [Fig. 5.12 (a)]. When flashtube or laser diode supplies light energy to the active medium (Nd:YAG crystal), the lower energy state (E_1) electrons in the neodymium ions gains sufficient energy and moves to the pump state or higher energy state E_4 [Fig. 5.12 (b)].

The lifetime of the pump state or any higher energy state E_4 is too small (230 μs). As a result, the electrons in the energy state E_4 cannot stay for sufficient time. In fact, after a very short time, the electrons fall into the next lower energy state E_3 releasing non-radiation energy [Fig. 5.12 (c)].

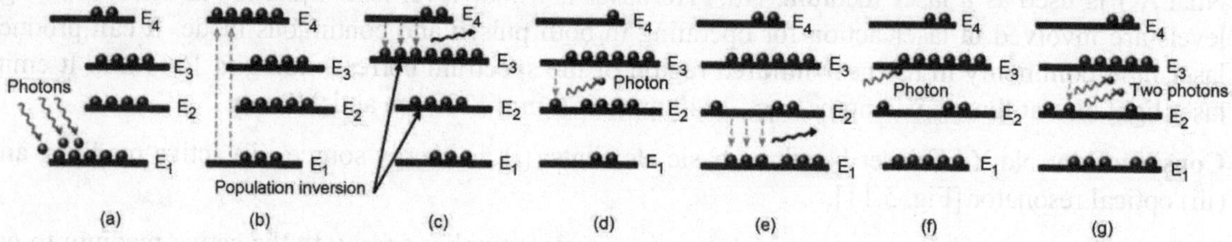

Fig. 5.12: Energy level diagram of Nd:YAG Laser

Lifetime of metastable state E_3 is large in comparison to the lifetime of pump state E_4. So the electrons go to the energy level E_3 faster than they leave E_3. This causes an increase in the electrons number in the metastable E_3 and hence population inversion is obtained. After a certain time, the electrons in the metastable state E_3 fall into the next lower energy state E_2 releasing photons [Fig. 5.12 (d)]. The photon emission in this way is known as spontaneous emission.

The lifetime of energy state E_2 is very small similar to the energy state E_4. So, after a short time, the electrons in the energy state E_2 will fall back to the ground state E_1 by releasing radiation-less energy [Fig. 5.12 (e)]. When photon emitted owing to spontaneous emission is interacted with the other metastable state electron, it stimulates that electron and makes it fall into the lower energy state by releasing the photon [Fig. 5.12 (f)]. Consequently, two photons are released [Fig. 5.12 (g)]. The photon emission is known as the stimulated emission of radiation. If two photons interact with the metastable state electrons, a release of four photons occurs. In this way, millions of photons are emitted. Thus, optical gain is obtained. Spontaneous emission is a natural process but stimulated emission is not a natural one. In order to get stimulated emission, external photons are required to supply in the active medium. The Nd:YAG active medium produces photons owing to spontaneous emission. Photons produced in the active medium will bounce back and forth within the two mirrors which stimulates other electrons to fall into lower energy state releasing photons. In this way, millions of electrons are stimulated to emit photons. The light produced in active medium is reflected multiple times within the mirrors before it escapes through the partly reflecting mirror.

Advantages of Nd:YAG laser

(i) Nd:YAG lasers consume low power

(ii) Nd:YAG laser offers high gain.

(iii) Nd:YAG laser has good thermal properties.

(iv) Nd:YAG laser has good mechanical properties.

(v) The efficiency of Nd:YAG laser is very high as compared to the ruby laser.

Applications of Nd:YAG laser

This laser has variety of applications particularly in medical and scientific fields, *e.g.*, laser surgery and laser spectroscopy. Some specified applications are in the following fields:

(i) **Military:** Nd:YAG lasers are used in laser designators and laser rangefinders. A laser designator (a laser light source) is utilized to target objects for the purpose of attacking. Laser range finder uses a laser light to find any distance object.

(ii) **Medicine:** Nd:YAG lasers are utilized for correcting posterior capsular opacification (PCO). It is a situation that may appear sometimes after cataract surgery. Nd:YAG lasers are utilized for removing skin cancers.

(iii) **Manufacturing:** Nd:YAG lasers can be utilized for etching plastics and metals. These lasers are also used for welding and cutting steel.

5.8 APPLICATIONS OF LASER

Laser beam is highly (a) monochromatic, (b) coherent and (c) intense. So, other than the definite applications already discussed above for different specific category of lasers, the lasers have numerous general applications:

(i) Holography; where radiating sources should be highly coherent and stable with respect to time.

(ii) Radio communication, in outer space.

(iii) Guiding satellites and rockets.

(iv) Medical fields (like treatment of detached retina or destruction of malignant tumor).

(v) Astronomical distance measurement.

(vi) Industry; for wielding, cutting or pieceing hole in hard materials.

(vii) Lasers play significant role in many research areas.

UNIT SUMMARY

- ***Laser and ordinary light***

  The following are the distinct features of laser:

  i) Monochromaticity
  ii) High directivity
  iii) High degree of coherence and
  iv) High brightness

- ***Einstein's theory of matter radiation interaction and A and B coefficients***

  $$\upsilon = \frac{E_2 - E_1}{h} \qquad \frac{A_{21}}{B_{21}} = \frac{8\pi h\upsilon^3}{c^3} \qquad \frac{B_{12}}{B_{21}} = 1$$

- ***Working principle of laser***

  Amplification of light by population inversion

  $$N_2 = N_1 e^{-\frac{h\upsilon}{kT}}$$

  As $E_2 > E_1$; so $N_1 > N_2$

- ***Optical resonator***

  $$\Delta\upsilon = \upsilon_{n+L} - \upsilon_n = \frac{c}{2L}$$

- ***Different parameters of laser***

  Energy of the emitted laser beam $E = \frac{hc}{\lambda}$

  Wavelength spread $d\lambda = \frac{\lambda^2}{c\tau_c}$

  Frequency difference between two consecutive modes $\Delta\nu = \frac{c}{2d}$

  Coherence time $\tau_c = \frac{1}{\Delta\upsilon}$

  Coherence length $l_c = c\tau_c = \frac{c}{\Delta\upsilon}$

- ***Beer's law***

  $$I = I_0 e^{-aNd}$$

- ***Applications of laser***

  Lasers are used in

  i) holography, ii) radio communication, iii) satellites and rockets, iv) medical fields, v) wielding, cutting industry, vi) spectroscopy *etc.*

EXERCISES

Multiple Choice Questions

5.1 In a ruby laser, population inversion is achieved by

(a) optical pumping
(b) inelastic atom-atom collision
(c) chemical reactions
(d) applying strong electric field

5.2 In the He-Ne laser, the laser lights emitted due to the transition from

(a) 3*s* to 2*p* (b) 3*s* to 3*p* (c) 2*s* to 2*p* (d) cannot be predicted

5.3 The probability of spontaneous emission in comparison to stimulated emission is proportional to which power of frequency of incident radiation?

(a) 1st (b) 2nd (c) 3rd (d) 4th

5.4 In ruby laser, the active medium is

(a) solid (b) liquid (c) gas (d) solid-gas mixture

5.5 For a laser action to occur, the medium used must have at least

(a) 4 energy levels (b) 2 energy levels (c) 3 energy levels (d) 1 energy level

5.6 *In He-Ne laser, Ne atoms obtain energy*

(a) by collision with He atom
(b) from chemical reaction
(c) from optical pumping
(d) from electrical pumping

5.7 Which of the following schemes does not produce lasing action?

(a) two level scheme
(b) three level scheme
(c) four level scheme
(d) five level scheme

5.8 In ruby laser, the host crystal is

(a) Al_2O_3 (b) MnO_2 (c) $CaCO_3$ (d) Al_2SO_4

5.9 Pulsed laser beam is generated from a

(a) semiconductor laser (b) He-Ne laser (c) CO_2 laser (d) ruby laser

5.10 The colour of the laser output in case of ruby laser is

(a) violet (b) blue (c) red (d) green

5.11 Emission of photon caused due to transition of an electron from higher to lower energy level with the help of external energy is known as

(a) stimulated absorption
(b) spontaneous emission
(c) stimulated emission
(d) population inversion

5.12 Ratio of He and Ne in He-Ne laser is

(a) 100 : 1 (b) 10 : 1 (c) 1 : 10 (d) 1 : 100

5.13 Life-time of metastable state is of the order of

(a) 10^{-15} s (b) 10^{6} s (c) 10^{-3} s (d) 10^{-10} s

5.14 Standard wavelength emitted from a He-Ne laser is

(a) 632.8 nm (b) 623.8 nm (c) 628.3 nm (d) 638.2 nm

5.15 Population inversion can be achieved
(a) when excited state is less populated than the ground state
(b) when excited state is more populated than the ground state
(c) when the population of excited state and ground state are equal
(d) cannot be predicted

5.16 Ruby laser system is an example of
(a) continuous laser (b) pulsed laser (c) both (a) and (b) (d) none of these

5.17 The condition for inverse population for n_1, n_2 number of atoms in energy states E_1, E_2 ($E_2 > E_1$) is
(a) $n_1 = n_2$ (b) $n_1 > n_2$ (c) $n_1 < n_2$ (d) none of these

5.18 Rate of stimulated absorption depends on
(a) number of atoms in the ground state (b) intensity of the incident radiation
(c) both (a) and (b) (d) none of these

5.19 The colour emitted from a ruby laser output is
(a) red (b) green (c) blue (d) violate

5.20 He-Ne laser system is an example of
(a) continuous laser (b) pulsed laser (c) both (a) and (b) (d) none of these

5.21 Rate of spontaneous emission depends on the
(a) number of atoms in the ground state (b) number of atoms in the excited state
(c) radiation intensity (d) none of these

5.22 In ruby laser, population inversion is achieved by which type of pumping?
(a) chemical (b) electrical (c) optical (d) none of these

5.23 Ratio of stimulated emission and spontaneous emission is given by
(a) $e^{h\upsilon/kT}+1$ (b) $\frac{1}{e^{h\upsilon/kT}-1}$ (c) $e^{h\upsilon/kT}-1$ (d) $\frac{1}{e^{h\upsilon/kT}+1}$

5.24 Light amplification may be achieved by
(a) stimulated emission (b) spontaneous emission
(c) spontaneous absorption (d) none of these

5.25 Intensity (I) of a laser beam passing through a distance x is given by (where μ is the amplification)
(a) $I = I_o e^{\mu x}$ (b) $I = I_o e^{-\mu x}$ (c) $I = I_o e^{-\mu x}$ (d) none of these

Answers of Multiple Choice Questions

5.1 (b), 5.2 (b), 5.3 (c), 5.4 (a), 5.5 (b), 5.6 (b), 5.7 (c), 5.8 (c), 5.9 (d), 5.10 (b), 5.11 (c), 5.12 (b), 5.13 (c), 5.14 (a), 5.15 (b), 5.16 (b), 5.17 (c), 5.18 (c), 5.19 (a), 5.20 (a), 5.21 (b), 5.22 (c), 5.23 (b), 5.24 (a), 5.25 (b)

Short and Long Answer Type Questions

Category I

5.1 Explain briefly the characteristics of laser beam.

5.2 What does the acronym LASER stands for?

5.3 Find the differences between laser light and ordinary light.

5.4 Explain the terms spontaneous emission and discuss its role in laser.

5.5 Distinguish between spontaneous and stimulated emission of radiation.

5.6 Explain population inversion.

5.7 "Higher probability of stimulated emission compared to that of spontaneous emission is achieved in laser." Give your comments.

5.8 Describe the essential parts of optical resonator.

5.9 Explain briefly the working principle of a laser.

5.10 Explain the role of an optical resonator.

5.11 Explain whether a laser is an amplifier or an oscillator.

5.12 Explain the working of CO_2 laser.

5.13 Mention some important uses of ruby laser.

5.14 Discuss the operation of a Nd:YAG laser with the help of energy level diagram.

5.15 Explain the principle of He-Ne laser with energy level diagram.

5.16 Discuss some advantages of He-Ne laser in comparison to ruby laser.

5.17 Give some applications of laser.

5.18 Describe the function of an optical resonator. What is the use of such a device in the context of laser generation?

5.19 Give the meaning of population inversion in the context of laser. Does it violate Maxwell-Boltzmann distribution law?

5.20 State some important applications of He-Ne laser and ruby laser.

5.21 What is a metastable state? Explain the importance of metastable state in the context of laser.

5.22 Find the ratio of spontaneous and stimulated emission as derived in the basic theory of laser.

5.23 Explain, spontaneous emission, stimulated absorption and stimulated emission of radiation with neat diagram and deduce an expression relating various Einstein's coefficients.

5.24 Illustrate spontaneous emission and stimulated emission. Describe in brief why stimulated emission generates highly intense coherent beam.

5.25 Distinguish clearly between spontaneous and stimulated emission.

5.26 Mention the role of optical pumping. Why optical resonator is necessary to produce laser?

5.27 Describe the construction of a ruby laser. Also, discuss how population inversion is achieved in ruby laser by clearly drawing the energy diagrams. Discuss whether this result is suitable for continuous or pulse mode operation.

5.28 Describe the different techniques to produce population inversion. Why it cannot be created in a two-level system?

5.29 Write a technical note on dye laser.

5.30 Explain why population inversion is needed for lasing action.

Category II

5.31 (a) Explain Einstein's A, B coefficients and deduce their mutual relation. Show that the ratio of spontaneous and stimulated emission is proportional to the cube of frequency.

(b) Explain the construction and action of the optical resonator in ruby laser.

5.32 Describe briefly the working principle of laser action.

5.33 (a) Discuss the operation of ruby laser with the help of energy level diagram.

(b) Find the role of optical resonator in laser production.

5.34 (a) Calculate the ratio of stimulated to the spontaneous emission at a temperature 300 K for sodium D-line (= 5890 Å).

(b) With the help of a neat diagram, briefly describe the construction of a ruby laser.

(c) Why is optical pumping not generally employed in case of a gas laser?

5.35 (a) Explain the terms absorption, spontaneous emission and stimulated emission by some suitable diagrams. Hence derive the relation for Einstein's A, B coefficient.

(b) Explain the working principle of He-Ne laser with energy level diagram.

(c) A relative population (Boltzman ratio) of 1/e is representative of the ratio of populations in two energy states at room temperatures ($T = 27^0C$). Determine the wavelength of the radiation emitted at the temperature.

(d) If the wavelength of radiation of ruby laser is 6943 Å what is the energy of a photon emitted?

5.36 (a) Relate spontaneous and stimulated emission probabilities and hence find out the relation between field energy and A, B coefficient.

(b) Show that the probability of spontaneous emission increases rapidly with the energy difference between the involved energy states.

5.37 Explain Einstein's A and B coefficients of absorption and emission. Find out the relation between them.

5.38 (a) Find the differences between spontaneous emission and stimulated emission of radiation.

(b) Why is stimulated radiation used for laser action instead of spontaneous emission of radiation?

5.39 Relate spontaneous and stimulate emission probabilities and hence find out the relation between them.

5.40 Explain the construction and action of the optical resonator in a ruby laser.

Numerical Problems

5.1 In a He-Ne laser transition from 3s-2p level gives a laser transition of wavelength 632.8 nm. If the energy of the 2p level is 15.2×10^{-19} J, how much pumping energy is required, if there is no energy loss in the He-Ne laser? [*Ans*: 11.4 eV]

5.2 Determine the relative populations of the two states in a ruby laser that produces a light beam of wavelength 694.3 nm at 400K and 600K.

5.3 Find the ratio of populations of the two states in a He-Ne laser that produce light of wavelength 632.8 nm at 300 K.

5.4 Calculate the ratio of stimulated to the spontaneous emission at a temperature 300 K for sodium D-line (= 5890 Å).

5.5 A ruby laser emits light of 693.95 nm wavelength. If 1 mole of Cr^{3+} ions are involved in population inversion process in a pulse, calculate the pulse energy in eV.

5.6 An optical source emits a radiation of wavelength 5000 Å at a temperature 800 K. Show that for the system the stimulated emission is negligible compared to the spontaneous emission.

5.7 Find the wavelength spread of a 1 nano-second pulse from a ruby LASER with a wavelength of 630 nm. [*Ans*: 1.323 pm]

5.8 Compute the coherence length of a light with 589 nm wavelength in 10^{-12} s pulse duration.

PRACTICAL

To determine the wavelength of a He-Ne laser

Aim

To measure the wavelength of a He-Ne laser using diffraction grating.

Apparatus required

Diffraction grating, He-Ne laser, viewing screen, meter stick, rubber band and binder clips

Theory

A grating is made by cutting parallel lines on a glass plate using a precision ruling machine. The space between any two lines in a transmission grating is transparent to the light and so acts as a separate slit. Diffraction grating consists of a large number of equally separated parallel slits.

In order to find where the maxima are created we have to know the value of $d \sin \theta$, as illustrated in Fig. (i). In the figure, d represents the difference in distance of the two light beams coming from the two slits to the screen. The maxima are found to occur when the difference in distances becomes equal to the multiple of the wavelength of the light. This difference will be equal to one multiple of the wavelength of light between the two beams and so we can write,

$$n\lambda = 2d \sin \theta \qquad (i)$$

Now we have,

$$\tan \theta = \frac{x}{L} \qquad (ii)$$

Hence by applying the Pythagorean theorem, we find the beam length using $L^2 + x^2 = h^2$ as

$$h = \sqrt{L^2 + x^2} \qquad (iii)$$

Again

$$\sin\theta = \frac{x}{h} = \frac{x}{\sqrt{L^2 + x^2}} \qquad (iv)$$

Substituting Eq. (*iv*) into Eq. (*i*) we have,

$$\lambda = d\frac{x}{n\sqrt{L^2 + x^2}} \qquad (v)$$

Using Eq. (*v*) we can calculate the wavelength of the used laser beam.

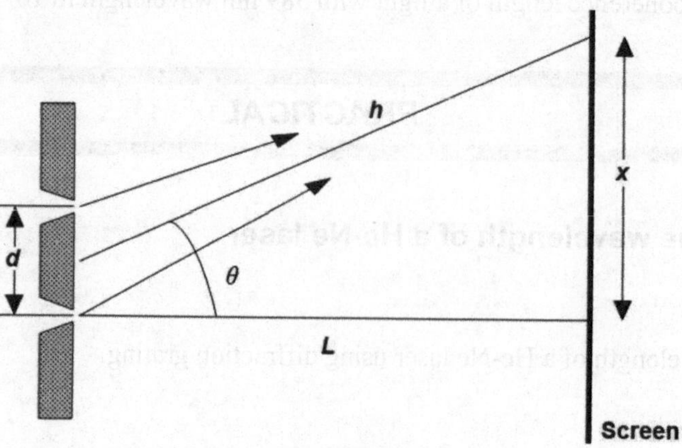

Fig. (i)

Procedure

1. Set up the experimental arrangement and properly mount the meter stick.
2. Line up the laser properly to hit the meter stick at a distance of 50 cm point, though the distance from the laser to the screen is not critical.
3. Set the diffraction grating with 100 lines/mm window in between the laser and the screen. Make the following geometric adjustments:

   i) Vertical - where the dot measurements are easy to read on the meter stick.

   ii) Horizontal - by increasing the value of *L* scatter the dots.
4. Record the distance values:

   i) The *L* distance set should be same for all of the *x* and is to be recorded.

   ii) Record the distances *x* of the light dots from the 50 cm mark on the meter stick.
5. Repeat the steps for the 300 and 600 line/mm windows on the diffraction grating. The distance *L* should be readjusted for each window.

6. It is to be mentioned that you will be limited in the number dots you can get with the denser gratings.
7. Draw graphs for each of the diffraction gratings. Then use a trend line to find the slope of the curves.
8. From the slope, calculate the percentage error.

Result

Table 1: Determination of the wavelength of the used laser beam

Grating Window	100 lines/mm (L = ...cm, d = ...mm)			300 lines/mm (L = ...cm, d = ...mm)			600 lines/mm (L = ...cm, d = ...mm)		
n	*x*	*λ*	*Average λ*	*x*	*λ*	*Average λ*	*x*	*λ*	*Average λ*
1									
2									
3									
4									
5									
6									
7									

Precautions

1. Do not directly observe the laser light as it has a high intensity and can be focused easily. A direct laser light on your eye will be focused by cornea onto a small spot on your retina and can burn or even can detach the retina.
2. When the light strikes any object, a reflection will occur. This can be nearly as strong as the incident light. Knowing the positions of reflections it should be blocked if required.
3. The laser light maintains a shutter in front of the beam. At the time, when not taking data the shutter is to be placed in front of the laser light.
4. Never hold any reflecting object by hand in front of the laser light. This will prevent the possibility of shining the light accidentally into your eyes.
5. Always put your head above the plane of the laser light.

KNOW MORE

A laser is an unusual source of light which produces a very narrow beam of light. This light is very useful for a number of technologies and instruments. Laser is a device that emits light by implementing a process of optical amplification based on the stimulated emission of electromagnetic radiation.

Activity

Lasers do not occur in nature but we have figured ways to artificially create them. Lasers produce a narrow beam of light where all of the light waves have similar wavelengths. The light waves of the laser travel together with their peaks all lined up or in phase. That is the reason why laser beams are very narrow, very bright and can remain focused into a very tiny spot.

After researching the assigned laser type, one can create a PowerPoint presentation, a pamphlet or a small film clip to the class. After the presentations, one can be able to select the most appropriate laser for using in the mummified troll security system. This can clarify how properties of light in different media enable the system for detecting movement near the troll.

Interesting facts

Since laser light remained focused and does not spread out much like a flashlight, laser beams can travel a very long distance. They can also concentrate a huge amount of energy on a very small area.

Analogy

Laser is a device which stimulates atoms or molecules to emit light at any particular wavelength and amplifies that light, producing a narrow beam of radiation. In general, the emission covers an extremely limited range of visible, infrared or ultraviolet wavelengths. Different types of lasers have been developed with varying characteristics.

History

The laser is an outgrowth of a suggestion given by Einstein in 1917 that under suitable circumstances atoms could release excess energy as light. This can be done either spontaneously or when stimulated by light. German physicist Rudolf Walther Ladenburg first found stimulated emission in 1928, although at that time it seemed to have no practical importance.

Timelines

1917: Einstein proposed the theory of "stimulated emission," by which a photon or light particle induces an atom to emit an identical photon.

1953: American physicist Charles Townes builts forerunner of the laser; a "maser," for Microwave Amplification by Stimulated Emission of Radiation.

1957: Gordon Gould, a doctoral student under Townes, coined the term "laser," theorizing that light could be used to excite atoms into making a coherent beam of light. Later he filed a patent but legal dispute lasted almost for three decades.

1960: First operational laser, built by Theodore Maiman of Hughes Research Laboratories in California.

1961: Laser used for the first time in surgery, to destroy retinal tumor.

1962: Invention of semi-conducting diode laser, the mainstay of small commercial lasers today.

1969: Laser used in telemetry. A beam bounced back by a mirror deployed by the Apollo 11 crew to measure the distance between Earth and the Moon to within a few meters.

1971: Laser enters the arts. Awarded the Nobel Prize to Dennis Gabor, a British-Hungarian, for the invention of holography.

1974: First supermarket bar-code scanner introduced.

1975: IBM introduced first commercial laser printer.

1978: First laser disc player, made by Philips, but its high cost was a barrier to success.

1982: First Compact Disc player; the first CD to be pressed is "52nd Street" by Billy Joel.

1983: President Ronald Reagan gave "Star Wars" speech, sketching vision of space-based laser weapons.

1988: North America and Europe were linked by first fiber-optic cable, which used laser pulses to transport data.

1990s: Lasers become established in different manufacturing processes, including integrated circuits and car manufacturing.

Applications (Real Life / Industrial)

Lasers have a number of uses today. They are used in precision tools and can cut through diamonds or thick metal. They can be used to help in delicate surgeries. Lasers are widely used for recording and retrieving information. They are also used in communications and in carrying TV and internet signals. We find them in laser printers, bar code scanners and DVD players. They are also used as parts for computers and other electronic devices. Lasers are also used in instruments called spectrometers which help scientists to figure out what things are made of. For example, the Curiosity rover used a laser spectrometer to observe what kinds of chemicals are in certain rocks on Mars.

A list of both industrial and commercial laser applications includes: (i) Laser cutting (ii) Laser welding (iii) Laser drilling (iv) Laser cladding, a surface engineering process which can be applied to mechanical components for reconditioning, repair work and also for hard facing (v) Photolithography (vi) Laser marking (vii) Laser cleaning. As far as the medical applications are concerned Lasers are used for bloodless surgery, for eye lens curvature corrections, in fiber-optic endoscope to detect ulcers in the intestines, to destroy kidney stones, in cancer diagnosis and therapy

Case Study (Environmental / Sustainability / Social / Ethical Issues)

NASA missions have suitably used lasers for investigating the gases in Earth's atmosphere. Lasers have also been used in some of the instruments that map the surfaces of planets, moons, and asteroids. Scientists have been able to measure the distance between the moon and Earth using lasers.

Picture of Martian soil before (left) and after (right) it was zapped by the Curiosity rover's laser instrument called ChemCam (Image credit: NASA/JPL-Caltech/LANL/ CNES/IRAP/LPGN/CNRS)

Calculating the amount of time it takes for a laser beam to travel to the moon and back, astronomers can tell correctly how far away it is. The picture shows the Martian soil before (left) and after (right) it was zapped by the Curiosity rover's laser instrument called ChemCam. By zapping tiny holes inside the Martian soil and rock, ChemCam was able to find what the material is made of.

Inquisitiveness and Curiosity Topics

Lasers caught the public imagination very fast, for their similarity to the "heat rays" of science fiction. But the practical applications took years to develop. A young physicist, Irnee D'Haenens, while working with Maiman on the ruby laser, joked that the device was "a solution looking for a problem," and the line lingered in the laser community for many years. Townes and Schawlow used laser beams in basic research and to send signals through air or space. A key early success came in late 1963 when two researchers at the University of Michigan, Emmett Leith and Juris Upatnieks, used lasers to make the first three-dimensional holograms.

REFERENCES AND SUGGESTED READINGS

1. A. E. Siegman, *Lasers*, University Science Books, 1986.
2. T. Pearsall, *Quantum Photonics*, 2nd Edition, Graduate Texts in Physics, Springer, 2020.
3. N. Taylor, *Laser: The Inventor, The Nobel Laureate, and The Thirty-Year Patent War*, Simon and Schuster, 2000.
4. T. Pearsall, *Photonics Essentials*, 2nd Edition, McGraw-Hill, 2010.
5. C. H. Townes, *How the Laser Happened: Adventures of a Scientist*, Oxford University Press, 1999.
6. C. H. Townes, *The First Laser*, University of Chicago, Retrieved May 15, 2008.
7. J. Hecht, *Beam: The Race to Make the Laser*, Oxford University Press, 2005.
8. D. Schuocker, *Handbook of the Eurolaser Academy*, Springer, 1998.
9. https://application.wiley-vch.de/books/sample/3527327150_c01.pdf
10. https://www.bgu.ac.il/~glevi/website/Guides/Lasers.pdf
11. https://arxiv.org/ftp/arxiv/papers/2008/2008.03940.pdf
12. http://staff.ustc.edu.cn/~litao/ORS5_v2.ppt.pdf
13. https://nptel.ac.in/courses/104/104/104104085/
14. https://nptel.ac.in/courses/115/102/115102124/

TABLE OF PHYSICAL CONSTANTS

Table I: Refractive indices of different materials

Material	Value of RI	Material	Value of RI
Acetone	1.00109	Ethanol	1.36
Air (-15°C)	1.00030942	Methanol	1.33
Air (0°C)	1.00029238	Propyl Alcohol	1.38
Air (15°C)	1.00027712	Aniline	1.586
Air (30°C)	1.00026337	Benzene	1.501
Air (60°C)	1.00023958	Benzyl benzoate	1.568
Alcohol, ethyl	1.000878	Carbon disulfide	1.63
Alcohol, methyl	1.000586	Carbon tetrachloride	1.46
Ammonia	1.000376	Chlorobenzene	1.525
Argon	1.000281	Chloroform	1.44
Benzene	1.001762	Decane	1.41
Bromine	1.001132	Dodecane	1.41
Carbon dioxide	1.000449	Ether	1.35
Carbon disulphide	1.001481	Ethylene glycol	1.43
Carbon monoxide	1.000338	Ethyl salicylate	1.523
Chlorine	1.000773	Toluene	1.497
Chloroform	1.00145	Turpentine (wood)	1.47
Ether, ethyl	1.001533	Glycerin (Glycerol)	1.47
Ether, methyl	1.000891	Milk	1.35
Helium	1.000035	Octane	1.4
Hydrochloric acid	1.000447	Paraffin, liquid	1.48
Hydrogen	1.000132	Propane	1.34
Methane	1.000444	Quinoline	1.627
Nitric oxide	1.000297	Water 0°C	1.33346
Nitrogen	1.000298	Water 20°C	1.33283
Nitrous oxide	1.000516	Water 100°C	1.31766
Oxygen	1.000271	Crown glass	1.5 - 1.62
Pentane	1.000686	Diamond	2.417
Sulphur dioxide	1.000686	Flint glass	1.57 - 1.75
Water vapor	1.000261	Ice	1.31
Acetic Acid	1.37	Plexiglas	1.49
Acetone	1.36	Salt	1.52

Table II: Refractive indices of crown and flint glasses for different wavelengths

Color	Wave length (nm)	Crown glass	Flint glass
Violet	396.9	1.533	1.663
Blue	486.1	1.523	1.639
Yellow	589.3	1.517	1.627
Red	656.3	1.515	1.622
Dispersive power		0.02	0.04

Table III: Values of critical angles for different medium

Medium	Critical angle
Water	48.75°
Crown glass	41.14°
Dense flint glass	37.31°
Diamond	24.41°

Table IV: Numerical aperture with magnification

Magnification	Depth of field	Numerical Aperture
4x	50	0.10
10x	7.7	0.25
20x	2.9	0.40
40x	0.9	0.65

APPENDICES

APPENDIX-A: Suggestive Template for Practicals

- **Aim**
  Explain briefly about the aim of the experiment.
- **Relevance**
  Explain about the relevancy of the experiment in your own words.
- **Requirements**
  List out all the required apparatus along with their proper specifications.
- **Procedure, Observations and Inference**
  Explain the procedure of the experiment step-wise and note the observations properly to infer accurately. You can use a table similar to that given below:

Step No.	Procedure	Observation	Inference
1			
2			
3			

- **Video/animation**
  If possible, you can go through some video/animation to visualize the steps physically.
- **Calculations**
  Properly calculate all the required physical quantities essential for your experiment.
- **Result and Discussion (Error measurement)**
  Obtain the final result and discuss about it with proper considerations of errors which can be introduced during your experiment.
- **Conclusions**
  Finally make a conclusion based on the obtained results.
- **Validation of the topics in Experiment**
  Try to validate the result of the experiment in real life scenario.
- **Use of ICT**
  You can also study using the available online resources. These are useful as there is no time constraint at all. Some of which are listed (not limited to) below:
  https://swayam.gov.in/
  https://nptel.ac.in/
  https://www.swayamprabha.gov.in/

Note for Instructor and lab-Technicians

Some general and specific instructions are listed separately [see Annexure V] for laboratory preparation, maintenance, safety aspects, etc. Laboratory Instructor and lab-Technicians can follow those instructions properly to run the laboratory smoothly without any hazard.

APPENDIX-B: Indicative Guidelines for evaluation of Group presentation for New Practicals/ Projects/ Activities

Process Related Skills

Criteria and Level	Developing	Competent	Proficient
Handling the Set-up			
Recording of Data			
Time management			
Team Work			
Individual Work			
Safety Precautions			

Product Related Skills

Criteria and Level	Developing	Competent	Proficient
Content			
Research/Survey			
Use of latest Technology			
Stays on Topic			
Preparedness			
Confidence of Presentation			
ICT usage including ppt making skill			
Time management			
Group Efforts			
Individual Efforts			

APPENDIX-C: Assessments Aligned to Bloom's Level

- Revised Bloom's Taxonomy – It has been coupled into following two categories for development of Questions for this Quadrant as given below:

Category I Questions	Category II Questions *- Higher Order Thinking Skills*
Bloom's Level 1: Remember Bloom's Level 2: Understand Bloom's Level 3: Apply	Bloom's Level 4: Analyze Bloom's Level 5: Evaluate Bloom's Level 6: Create

APPENDIX-D: Index for Practicals

S. No	Page No	Name of the Practical	Date			Marks	Signature
			Actual	Repeat	Record		
1		To find the value of acceleration due to gravity using angular oscillations of a long bar					
2		Measuring the speed of light using PC-based oscilloscope					
3		To verify the laws of reflection					
4		Determination of angle of minimum deviation for a given prism					
5		Young's double-slit experiment for interference of light					
6		Combined interference and diffraction experiment					
7		To establish whether a light source is polychromatic or not using a CD as a reflection grating					
8		To determine the wavelength of a He-Ne laser					

ANNEXURES

Annexure- I Diffraction limits and Rayleigh criterion

Physical limits of the wavelength of light cause resolution restrictions which are generally referred to as the diffraction limit. This limit does not permit optical instrument for differentiating between two objects which are separated by a distance of about half the wavelength utilized. Airy diffraction patterns are formed by light from two different points going through a circular aperture. The Fig. A.1 below reveals the following patterns: (a) points far apart, (b) points meeting the Rayleigh criterion and (c) points closer than the Rayleigh criterion. If we look at the figures we find that in Fig. A.1 (b) the diffraction pattern is produced by two point light sources close to each other. This pattern appears to be similar to that for a single point source and it is hardly possible to say that there are two light sources instead of one. When these light sources come closer together as presented in Fig. A.1 (c), it becomes difficult to distinguish them and so making a limit which is an inescapable result of the wave nature of light.

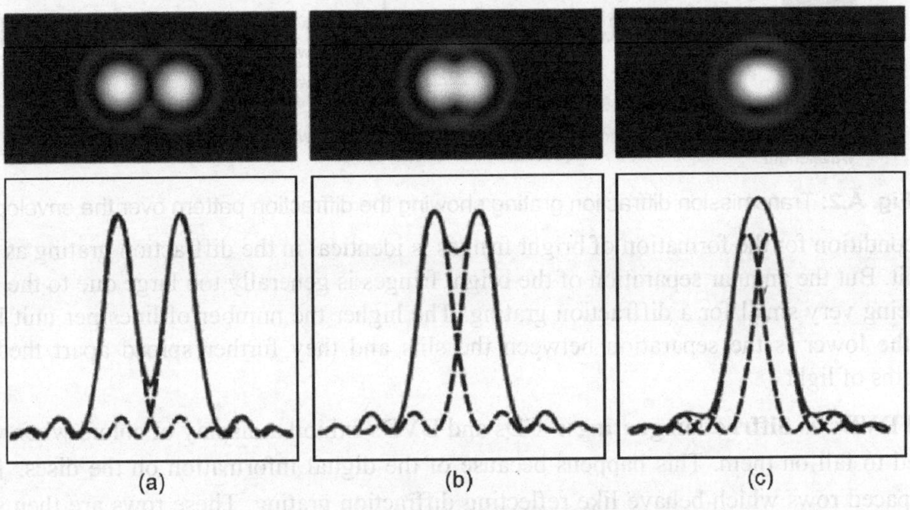

(a) (b) (c)

Fig. A.1: (a) Points far apart, (b) points meeting the Rayleigh criterion and (c) points closer than the Rayleigh criterion

Many situations are there when diffraction limits the resolution. The acuity of vision is limited as the light goes through the pupil of our eye. This diffraction-like spreading of light is for the limited diameter of a light beam and is never due to the interaction with the aperture. Thus diffraction limits the resolution of a lens or mirror. Even the telescopes are limited by diffraction due to the finite diameter of their primary mirror. It is simpler to distinguish between points far apart [Fig. A.1 (a)] and meeting the Rayleigh criterion [Fig. A. 1 (b)]. However, if the points are closer than the Rayleigh criterion as shown in Fig. A. 1 (c), it becomes difficult to distinguish.

Annexure- II Diffraction grating

Diffraction grating uses the properties of diffraction and interference for separating light into its component colors. This produces an effect which is comparable to that of a prism. Diffraction gratings are usually constructed to either transmit or reflect light. A transmission grating is that where light is able to pass through the surface of the grating. It is comprised of a number of equally spaced parallel slits. Reflection gratings, on the other hand, are either plane or concave. These gratings reflect the incident light at various angles for producing a diffraction pattern. Fig. A.2 below reveals a transmission diffraction grating where monochromatic light passes through each of the slits of the grating for constructively interfering at a particular angle.

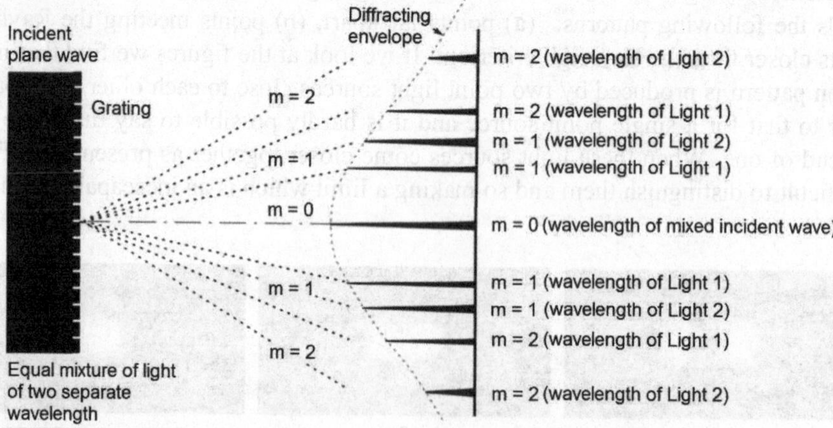

Fig. A.2: Transmission diffraction grating showing the diffraction pattern over the envelope

The condition for the formation of bright fringes is identical in the diffraction grating as that of the double-slit. But the angular separation of the bright fringes is generally too large due to the spacing of the slit being very small for a diffraction grating. The higher the number of lines per unit length in a grating, the lower is the separation between the slits and they further spread apart the individual wavelengths of light.

CDs and DVDs as diffraction gratings: CDs and DVDs exhibit a display of colors when white light is allowed to fall on them. This happens because of the digital information on the discs. This forms closely spaced rows which behave like reflecting diffraction grating. These rows are then spaced 1.6 μm from one another on CD while 0.74 μm on DVD. These rows do not reflect light as the part of the disc that separates them. In these parts the reflected light undergoes constructive interference in some directions. So if white light is reflected from the disc each wavelength of light can be found at a certain angle with respect to the surface of the disc for the formation of light spectra. Both CDs and DVDs have a protective polycarbonate coating. This coating does not interfere in the reflective diffraction grating but creating a transmission grating. Electronic circuits linked with the detector translate the available binary data into electrical signal which is then amplified to reconstruct the original recorded sample. The working principle of a DVD player is also similar except it uses a low wavelength laser and so can detect information coded in much smaller pits. Compared to a CD, a DVD can hold more than seven times information.

Annexure- III Three and four level Laser media

Inversion cannot be obtained in a two-level Laser system by using optical pumping. In fact, the coherent regime is inaccessible by making use of typical optical pumping sources. Inversion by optical pumping can be obtained if we use a three-level (Fig. A.3) or four-level (Fig. A. 4) laser systems.

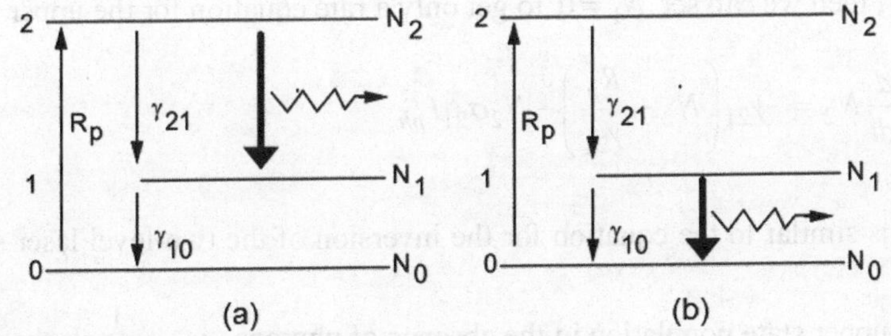

Fig. A. 3: Three-level laser medium

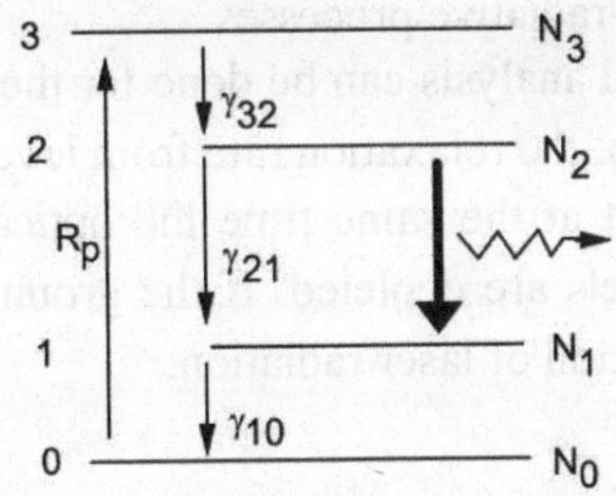

Fig. A. 4: Four-level laser medium

When the medium is under thermal equilibrium, only the ground state is typically occupied. By using optical pumping with an intense flash lamp or another laser we can pump a considerable fraction of the atoms from the ground state with population N_0 to the excited state with population N_3 for the three level lasers operating following Fig. A.3. This is true with population N_4 for four-level laser systems as shown in Fig. A.4. The relaxation rate γ_{10} is very fast compared to γ_{21}, where the laser action occurs to achieve inversion under the condition $N_2 > N_1$.

In the case of four level lasers the relaxation rate γ_{32} should be also fast compared to γ_{21}. These systems can be analyzed in rate equation approximation, adiabatically eliminating the dipole moments. In the three-level system, as in Fig. A.3 (a), we get the rate equations using the analogy to the two-level system as follows:

$$\frac{d}{dt}N_2 = -\gamma_{21}N_2 - \sigma_{21}(N_2 - N_1)I_{ph} + R_p$$

$$\frac{d}{dt}N_1 = -\gamma_{10}N_1 + \Upsilon_{21}N_2 + \sigma_{21}(N_2 - N_1)I_{ph}$$

$$\frac{d}{dt}N_0 = \gamma_{10}N_1 - R_p$$

Here, σ_{21} represents the cross section for stimulated emission between levels 2 and 1 and I_{ph} is the photon flux at the transition frequency f_{21}. In many cases, there are atoms available in the ground state so that optical pumping cannot deplete the number of atoms N_0 in the ground state. That is the reason why one can assume a constant pump rate R_p. If the relaxation rate γ_{10} is highly faster than γ_{21} and the number of possible stimulated emission events that can occur be equal to $\sigma_{21}(N_2 - N_1)I_{ph}$, then we can set $N_1 = 0$ to get only a rate equation for the upper laser level, *i.e.*,

$$\frac{d}{dt}N_2 = -\gamma_{21}\left(N_2 - \frac{R_p}{\gamma_{21}}\right) - N_2\sigma_{21}I_{ph}$$

This equation is similar to the equation for the inversion of the two-level laser system. Here, $\frac{R_p}{\gamma_{21}}$ is the equilibrium upper state population in the absence of photons, $\gamma_{21} = \frac{1}{\tau_L}$ is the inverse upper state lifetime owing to radiative and non-radiative processes.

It may be noted that an identical analysis can be done for the three-level laser operating according to Fig. A. 3 (b). Under that situation, the relaxation rate from level 3 to level 2, which is now the upper laser level has to be very fast. But at the same time the optical pumping should be so strong that essentially all the ground state levels are depleted. If the ground state populations are not depleted, then it would always lead to absorption of laser radiation.

Annexure- IV Different Types of Error in Measurements

Measurement and its importance in Science and Technology

Measurements represent quantities relating to a real time system by making use of numerical values. Requirements for accurate measurements are:

i) Apparatus must be accurate,

ii) Method used ought to be provable and

iii) Standard used should be defined correctly.

In science and technology, advancement is of little significance without the availability of actual measured values with practical proofs. A device to determine variable is known as an instrument. It serves an aid for humans to measure values of unknown quantities. An instrument can be electronic, mechanical or electrical. Based on the degree of variation of the measured quantity with respect to time, an instrument can have static or dynamic characteristics.

Errors in Measurement

If ideal conditions are applied for measuring any parameter, the average deviations of different factors tend to be zero. Average of these infinite numbers of measured values is known as True Value. But this is hypothetical as the negative and positive deviations do not cancel each other in practice. The measured value obtained under ideal conditions is considered as the true value or the best-measured value. A difference between the actual value and the true value is called an Error.

Types of Error

Systematic errors are broadly of three categories;

i) Instrumental Error,

ii) Environmental Error and

iii) Observational Error.

Instrumental errors occur owing to shortcomings in the instruments, improper use of instruments or loading effect of the instrument. Improper construction, calibration or operation of an instrument might result in some inherent errors. Environmental errors occur due to external ambient conditions of the instrument which include changes in temperature, humidity, availability of dust, vibrations or effects of external magnetic or electrostatic fields. Random errors occur due to some small factors which fluctuate from one measurement to another.

The manufacturer of any instrument defines a certain accuracy that depends on the type of material and the effort to manufacture the instrument. This accuracy is defined within a certain deviations from the nominal value. The limits of these deviations are termed as Limiting Errors. The ratio of error to the specified nominal value is called relative Limiting Error. Manual errors in reading an instrument or recording any measurement are known as Gross errors which occur during the experiments. The errors in any scientific measurement may happen from different sources which are presented in Fig. A.5. Observational errors may occur due to the fault study of the instrument reading for many sources. Environmental errors, on the other hand, happen due to the outside situation of the measuring instruments, mostly due to the temperature result, force, moisture, dirt, vibration.

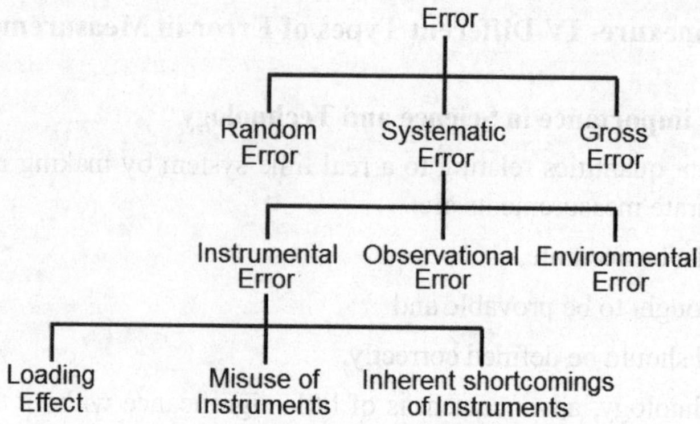

Fig. A.5: Error in measurements

There is an inherent limitation of devices due to their mechanical arrangement.

Calculation of Error in Measurement

These errors are categorized into three types:

i) absolute error,
ii) relative error, and
iii) percentage error.

The absolute error is defined as the variation between the values of actual and measured quantities.

If we denote the measured value as V_A, and the exact value as V_E, then we have

$$\text{Absolute error} = |V_A - V_E|$$

$$\text{Relative Error} = \frac{\text{Absolute error}}{\text{Actual error}}$$

$$= \frac{|V_A - V_E|}{V_E}$$

$$\text{Percentage error (\%)} = \frac{|V_A - V_E|}{V_E} \times 100$$

Example: A length was calculated as 6.8 cm but the absolute length was 6.74 cm. Find the absolute, relative and percentage errors.

Ans: Given that $V_A = 6.8$ cm and $V_E = 6.74$ cm

$$\text{Absolute error} = |V_A - V_E| = |6.8\text{-}6.74| = 0.06 \text{ cm}$$

$$\text{Relative Error} = \frac{|V_A - V_E|}{V_E} = \frac{0.06}{6.74} = 0.0089$$

$$\text{Percentage error (\%)} = \frac{|V_A - V_E|}{V_E} \times 100 = \frac{0.06}{6.74} \times 100 = 0.89\%$$

Arithmetic Mean Value

To minimize random errors, the measurements are repeated and the average value is taken as the correct value of the measured quantity. The arithmetic mean value would be very close to the most accurate reading. If the number of observation is taken '*n*' times, the random error reduces to $\frac{1}{n}$ times.

Let $a_1, a_2, a_3..., a_n$ are the *n* different measured readings of a physical quantity. The most accurate value is its arithmetic mean value which can be obtained from,

$$a_{mean} = \frac{a_1 + a_2 + ... + a_n}{n} = \frac{1}{n}\sum_{i=1}^{n} a_i$$

Absolute Error

The magnitude of the difference between the true value of the quantity and the measured value is known as the absolute error in the measurement. Since the true value of the quantity is not known, the arithmetic mean of the measured values is taken as the true value.

If $a_1, a_2,$ ..... are the measured values of a certain quantity, the errors $\Delta a_1, \Delta a_2,$ .... in the measurements are

$$\Delta a_1 = a_{\text{mean}} - a_1$$

$$\Delta a_2 = a_{\text{mean}} - a_2$$

$$\Delta a_3 = a_{\text{mean}} - a_3$$

$$\Delta a_4 = a_{\text{mean}} - a_4 \qquad \text{and so on}$$

The arithmetic mean of all the absolute errors is considered as the final absolute error in the measurement and is called mean absolute error. The value obtained in a single measurement may be in the range: $a_{\text{mean}} \pm \Delta a_{\text{mean}}$

Relative Error

The ratio of the absolute error to the true value of the measured quantity is known as the relative error or fractional error. As the arithmetic mean value is taken as the true value, the relative error can be expressed as,

$$\text{Relative error, } \delta a = \frac{\Delta a_{mean}}{a_{mean}}$$

Percentage Error

It is the relative error when expressed in percentage. Thus, we have,

$$\text{Percentage error } = \frac{\Delta a_{mean}}{a_{mean}} \times 100\%$$

Example: Using a screw gauge successive measured readings of the radius of a wire are 1.21 mm, 1.19 mm, 1.20 mm, 1.18 mm and 1.17 mm respectively. Determine the absolute errors and the relative error in the measurement.

Ans: Arithmetic mean value is, $a_{mean} = \dfrac{1.21+1.19+1.20+1.18+1.17}{5} = 1.19\,\text{mm}$

Table A.1: Error in measurements

Difference between a_{mean} and measured value in mm	Magnitude of error in mm
1.21–1.19 = 0.02	0.02
1.19–1.19 = 0.00	0.00
1.20–1.19 = 0.01	0.01
1.18–1.19 = –0.01	0.01
1.17–1.19 = –0.02	0.02

The arithmetic mean of the absolute errors is, $\Delta a_{mean} = \dfrac{0.02+0.00+0.01+0.01+0.02}{5} = 0.016\,\text{mm}$

$$\text{Relative error, } \delta a = \frac{\Delta a_{mean}}{a_{mean}} = \frac{0.016}{1.19} = 0.0134$$

$$\text{Percentage error } = \frac{\Delta a_{mean}}{a_{mean}} \times 100\% = \pm 1.34\%$$

Combination of errors

If a quantity is obtained by combining a few measurements, the errors in those measurements can be combined in some way or other.

Error in the sum of the quantities

We consider two quantities A and B which have measured values $A \pm \Delta A$ and $B \pm \Delta B$ respectively; ΔA and ΔB are the absolute errors in the measurements.

To determine the error ΔZ which may occur in the sum $Z = A + B$, we consider

$$Z \pm \Delta Z = (A \pm \Delta A) + (B \pm \Delta B) = A + B \pm \Delta A \pm \Delta B$$

Maximum possible error in the value of Z is,

$$\Delta Z = \Delta A + \Delta B.$$

Hence when two quantities are added, the absolute error in the result is the sum of the absolute errors in the measured quantities.

Error in the difference of the quantities

We consider two quantities, A and B which have measured values $A \pm \Delta A$ and $B \pm \Delta B$ respectively; ΔA and ΔB are the absolute errors in their measurements.

To determine the error ΔZ which may occur in the difference $Z = A - B$, we consider

$$Z \pm \Delta Z = (A \pm \Delta A) - (B \pm \Delta B) = A - B \pm \Delta A \pm \Delta B$$

Hence the maximum possible error in the value of Z is given by

$$\Delta Z = \Delta A + \Delta B.$$

So when two quantities are subtracted, the absolute error in the result is the sum of the absolute errors in the measured quantities.

Error in the product of the quantities

We consider two quantities, A and B which have measured values $A \pm \Delta A$ and $B \pm \Delta B$ respectively; ΔA and ΔB are the absolute errors in their measurements.

To get the error ΔZ which may occur in the product $Z = AB$, we consider

$$Z \pm \Delta Z = (A \pm \Delta A)(B \pm \Delta B) = AB \pm A\Delta B \pm B\Delta A \pm \Delta A\Delta B$$

Dividing left side by Z and right side by AB we have,

$$1 \pm \frac{\Delta Z}{Z} = 1 \pm \frac{\Delta A}{A} \pm \frac{\Delta B}{B} \pm \frac{\Delta A\Delta B}{AB}$$

The maximum fractional error in Z is,

$$\frac{\Delta Z}{Z} = \frac{\Delta A}{A} \pm \frac{\Delta B}{B}$$

(As ΔA and ΔB are small, their products $\frac{\Delta A\Delta B}{AB}$ are too small and can be ignored)

Hence when two quantities are multiplied, the fractional error in the result is the sum of the fractional errors in the measured quantities.

Error in the quotient of the quantities

We consider two quantities, A and B which have measured values $A \pm \Delta A$ and $B \pm \Delta B$ respectively; ΔA and ΔB are the absolute errors in their measurements.

To find the error ΔZ that may occur in the quotient $Z = \frac{A}{B}$, we consider

$$Z \pm \Delta Z = \frac{A \pm \Delta A}{B \pm \Delta B}$$

In a similar way, solving we have, the maximum fractional error in Z as,

$$\frac{\Delta Z}{Z} = \frac{\Delta A}{A} \pm \frac{\Delta B}{B}.$$

Hence when two quantities are divided, the fractional error in the result is the sum of the fractional errors in the measured quantities.

It shows that the maximum percentage error in Z is the sum of the maximum percentage error in A and maximum percentage error in B *i.e.*,

$$\frac{\Delta Z}{Z} \times 100 = \frac{\Delta A}{A} \times 100 + \frac{\Delta B}{B} \times 100$$

Error when a quantity is raised to a power

The error ΔZ which may occur when a quantity is raised to its nth power is n times the fractional error in the quantity itself *i.e.*, if $Z = A^n$,

$$\frac{\Delta Z}{Z} = n\frac{\Delta A}{A}$$

The maximum percentage error in Z can be written as,

$$\frac{\Delta Z}{Z} \times 100 = n \times \frac{\Delta A}{A}$$

Systematic Errors

Systematic errors occur due to fault in the measuring device. They are also called as Zero Error – a positive or negative error. These errors can be detached after correction of the measurement device. These errors are classified into different categories.

Systematic errors are categorized as:

- Instrumental error
- Environmental error
- Observational error
- Theoretical error

Parallax Error

This error occurs due to wrong observations of reading in the instruments. The wrong observations may be due to parallax. To minimize the parallax error highly accurate meters provided with mirror scales are essential.

Estimating Random Errors

There are many ways to make an estimate of the random error in a particular measurement. The simplest way is to make a series of measurements of a given quantity (say, *x*) and calculate the mean and standard deviation ($\overline{x}$ and σ_x) from this data.

The mean $\overline{x}$ is defined as,

$$\overline{x} = \frac{1}{N}\sum_{i=1}^{n} x_i$$

where, x_i is the result of the ith measurements, *N* is the number of measurements

The standard variation is given by,

$$\sigma_x = \frac{1}{N}\sum_{i=1}^{n}(x_i - \overline{x})^2$$

Annexure-V: Some general and specific instructions when working in the laboratory

General Instructions

1. In the laboratory, work quietly and cautiously. Remember the main purpose of doing any experiment is to make faithful measurements.
2. Always share equally all the steps of the work with your partner.
3. Presentations of data in tabular form, graphs and calculations should be done correctly and sincerely.
4. Be always honest at the time of recording and representing the experimental data.
5. It is very important to keep in mind that never make up readings or doctor them to get a better fit of the graph as per theory. If any reading appears incorrect, you have to repeat the measurement again and again to find the source of error.
6. At the time of drawing the graph all the data obtained from experiment are to be properly plotted.
7. It is a fact that the objective of the laboratory is learning and also a verification of the knowledge that you have gathered. The experiments are designed properly for the purpose of illustrating different phenomena in all the important areas of Physics.
8. By doing the experiment with your own interest only it is possible to be familiar with all the fine points and to expose you to measuring instruments.
9. Always perform the experiment with an attitude of learning and with your interest to verify the theoretical knowledge that you have gathered.
10. Be very particular to arrive in time in the laboratory and always with proper preparation with a clear knowledge about the experiment.

Specific Instructions

1. When working in the laboratory for collecting data of your experiment, it is important to note all the measured data neatly in the notebook.
2. The recorded data entered in the notebook have to confirm by your instructor before leaving the laboratory.
3. All the students doing the same experiment have to maintain individual copy of the recorded data. The laboratory notebook is required to bring in the laboratory regularly when you come for doing the experiment.
4. Graphs are to be drawn properly at the end of each of experiment.
5. For this you need to know how to optimize on usage of graph paper. Remember all the repeated data are to be accommodated on a single graph sheet.
6. Graphs are to be labeled properly along with the axes showing the corresponding units.
7. During the working hours in the laboratory you are supposed to fully utilize the duration and do not leave the laboratory before the completion of the working hours. If you finish early, you may spend the remaining time to complete the calculations and graphs drawing and for that in the laboratory you are supposed to come equipped with calculators, pencils and scale.

REFERENCES FOR FURTHER LEARNING

1. Fowles, Grant R.; Cassiday, George L., Analytic Mechanics (5th ed.), Fort Worth: Saunders College Publishing, 1986.
2. Hayek, Sabih I., "Mechanical Vibration and Damping". Encyclopedia of Applied Physics. WILEY-VCH Verlag GmbH and Co KGaA, 2003.
3. Kreyszig, Erwin, Advanced Engineering Mathematics (3rd ed.), New York: Wiley, 1972.
4. Serway, Raymond A.; Jewett, John W., Physics for Scientists and Engineers. Brooks/Cole. 2003.
5. Tipler, Paul, Physics for Scientists and Engineers: Vol. 1 (4th ed.). W. H. Freeman, 1998.
6. Wylie, C. R., Advanced Engineering Mathematics (4th ed.). McGraw-Hill, 1975.
7. Craik, A.D.D., "The origins of water wave theory", Annual Review of Fluid Mechanics, 36: 1-28, 2004.
8. Dean, R.G.; Dalrymple, R.A., "Water wave mechanics for engineers and scientists", Eos Transactions, Advanced Series on Ocean Engineering, 2 (24): 490, 1991.
9. Dingemans, M.W., "Water wave propagation over uneven bottoms", NASA Sti/Recon Technical Report N, Advanced Series on Ocean Engineering, 13, 1997.
10. Lamb, H., Hydrodynamics (6th ed.), Cambridge University Press, 1994.
11. Landau, L.D.; Lifshitz, E.M., Fluid Mechanics, Course of theoretical physics, 6 (2nd ed.), Pergamon Press, 1987
12. Lighthill, M.J., Waves in fluids, Cambridge University Press, 504 pp, 1978.
13. Phillips, O.M., The dynamics of the upper ocean (2nd ed.), Cambridge University Press, 1977.
14. Whitham, G. B., Linear and nonlinear waves, Wiley-Interscience, 1974.
15. Arthur Schuster, An Introduction to the Theory of Optics, London: Edward Arnold, 1904.
16. Greivenkamp, John E., Field Guide to Geometrical Optics. SPIE Field Guides. 1. SPIE. pp. 19–20. 2004.
17. Hugh D. Young, University Physics 8e. Addison-Wesley, 1992.
18. E. W. Marchand, Gradient Index Optics, New York, NY, Academic Press, 1978.
19. Hecht, Eugene, Optics (2nd ed.). Addison Wesley, 1987
20. Sommerfeld, A., and Runge, J., Anwendung der Vektorrechnung auf die Grundlagen der geometrischen Optik. Annalen der Physik, 340(7), 277-298. 1911.
21. Born, M. and Wolf, E., Principles of optics: electromagnetic theory of propagation, interference and diffraction of light. Elsevier, 2013.
22. Borowitz, S., Fundamentals of quantum mechanics, particles, waves, and wave mechanics, 1967.
23. Luneburg, R. K., Methematical Theory of Optics, Brown University Press 1944 [mimeographed notes], University of California Press 1964
24. Kline, M., Kay, I. W., Electromagnetic Theory and Geometrical Optics, Interscience Publishers 1965.
25. Fleisch, D.; Kinnaman, L., A student's guide to waves. Cambridge: Cambridge University Press. 2015.
26. French, A.P., Vibrations and Waves (M.I.T. Introductory physics series). Nelson Thornes1971.
27. Campbell, Murray; Greated, Clive, The musician's guide to acoustics (Repr. ed.). Oxford: Oxford University Press, 2001.

28. Hall, D.E., Musical Acoustics: An Introduction. Belmont, CA: Wadsworth Publishing Company, 1980.
29. Hunt, Frederick Vinton, Origins in acoustics. Woodbury, NY: Published for the Acoustical Society of America through the American Institute of Physics, 1978.
30. Ostrovsky, L.A.; Potapov, A.S., Modulated Waves, Theory and Applications. Baltimore: The Johns Hopkins University Press, 1999.
31. Griffiths, G.; Schiesser, W.E., Traveling Wave Analysis of Partial Differential Equations: Numerical and Analytical Methods with Matlab and Maple. Academic Press, 2010.
32. Bertolotti, Mario, *The History of the Laser*. Institute of Physics, 2004.
33. Bromberg, Joan Lisa, *The Laser in America, 1950–1970*. MIT Press, 1991.
34. Csele, Mark, Fundamentals *of Light Sources and Lasers*. Wiley, 2004.
35. Koechner, Walter, *Solid-State Laser Engineering*. 3rd ed. Springer-Verlag, 1992.
36. Siegman, Anthony E., *Lasers*. University Science Books, 1986.

CO AND PO ATTAINMENT TABLE

Course outcomes (COs) for this course can be mapped with the programme outcomes (POs) after the completion of the course and a correlation can be made for the attainment of POs to analyze the gap. After proper analysis of the gap in the attainment of POs necessary measures can be taken to overcome the gaps.

Table for CO and PO attainment

Course Outcomes	Attainment of Programme Outcomes (1- Weak Correlation; 2- Medium Correlation; 3- Strong Correlation)											
	PO-1	PO-2	PO-3	PO-4	PO-5	PO-6	PO-7	PO-8	PO-9	PO-10	PO-11	PO-12
CO-1												
CO-2												
CO-3												
CO-4												
CO-5												
CO-6												

The data filled in the above table can be used for gap analysis.

INDEX